**Approximate Methods
in
Engineering Design**

This is Volume 155 in
MATHEMATICS IN SCIENCE AND ENGINEERING
A Series of Monographs and Textbooks
Edited by RICHARD BELLMAN, *University of Southern California*

The complete listing of books in this series is available from the Publisher upon request.

Approximate Methods in Engineering Design

T. T. FURMAN

*Department of Aeronautical and Mechanical Engineering,
University of Salford,
Salford, England*

1981

ACADEMIC PRESS
A Subsidiary of Harcourt Brace Jovanovich, Publishers
London   New York   Toronto   Sydney   San Francisco

ACADEMIC PRESS INC. (LONDON) LTD.
24/28 Oval Road
London NW1

United States Edition published by
ACADEMIC PRESS INC.
111 Fifth Avenue
New York, New York 10003

Copyright © 1981 by
ACADEMIC PRESS INC. (LONDON) LTD.

All Rights Reserved
No part of this book may be reproduced in any form by photostat, microfilm, or any other means, without written permission from the publishers

British Library Cataloguing in Publication Data

Furman, Tobias Towie
  Approximate methods in engineering design.—
(Mathematics in science and engineering).
  1. Engineering design—Approximation methods
  2. Engineering mathematics
  I. Title   II. Series
  620′.0042   TA174   80–40891
  ISBN 0-12-269960-2

Printed in Great Britain by John Wright and Sons Ltd. at The Stonebridge Press, Bristol BS4 5NU

Preface

Many problems which arise in engineering defy exact analysis. It is often not even necessary to obtain rigorously accurate results to practical engineering problems, since data, variables and factors are themselves usually only approximations.

In design, as one proceeds from a brief to the manufacturing details, one needs to start with approximate assessment of forces, stresses, speeds, power etc., so as to be able to lay out principal geometries. The iterative process leading to a final design usually involves a gradual enhancement of the accuracy of calculated predictions. The value of mathematical precision is often reduced, and even negated, because one invariably has to use idealized simplifications and models to represent real physical situations. Thus approximate methods of calculations with a reasonable knowledge of the degree of accuracy involved can, in general, be much more valuable than complex and time-consuming precise and rigorous mathematical analyses. The expression "degree of accuracy" implies, *per se*, the need to assess the magnitude and direction of the error invoved.

This book has been written with this underlying philosophy. It is hoped that the information collated, the techniques and methods elucidated, and the case studies described will be of help to engineering designers in their various and demanding tasks.

December, 1980 *T. T. Furman*

*I dedicate this book
to my cousin Abbe Obstbaum.*

*But for his belief in me
and his support this book
might never have been written.*

T.T.F.

Contents

Preface   v

Chapter 1   Approximate Correlations   1

1.1	Taylor's Series	1
1.2	Linear Interpolation	4
1.3	Newton's Method	4
1.4	Iterative Method	8
1.5	Checking for Convergence	9
1.6	Inversion Technique	12
1.7	Estimation of Errors	14
1.8	Least Square Method	19
1.9	Finite Difference Equations	22

Chapter 2   Theory of Errors   31

2.1	Definition of Errors	31
2.2	Rounding off Last Figures	32
2.3	Operations with Approximate Numbers	32
2.4	Error of a Function	36
2.5	The Inverse Problem	38
2.6	Influence on Design	44

Chapter 3   Statistical Errors—Empirical Correlations   50

3.1	Statistical Design Factors	50
3.2	Tolerance and Standard Deviation	50
3.3	Reliability of Fatigue Calculations	52
3.4	Dynamic Life/Load Capacity of Rolling Element Bearings	53
3.5	Effect of Misalignment of the Life of Rolling Element Bearings	56

Chapter 4   Types of Distributions   69

4.1	The Binomial Distribution	69
4.2	Poisson's Distribution	71
4.3	Continuous Probability Distributions	73
4.4	Pattern of Failure Rates of Mechanical Equipment	75
4.5	Weibull Probability Distributions	76
4.6	The Normal (Gaussian) Distribution	80
4.7	Grouped Frequencies	90
4.8	Distribution of Errors	96
4.9	Effects of Sample Size	98

CONTENTS

Chapter 5	The Algebra of Random Numbers	101
5.1	Definitions	101
5.2	Basic Statistical Theory	101
5.3	Partial Derivative Method	106
5.4	Examples of Algebraic Derivations	107
5.5	Correlation Coefficients	111
5.6	Confidence Limits	118
5.7	Examples	120

Chapter 6	Variabilities of Material Properties	123

Chapter 7	Reliability	133
7.1	Component Reliability	133
7.2	Reliability and Safety Factors	147
7.3	Reliability of Assemblies and Systems	151
7.4	The Role of Design in Achieving Reliability	162
7.5	Case Studies	165

Chapter 8	Machine Failure	198
8.1	Definition of Machine Failure	198
8.2	Origins of Machine Failures	198
8.3	Causes of Machine Failures	200
8.4	Failure Diagnosis	203
8.5	Mechanisms of Component Failures	215

Chapter 9	Modelling	294
9.1	Shear Stress Distribution in a Thin Tubular Cylinder	295
9.2	Shear Stress Distribution across Diamond-shaped Section	297
9.3	Shear Centre of Channel Section Beam	300
9.4	Estimating the Interference Pressure between a Spring-clip and Shaft	303
9.5	Reversibility or Non-reversibility of Machines	308
9.6	Fatigue Failures of Burnishing Rollers	311
9.7	Base Plate Bolt Loads	313
9.8	Assessment of Spring force for Poppet Valve Assembly	321
9.9	Inequality of Forced and Natural Frequencies	326
9.10	Manufacturing Tolerances and Probability of Performance Achievement	332

Solutions		345
Subject Index		385

Chapter 1

Approximate Correlations

1.0 INTRODUCTION

In design we mostly handle approximate data. Complex functions can often be replaced by simpler approximate ones yielding results within acceptable levels of errors. We must, however, have some idea of the magnitude and direction of the errors thus introduced. A common technique is to replace a complex function by a series which can be made to converge on the true value of the function. Using the first few terms of such an expansion enables us to obtain approximate answers good enough for practical purposes. (See Table 1.1, p. 3, and Tables 1.2 and 1.3, pp. 29, 30.)

1.1 TAYLOR'S SERIES

These well-known series are often used to evaluate complex functions. If

$$y = f(x) \tag{1.1}$$

is a continuous function between values of (a) and $(a+h)$, and it has continuous derivatives from the 1st to the nth derivative, then

$$f(x) = f(a) + \frac{(x-a)}{1!}f'(a) + \frac{(x-a)^2}{2!}f''(a) + \frac{(x-a)^n}{n!}f^n(a) \tag{1.2}$$

or, alternatively, putting $(x-a) = h$;

$$f(a+h) = f(a) + \frac{h}{1!}f'(a) + \frac{h^2}{2!}f''(a) + \frac{h^n}{n!}f^n(a) \tag{1.3}$$

These and similar series can be used instead of the exact functions allowing simple and quick estimation of good approximations, provided, of course, the series converge rapidly. In order to express a function $f(x)$ as a series of powers of x, we can put $a = 0$ in Taylor's equation (1.2). Thus,

$$f(x) = f(0) + \frac{x}{1!}f'(0) + \frac{x^2}{2!}f''(0) + \frac{x^3}{3!} \cdot f'''(0) + \ldots + \frac{x^n}{n!} \cdot f^n(0) \tag{1.4}$$

Example 1.1.1

$$f(x) = e^x = e^0 + \frac{x}{1!} \cdot e^0 + \frac{x^2}{2!} \cdot e^0 + \ldots$$

$$= 1 + \frac{x}{1!} + \frac{x^2}{2!} + \frac{x^3}{3!} + \ldots$$

Example 1.1.2

$$f(x) = (1+x)^{1/3} = (1+0)^{1/3} + \frac{x}{1!}\tfrac{1}{3}(1+0)^{-2/3} + \frac{x^2}{2!} \cdot \tfrac{1}{3}(-\tfrac{2}{3})(1+0)^{-5/3}$$

$$+ \frac{x^3}{3!} \cdot \tfrac{1}{3}(-\tfrac{2}{3})(-\tfrac{5}{3})(1+0)^{-8/3} + \ldots$$

or $\quad = 1 + \tfrac{1}{3}x - \tfrac{1}{3} \cdot \tfrac{2}{6}x^2 + \tfrac{1}{3} \cdot \tfrac{2}{6} \cdot \tfrac{5}{9}x^3 - \ldots$

(see Table 1.1)

Example 1.1.3

$$f(x) = \tan x = \tan(0) + \frac{x}{1!}[1 + \tan^2(0)]$$

$$+ \frac{x^2}{2!}\{2\tan(0) \cdot [1 + \tan^2(0)]\}$$

$$+ \frac{x^3}{3!}[[2\{[1+\tan^2(0)]^2 + \tan(0)(1+\tan^2(0))\}]] \quad \text{etc.}$$

or $\quad \tan x = x + \dfrac{x^3}{3} + \ldots \quad$ etc.

(see Table 1.1)

Example 1.1.4

Consider function $f(x) = \cos x - \sqrt{1-x}$. Find the value of $f(x)$ for $x = 0 \cdot 10$. Using six-figure trigonometrical and log tables one obtains, at best, a value for

$$f(0 \cdot 1) = 0 \cdot 000\,017$$

Using Taylor's Series (Table 1.1) we obtain

$$\cos x = 1 - \tfrac{1}{2}x^2 + \tfrac{1}{24}x^4 - \tfrac{1}{720}x^6 + \tfrac{1}{40\,30}x^8 \ldots \quad \text{(i)}$$

and

$$\sqrt{1-x} = 1 - \tfrac{1}{2}x^2 - \tfrac{1}{8}x^4 - \tfrac{1}{6}x^6 - \tfrac{5}{128}x^8 \ldots \quad \text{(ii)}$$

Adding the two series, we get

$$f(x) = \tfrac{1}{6}x^4 - \tfrac{11}{180}x^6 + \tfrac{197}{5040}x^8 \ldots, \quad \text{(iii)}$$

Table 1.1. Power series of functions.

Function	Equivalent power series	Domain of convergence
$\sin x$	$x - \dfrac{x^3}{3!} + \dfrac{x^5}{5!} - \dfrac{x^7}{7!} + \ldots - + - \ldots$	$|x| < \infty$
$\cos x$	$1 - \dfrac{x^2}{2!} + \dfrac{x^4}{4!} - \dfrac{x^6}{6!} + \ldots - \ldots +$	$|x| < \infty$
$\tan x$	$x + \tfrac{1}{3}x^3 + \tfrac{2}{15}x^5 + \tfrac{17}{315}x^7 + \tfrac{62}{2835}x^9 +$	$|x| < \dfrac{\pi}{2}$
$\cot x$	$\left[\dfrac{1}{x} - \dfrac{x}{3} + \dfrac{x^3}{45} + \dfrac{2x^5}{945} + \dfrac{x^7}{4725} + \ldots\right]$	$0 < |x| < \pi$
e^x	$1 + \dfrac{x}{1!} + \dfrac{x^2}{2!} + \dfrac{x^3}{3!} + \ldots$	$|x| < \infty$
$\ln x$	$2\left[\dfrac{x-1}{x+1} + \dfrac{(x-1)^3}{3(x+1)^3} + \dfrac{(x-1)^5}{5(x+1)^5} + \ldots\right]$	$x > 0$
$\ln(1+x)$	$x - \dfrac{x^2}{2} + \dfrac{x^3}{3} - \dfrac{x^4}{4} + \ldots$	$-1 < x \leqslant 1$
$\ln(1-x)$	$-\left[x + \dfrac{x^2}{2} + \dfrac{x^3}{3} + \dfrac{x^4}{4} + \ldots\right]$	$1 \leqslant x < 1$
$(a \pm x)^m$	Rewrite as $a^m\left(1 \pm \dfrac{x}{a}\right)^m$ as below	
$(1 \pm x)^m$ for $m > 0$	$1 \pm mx + \dfrac{m(m-1)x^2}{2!} \pm \dfrac{m(m-1)(m-2)x^3}{3!} + $	$|x| \leqslant 1$
$\arcsin x$	$x + \dfrac{x^3}{2.3} + \dfrac{1.3.x^5}{2.4.5} + \dfrac{1.3.5.x^7}{2.4.6.7} + \ldots$	$|x| < 1$
$\arccos x$	$\dfrac{x}{2} - \left[x + \dfrac{x^3}{2.3} + \dfrac{1.3.x^5}{2.4.5} + \dfrac{1.3.5.x^7}{2.4.6.7} + \right]$	$|x| < 1$
$\arctan x$	$x - \dfrac{x^3}{3} + \dfrac{x^5}{5} - \dfrac{x^7}{7} + \ldots$	$|x| < 1$
$\sinh x$	$x + \dfrac{x^3}{3!} + \dfrac{x^5}{5!} + \dfrac{x^7}{7!} + \ldots$	$|x| < \infty$
$\cosh x$	$1 + \dfrac{x^2}{2!} + \dfrac{x^4}{4!} + \dfrac{x^6}{6!} + \ldots$	$|x| < \infty$
$\tanh x$	$x - \tfrac{1}{3}x^3 + \tfrac{2}{15}x^5 - \tfrac{17}{315}x^7 + \tfrac{62}{2835}x^9 + \ldots$	$|x| < \dfrac{\pi}{2}$

which for $x = 0\cdot01$ yields

$$f(0\cdot 1) = 0\cdot 000\,016\,7 - 0\cdot 000\,000\,061 + \ldots$$
$$= 0\cdot 000\,016\,61 \text{ correct to eight significant digits.}$$

1.2 LINEAR INTERPOLATION
(Balfour and McTernan, 1967)

If the root of a function f(x) = 0 lies between the interval $a < x_0 < b$, it can be shown that a closer approximation can be found by linearly interpolating between the two values of *a* and *b* (see Fig. 1.1).

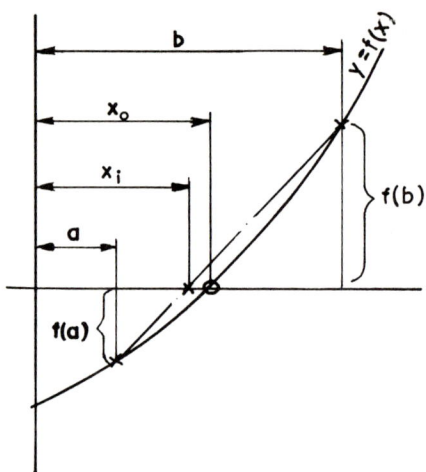

Fig. 1.1. Linear interpolation.

From similar triangles we have

$$\frac{x_i - a}{-f(a)} = \frac{b - x_i}{f(b)}$$

$$\therefore \quad x_i \cdot f(b) - a \cdot f(b) = x_i \cdot f(a) - b \cdot f(a)$$

$$\text{or} \quad x_i = \frac{a \cdot f(b) - b \cdot f(a)}{f(b) - f(a)} \tag{1.5}$$

The error of $(x_0 - x_i)$ will, of course, depend on the closeness of the values of *a* and *b* to x_0.

1.3. NEWTON'S METHOD

Newton showed how, if x_0 is the true root of a function f(x) = 0 and only one value close to x_0 is known or can be guesstimated, this first value can be iteratively converged closer to the true value of x_0 (see Fig. 1.2). If the first value of $x = x_1$, then from the triangle ABC, we have

$$f'(x_1) = \frac{f(x_1)}{x_1 - x_2}$$

1. APPROXIMATE CORRELATIONS

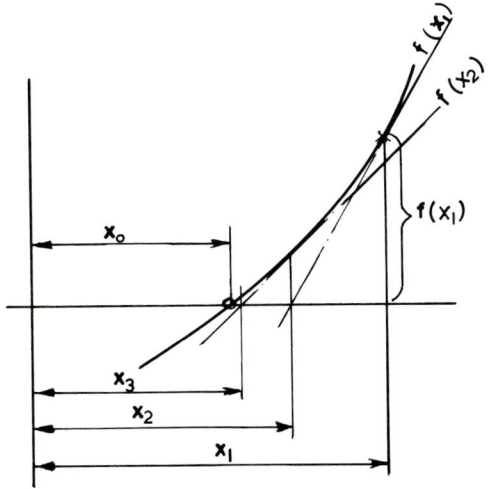

Fig. 1.2. Newton's method.

i.e.
$$x_2 = x_1 - \frac{f(x_1)}{f'(x_1)}$$

and
$$x_3 = x_2 - \frac{f(x_2)}{f'(x_2)}$$

$$x_{n+1} = x_n - \frac{f(x_n)}{f'(x_n)} \tag{1.6}$$

If the function is convergent at or near the point $x = x_0$, this sequence of steps will reduce the error and make

$$x_n \to x_0.$$

Newton's method can also be derived from Taylor's series, thus
Let $x_1 = x_0 + \varepsilon$, where

$$\varepsilon = \text{small initial error}$$
$$x_0 = \text{true root}, \quad \therefore \quad f(x_0) = 0$$

From equation (1.3) then

$$0 = f(x_0) = f(x_1 - \varepsilon) = f(x_1) - \varepsilon f'(x_1) + \tfrac{1}{2}\varepsilon^2 f''(x_1) + - \ldots$$

or
$$\varepsilon = \frac{f(x_1)}{f'(x_1)} + \tfrac{1}{2}\varepsilon^2 \frac{f''(x_1)}{f'(x_1)} + \ldots$$

Thus

$$x_1 = x_0 + \varepsilon = x_0 + \frac{f(x_1)}{f'(x_1)} + \tfrac{1}{2}\varepsilon^2 \cdot \frac{f''(x_1)}{f'(x_1)} + \ldots$$

or

$$x_0 = x_1 - \frac{f(x_1)}{f'(x_1)} - \tfrac{1}{2}\varepsilon^2 \cdot \frac{f''(x_1)}{f'(x_1)} + \ldots \quad (1.7)$$

The first approximation of equation (1.7) therefore is

$$x_2 = x_1 - \frac{f(x_1)}{f'(x_1)}$$

and subsequent iterations become

$$x_{n+1} = x_n - \frac{f(x_n)}{f'(x_n)}$$

as for equation (1.6).

N.B. Whereas Newton's method continues to give an error on the same side as the initial error, the Linear Interpolation method causes the errors to alternate on either side of the root value. Thus using both methods, where applicable, allows an estimate to be made of the magnitude of the error at any stage in the calculations.

Example 1.3.1

Find the smallest positive root of the equation

$$\sin x - x \cdot \cos x = 0 \quad (i)$$

Rewrite this equation in the form

$$x = \tan x \quad (ii)$$

Plot approximate graph of $y = \tan x$, and draw a line through the origin with 45° slope. This line will intersect the $y = \tan x$ curves indicating an approximate value for x, such that $\tan x \simeq x$ (see Fig. 1.3).

It will be seen from Fig. 1.3 that equation (i) will be satisfied if $x \simeq \tfrac{3}{2}\pi \simeq 4\cdot 71$.

We can thus start with $x_1 = (\tfrac{3}{2}\pi)$, and being a single starting point, we may use Newton's method to move closer to the true root value.

Now generally from (i)

$$f(x) = \sin x - x \cos x \quad (iii)$$

$$\therefore f'(x) = x \sin x \quad (iv)$$

Thus for $x_1 = 4\cdot 71$, we get

$$f(\tfrac{3}{2}\pi) = \sin(\tfrac{3}{2}\pi) - \tfrac{3}{2}\pi \cdot \cos(\tfrac{3}{2}\pi)$$

$$f(x_1) = -1 \quad (v)$$

1. APPROXIMATE CORRELATIONS

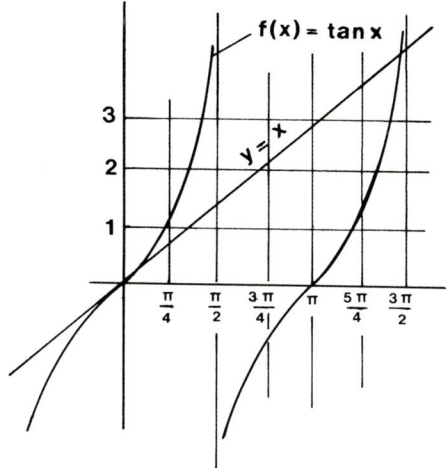

Fig. 1.3. Approximate graphical determination of $x = \tan x$.

and $f'(x_1) = \frac{3}{2}\pi(-1) \simeq -4.71$.

$$\therefore x_2 = 4.71 - \frac{-1}{-4.71} = 4.50 \qquad \text{(iv)}$$

and

$$f(x_2) = -0.9775 + 4.5 \cdot 0.2108$$

$$f(x_2) = -0.029 \qquad \text{(vii)}$$

We aim at $f(x_n) \to 0$; $f(x_1)$ and $f(x_2)$ are both negative, and are therefore to the same side of the true root. In order to get an idea of the accuracy of $x_2 = 4.50$, we want to use the Linear Interpolation method, but we require first to guess a third value x_3, which will make $f(x_3) = +ve$.

Try $x_3 = 4.45$, this gives

$$f(x_3) = -0.96577 - 4.45(-0.259)$$

$$f(x_3) = +0.189 \qquad \text{(viii)}$$

We can now use the Linear Interpolation method between x_2 and x_3 to get a closer approximation to x_0. Thus

$$x_i = \frac{4.5(0.189) - 4.45(-0.029)}{(0.189) - (-0.029)}$$

or

$$x_i = 4.4933$$

Thus the true root x_0 will be

$$4.4933 < x_0 < 4.50$$

This is reasonably close for all practical purposes.

1.4 ITERATIVE METHOD

In this method the standard form of equation $f(x) = 0$ is rewritten as a "recurrence equation", i.e.

$$x_{n+1} = \phi(x_n) \tag{1.8}$$

An initial near approximate value (x_1) is estimated and the recurrence equation is used iteratively to reduce the error. This will happen only if the function converges between (x_1) and (x_0). Thus, if

$x_1 = $ initial approximate value of root, then

$$x_2 = \phi(x_1)$$
$$x_3 = \phi(x_2)$$
$$x_{n+1} = \phi(x_n)$$

when $x_{n+1} \to x_n \to x_0$.

Example 1.4.1

$x^3 + x - b = 0$. Recurrence equation is $x_{n+1} = (b - x_n^3)$. Consider case for $b = 0.50$. Take first approximate value for root

$$x_1 = 0.42$$
$$x_2 = 0.5 - 0.0741 \qquad = 0.4259$$
$$x_3 = 0.5 - (0.4259)^3 = 0.422\,74$$
$$x_4 \qquad\qquad\qquad\qquad = 0.424\,453$$
$$x_5 \qquad\qquad\qquad\qquad = 0.423\,53$$
$$x_6 \qquad\qquad\qquad\qquad = 0.424$$
$$x_7 \qquad\qquad\qquad\qquad = 0.423\,76$$
$$x_8 \qquad\qquad\qquad\qquad = 0.4239$$
$$x_9 \qquad\qquad\qquad\qquad = 0.4238$$
$$x_{10} \qquad\qquad\qquad\qquad = 0.423\,87$$
$$x_{11} \qquad\qquad\qquad\qquad = 0.423\,85$$
$$x_{12} \qquad\qquad\qquad\qquad = 0.423\,856$$

Try also case when $b = 5.0$, say.

1. APPROXIMATE CORRELATIONS

1.5 CHECKING FOR CONVERGENCE

In the preceding sections we showed how by guesstimating a value for a function, one can use techniques to reduce the error between the true root and the approximate value. In all instances it was emphasized that this could be achieved only if the function at the guesstimated and true root value was convergent. Thus before attempting to find the true root by such processes one should ascertain convergence.

1.5.1 First order of approximation

We may start with the "recurrence equation" (1.8)

$$x_{n+1} = \phi(x_n) \tag{1.8}$$

Denote the true root to the function by x_0. If the first close approximate value is x_1, then

$$x_1 = x_0 + \varepsilon_1 \tag{a}$$

where $x_1 = \phi(x_0)$ and $\varepsilon_1 = $ first error relative to the true value. By successive approximations, we get

$$x_2 = x_1 + \varepsilon_2; \quad x_2 = \phi(x_1)$$
$$x_3 = x_2 + \varepsilon_3; \quad x_3 = \phi(x_2)$$
$$x_{n+1} = x_n + \varepsilon_{n+1}; \quad x_{n+1} = \phi(x_n) \tag{b}$$

If $\varepsilon_n \to 0$, then $x_n \to x_0$ and the series between x_1 and x_0 converges. Equations (b) can also be written in the form:

$$[x_0 + \varepsilon_{n+1}] = \phi[x_0 + \varepsilon_n] \tag{c}$$

and using Taylor's expansion, we get

$$[x_0 + \varepsilon_{n+1}] = \phi(x_0) + \varepsilon_n \phi'(x_0) + \frac{\varepsilon_n^2}{2!} \phi''(x_0) + \ldots \quad \text{etc.} \tag{d}$$

If the first value of x_1 was fairly close to x_0, then ε_1 would be small as would successive ε_n values. The "first order" of approximation from equation (d) then becomes

$$(x_0 + \varepsilon_{n+1}) \simeq \phi(x_0) + \varepsilon_n \cdot \phi'(x_0) \tag{e}$$

Since x_0 is a perfect root, then

$$x_0 = \phi(x_0)$$

thus

$$\varepsilon_{n+1} \simeq \varepsilon_n \cdot \phi'(x_0) \tag{f}$$

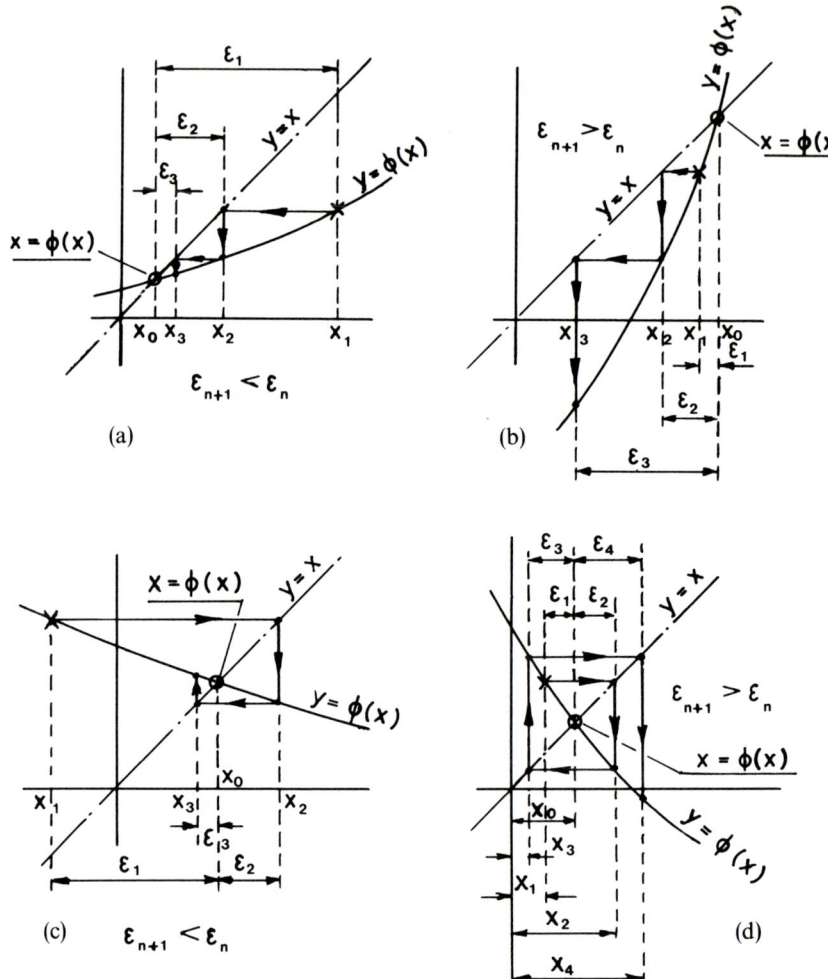

Fig. 1.4. First-order approximation. (a) $0 < \phi'(x) < 1$ convergence. (b) $\phi'(x) > 1$ non-convergence. (c) $-1 < \phi'(x) < 0$ convergence. (d) $\phi'(x) < (-1)$ non-convergence. (a) and (d) applicable; (c) and (d) non-applicable.

For convergence, therefore,

$$\varepsilon_{n+1} < \varepsilon_n$$

i.e.

$$0 < \phi'(x_0) < 1 \tag{g}$$

But initially we do not know the value of x_0. Therefore the conditions for convergence as per equation (a) have to be made for x_1 thus: first test for

convergence:
$$0 < \phi(x_1) < 1 \tag{h}$$
This means that the slope of the curve at $x = x_1$ should be between 0° and 45° (see Fig. 1.4).

Example 1.5.1
Given equation
$$e^x - \ln x - 20 = 0 \tag{i}$$

(A) Check for convergence:
Re-write equation (i) as a recurrence equation:
$$x_{n+1} = \ln(20 + \ln x_n) \tag{ii}$$
Guess a root value $x_1 = 3$ as near to the true root. For equation (ii)
$$\phi'(x_i) = \frac{1}{x_1(20 + \ln x_1)} = \frac{1}{3(20 + \ln 3)}$$
or
$$\phi'(x_1 = 3) = \frac{1}{3(20 + 1 \cdot 0986)} = \frac{1}{63 \cdot 3}$$
i.e.
$$\phi'(x_1 = 3) = \frac{1}{63 \cdot 3} < 1;$$
∴ function converges at $x_1 = 3$.

(B) Find true value of x_0:
$$x_1 = 3$$
$$x_2 = \ln(20 + \ln 3) = \ln(21 \cdot 0986) = 3 \cdot 0491$$
$$x_3 = \ln(20 + \ln 3.0491) = \ln(21 \cdot 114\,84) = 3 \cdot 049\,98$$
$$x_4 = \ln(20 + \ln 3 \cdot 049\,98) = \ln(21 \cdot 115\,13) = 3 \cdot 049\,99$$
$$x_5 = \ln(20 + \ln 3 \cdot 049\,99) = \ln(21 \cdot 115\,138) = 3 \cdot 049\,999\,025\,9$$
i.e. $x_5 = x_0 = 3 \cdot 049\,990\,3$

1.5.2 Second order of approximation
If $\phi(x_0)$ or $\phi(x_n) = 0$, we can use the second term in Taylor's expansion, equation (d) to check for convergence. Thus,
$$\varepsilon_{n+1} \simeq \varepsilon_n \cdot \phi'(x_0) + \frac{\varepsilon_n^2}{2!} \cdot \phi''(x_0)$$

and since in this case $\phi'(x_0)$ or $\phi'(x_n) = 0$, we have

$$\varepsilon_{n+1} \simeq \frac{\varepsilon_n^2}{2!} \phi''(x_0)$$

so for convergence $\varepsilon_{n+1} < \varepsilon_n$, therefore

(i) $\phi''(x_0)$ or $\phi''(x_n)$ must be finite, i.e. $\neq 0$

and

(ii) $\left[\frac{\varepsilon_{n+1}}{\varepsilon_n} \simeq \frac{1}{2}\phi''(x_n) \cdot \varepsilon_n\right] < 1$

Example 1.5.2 (see also Section 1.4):

$$x^3 + x - b = 0$$

Recurrence equation: $x = (b - x^3)$
$\phi(x) = (b - x^3)$
$\phi'(x) = (-3x^2)$
$\phi''(x) = (-6x)$
$\phi'''(x) = (-6)$

Convergence for "first order" approximation requires that

$$0 > (-3x_n^2) > (-1)$$

i.e. the

$$0 < x_n < 0.5774$$

i.e.

$$0 < b < 0.9699$$

Referring to the example in Section 1.4, we see that if $b = 5.0$ the function $x = 5 - x^3$ is not convergent because the root to this equation is $x_0 \simeq 1.5 > 0.5774$.

1.6 INVERSION TECHNIQUE

If on checking for convergence we find that the function does *not* converge, we cannot use the above iterative methods. The problem can often be overcome by "inverting" the function.

For example,

$$x = \tan x$$

1. APPROXIMATE CORRELATIONS

we have
$$\phi(x) = \tan x$$
and
$$\phi'(x) = (1 + \tan^2 x), \quad \text{which is always greater than 1.}$$

Thus the function $\phi(x) = \tan x$ does *not* converge for any value of x. Now, consider the inverted function

$$x = \arctan x.$$

i.e.
$$\psi(x) = \arctan x$$
and
$$\psi'(x) = \frac{1}{1+x^2}, \quad \text{which is less than 1, for any value of } x.$$

The function $\psi(x) = \arctan x$ is, therefore, always convergent, and iterative methods can be used to find a close approximation for the root to equation $x = \arctan x$, which will be the same as the root for $x = \tan x$.

Referring to the example discussed in Section 1.5, i.e. $x^3 + x - b = 0$, we recall that when using the recurrent equation $x = b - x^3$, we could find the root by iteration only if $0 < x_n < 0.5774$; which would be the case if $0 < b < 0.970$. Now, if we invert the recurrence equation, and write

$$x^3 = b - x,$$

we get
$$x = (b-x)^{1/3} \tag{i}$$

i.e.
$$\psi(x) = (b-x)^{1/3} \tag{ii}$$

$$\therefore \psi'(x) = +\tfrac{1}{3}(b-x)^{-2/3} \tag{iii}$$

For convergence we need

(A) $0 < \psi'(x) < 1$, or (B) $(-1) < \psi'(x) < 0$.

It can be shown that (B) leads to imaginary roots. We therefore consider case (A) only and note that

$$0 < \left[\frac{1}{3}\left(\frac{1}{b-x}\right)^{2/3}\right] < 1$$

or

$$0 < \left(\frac{1}{b-x}\right)^{2/3} < 3$$

i.e.
$$0 < \left(\frac{1}{b-x}\right) < 5\cdot 195$$

Thus for +ve values of b, x must be smaller than b, and
$$(b-x) > 0\cdot 1925$$
or
$$x < (b - 0\cdot 1925)$$
Now, we wanted to find the root for the equation when $b = 5\cdot 0$. Using the inverted recurrence equation, we get
$$x = (5-x)^{1/3}$$
Try $x_1 = 1\cdot 5$, which by iteration gives

$$x_2 = (5 - 1\cdot 5)^{1/3} = 1\cdot 5183$$
$$x_3 = \qquad\qquad = 1\cdot 5156$$
$$x_4 = \qquad\qquad = 1\cdot 516$$
$$x_5 = \qquad\qquad = 1\cdot 515\,973$$
$$x_6 = \qquad\qquad = 1\cdot 515\,98$$
$$x_7 = \qquad\qquad = 1\cdot 515\,98 = x_0$$

1.7 ESTIMATION OF ERRORS

We saw in the example in Section 1.3 that by using both "linear interpolation" and Newton's method an estimate can be obtained of the order of error at any point in the iteration procedure. An alternative method is shown by Bruckheimer et al. (1968) using specific examples to demonstrate their procedure.

Case 1
$$x^2 = b$$
Step 1. Guess a near approximate value, say x_1, with a small error ε_1. Then
and
$$x_0 = x_1 + \varepsilon_1$$
$$x_0^2 = (x_1 + \varepsilon_1)^2 = (x_1^2 + 2x_1\varepsilon_1 + \varepsilon_1^2)$$

Step. 2. For a first approximation neglect ε_1^2. Then
$$x_0^2 = b \simeq x_1^2 + 2\varepsilon_1 x_1$$

1. APPROXIMATE CORRELATIONS

and

$$\varepsilon_1 \simeq \frac{1}{2}\left(\frac{b}{x_1} - x_1\right)$$

Thus at any stage in the iteration

$$\varepsilon_n \simeq \frac{1}{2}\left(\frac{b}{x_n} - x_n\right) \tag{1.9}$$

Step 3. For a second approximation, we may add the error of the first approximation to the first guessed value. In this case, we get

$$x_2 \simeq x_1 + \varepsilon_1 = x_1 + \frac{1}{2}\left(\frac{b}{x_1} - x_1\right)$$

or

$$x_2 \simeq \frac{1}{2}\left(x_1 + \frac{b}{x_1}\right)$$

Thus generally at any stage in an iteration

$$x_{n+1} \simeq \frac{1}{2}\left(x_n + \frac{b}{x_n}\right) \tag{1.10}$$

In the above derivations two implicit assumptions have been made:
 (A) ε_n reduces as the iteration proceeds,
 (B) x_{n+1} converges on $x_0 = b^{\frac{1}{2}}$, as $x_{n+1} \to x_n$.
Criterion (B) is clearly seen to be valid from equation (1.10); Criterion (A) implies convergence of the function between x_1 and $x_n \to x_0$:

Proof. Let

$$\text{relative error } \delta_n = \frac{\varepsilon_n}{x_0} \tag{i}$$

Since

$$x_n = x_0 + \varepsilon_n$$

or

$$x_n = x_0(1 + \delta_n) = b^{1/2}(1 + \delta_n) \tag{ii}$$

Inserting (ii) into equation (1.10), we get

$$x_{n+1} \simeq \frac{1}{2}\left[b^{1/2}(1 + \delta_n) + \frac{b}{b^{1/2}(1 + \delta_n)}\right],$$

which reduces to

$$x_{n+1} \simeq b^{1/2}\left[1 + \frac{1}{2}\left(\frac{\delta_n^2}{1 + \delta_n}\right)\right] \tag{iii}$$

Thus the relative error at the $(n+1)$ iteration is

$$\delta_{n+1} = \frac{1}{2}\left(\frac{\delta_n^2}{1+\delta_n}\right) \qquad (iv)$$

which for small values of δ_n is clearly convergent.

Example 1.7.1 Find the value of $\sqrt{2}$ by successive iterations and estimate the error at each stage

$$x = \sqrt{2}, \quad \text{or } x^2 = 2$$

Let
$$x_1 = 1$$

then

$$x_2 = \frac{1}{2}\left(1 + \frac{2}{1}\right) = 1 \cdot 5$$

$$x_3 = \frac{1}{2}\left(1 \cdot 5 + \frac{2}{1 \cdot 5}\right) = 1 \cdot 4166$$

$$x_4 = \frac{1}{2}\left(1 \cdot 4166 + \frac{2}{1 \cdot 4166}\right) = 1 \cdot 414\,216$$

$$x_5 = \frac{1}{2}\left(1 \cdot 414\,216 + \frac{2}{1 \cdot 414\,216}\right) = 1 \cdot 414\,211\,4$$

etc.

Errors

$$\varepsilon_1 \simeq \frac{1}{2}\left[\frac{2}{1} - 1\right] \simeq 0 \cdot 5$$

$$\varepsilon_2 \simeq \frac{1}{2}\left[\frac{2}{1 \cdot 5} - 1 \cdot 5\right] \simeq -0 \cdot 0833$$

$$\varepsilon_3 \simeq \frac{1}{2}\left[\frac{2}{1 \cdot 4166} - 1 \cdot 4166\right] \simeq -0 \cdot 002\,41$$

$$\varepsilon_4 \simeq \frac{1}{2}\left[\frac{2}{1 \cdot 414\,216} - 1 \cdot 414\,216\right] \simeq 0 \cdot 000\,003\,56$$

etc.

Try this procedure to find the square roots of 5, 17·5, 33, etc., say to four decimals, and estimate the level of errors.

Case 2 Consider $x = \sqrt[3]{b}$; i.e. $x^3 = b$.

1. APPROXIMATE CORRELATIONS

Step 1. Guesstimate a first value x_1 as close to root as practical. Then

$$x_0 = x_1 + \varepsilon_1$$

and

$$x_0^3 = (x_1 + \varepsilon_1)^3 = b$$

i.e.

$$x_1^3 + 3x_1^2 \varepsilon_1 + 3x_1 \varepsilon_1^2 + \varepsilon_1^3 = b$$

Step 2. First approximation is

$$x_1^3 + 3x_1^2 \varepsilon_1 \simeq b$$

which yields

$$\varepsilon_1 \simeq \frac{1}{3}\left(\frac{b}{x_1^2} - x_1\right)$$

and subsequently

$$\varepsilon_n \simeq \frac{1}{3}\left(\frac{b}{x_n^2} - x_n\right) \tag{1.11}$$

Step 3. Second approximation is

$$x_2 \simeq x_1 + \varepsilon_1$$

i.e.

$$x_2 \simeq \frac{1}{3}\left(2x_1 + \frac{b}{x_1^2}\right)$$

or generally

$$x_{n+1} \simeq \frac{1}{3}\left(2x_n + \frac{b}{x_n^2}\right) \tag{1.12}$$

Again it is necessary to establish
 (A) that the limiting value $\to (b^{1/3})$
 (B) that the function $x_{n+1} = \phi(x_n)$ converges, i.e. that

$$\varepsilon_n > \varepsilon_{n+1} \to 0.$$

If (B) is proven, then

$$x_n \to x_{n+1} \to x_0$$

and equation (1.12) becomes

$$x_0 \to \frac{1}{3}\cdot\left(2x_0 + \frac{b}{x_0^2}\right)$$

and therefore

$$3x_0^3 \to (2x^3 + b)$$

i.e.
$$x_0^3 \to b$$
or
$$x_0 \to \sqrt[3]{b}$$

To prove (B), we proceed as for Case 1.

Let relative error $\delta_n = \dfrac{\varepsilon_n}{x_n}$

$$\therefore x_n = b^{1/3}(1+\delta_n) \tag{v}$$

Inserting (v) into (1.12), we get

$$x_{n+1} \simeq \frac{1}{3}\left\{2 \cdot b^{1/3}(1+\delta_n) + \frac{b}{\{b^{1/3}(1+\delta_n)\}^2}\right\}$$

and this simplifies to

$$x_{n+1} \simeq b^{1/3}\left\{1 + \frac{\delta_n^2(3+2\delta_n)}{3(1+\delta_n)^2}\right\}$$

Thus, the relative error is

$$\delta_{n+1} \simeq \frac{\delta_n^2(3+2\delta_n)}{3(1+\delta_n)^2} \tag{vi}$$

written as

$$\delta_{n+1} \simeq \frac{\delta_n^2}{(1+\delta_n)^2} + \frac{2}{3}\frac{\delta_n^3}{(1+\delta_n)^2}$$

we see that the first term is $< \delta_n^2$ and the second term is less than δ_n^3. Thus for small values of δ_n

$$\delta_{n+1} \ll \delta_n$$

i.e. equation (1.12) is converging.

Example 1.7.2 Apply Bruckheimer's method (Bruckheimer et al., 1968) to establish an iteration equation for the solution of equation

$$x^3 + x - b = 0$$

Estimate error levels at the iteration stages. Note that conditions for convergence have been already established at the end of Section 1.5 and also in Section 1.6 (For solution see p. 347.)

1. APPROXIMATE CORRELATIONS

Further examples for readers to attempt. Find root, check convergence and evaluate errors for

1.7.3 $x = \sin x - \cos x$

1.7.4 $\sin x = 5x - 2$

1.7.5 $3 \sin x = x + \dfrac{1}{x}$

1.7.6 $x = \sqrt{5}$ and $x = \sqrt[3]{3}$

Example 1.7.7 Find the depth of indentation of a relatively soft roller on a hard flat surface if $D = 200$ mm diameter and $a = 1$ mm $= \frac{1}{2}$ of width of contact.

Example 1.7.8 Check for convergence and give criteria for values of x relative to constants, a_1 to a_5 given:

$$a_0 \left[1 + \left(\frac{a_1}{a_2 + x} \right) \right]^{1/2} \cdot \left(a_3 - \frac{a_1}{a_2 \cdot x} \right) = a_4 (a_5 - x^2)^{1/2}$$

1.8 LEAST SQUARE METHOD

The least square method is a useful technique for determining approximate continuous functions for scattered empirical values or irregular graphs.

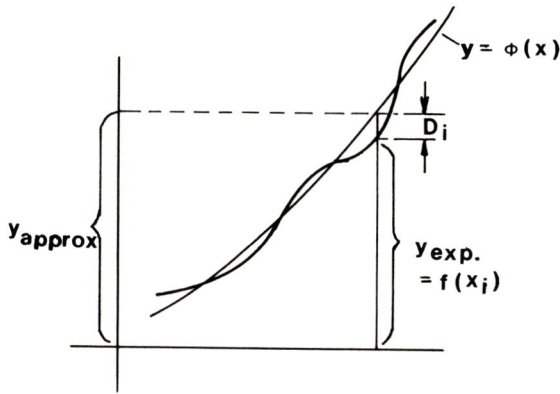

Fig. 1.5. Fitting smooth approximate curve to experimental values.

It is desired to find a smooth "best fit" curve to represent experimental results (see Fig. 1.5) of the following form:

$$y_{\text{approx}} = \phi(x) = a_0 + a_1 x + a_2 x^2 + a_3 x^3 + \ldots + a_n x^n$$

or
$$\phi(\log x) = a_0 + a_1 \log x + a_2\, 2\log x + \ldots + a_n\, n \log x \qquad (1.13)$$

(i) We proceed by selecting a number of x_i values and note the corresponding $f(x_i)$ values, i.e. from experimental graph or table.

(ii) The difference between any true and approximate value of y from the correlated function is

$$D_i = f(x_i) - \phi(x_i)$$

or

$$D_i = f(x_i) - (a_0 + a_1 x_i + a_2 x_i^2 + a_3 x_i^3 + \ldots + a_n x_i^n)]^2 \qquad (1.14)$$

(iii) Next we square all differences, D_i and obtain

$$(D_i)^2 = [f(x_i) - (a_0 + a_1 x_i + a_2 x_i^2 + \ldots + a_n x_i^n)]^2 \qquad (1.15)$$

The best fit is now defined as a curve which will make the sum of all $(D_i)^2$ a minimum, i.e.

$$S = \sum_1^n (D_i)^2 = \text{minimum} \qquad (1.16)$$

Such a minimum will be obtained when

$$\frac{\partial S}{\partial a_0} = \frac{\partial S}{\partial a_1} = \frac{\partial S}{\partial a_2} \ldots = \frac{\partial S}{\partial a_n} = 0 \qquad (1.17)$$

Thus,

$$\left.\begin{aligned}
\frac{\partial S}{\partial a_0} &= -2 \sum_1^n \{[f(x_i) - a_0 - a_1 x_i - a_2 x_i^2 - a_n x_i^n] \cdot 1\} = 0 \\
\frac{\partial S}{\partial a_1} &= -2 \sum_1^n \{[f(x_i) - a_0 - a_1 x_i - a_2 x_i^2 - a_n x_i^n] \cdot x_i\} = 0 \\
\frac{\partial S}{\partial a_2} &= -2 \sum_1^n \{[f(x_i) - a_0 - a_1 x_i - a_2 x_i^2 - a_n x_i^n] \cdot x_i^2\} = 0
\end{aligned}\right\} \qquad (1.18)$$

etc.

The above equations may be rewritten in the form

$$\left.\begin{aligned}
a_0 n + a_1 \sum_1^n x_i + a_2 \sum_1^n x_i^2 + \ldots &= \sum_1^n f(x_i) \\
a_0 \sum_1^n x_i + a_1 \sum_1^n x_i^2 + a_2 \sum_1^n x_i^3 + \ldots &= \sum_1^n [f(x_i) x_i] \\
a_0 \sum_1^n x_i^2 + a_1 \sum_1^n x_i^3 + a_2 \sum_1^n x_i^4 \ldots &= \sum_1^n [f(x_i) x_i^2]
\end{aligned}\right\} \qquad (1.18a)$$

etc.

1. APPROXIMATE CORRELATIONS

(iv) Next it is convenient to produce a table (or computer program):

Point	x_i	x_i^2	$x_i^3 ...$	$f(x_i)$	$x_i \cdot f(x_i)$	$x_i^2 \cdot f(x_i)$
1	x_1	x_1^2	$x_1^3 ...$	y_1	$x_1 \cdot y_1$	$x_1^2 \cdot y_1$
2	x_2	x_2^2	$x_2^3 ...$	y_2	$x_2 \cdot y_2$	$x_2^2 \cdot y_2$
3	x_3	x_3^2	$x_3^3 ...$	y_3	$x_3 \cdot y_3$	$x_3^2 \cdot y_3$
4	x_4	x_4^2	$x_4^3 ...$	y_4	$x_4 \cdot y_4$	$x_4^2 \cdot y_4$
n	x_n	x_n^2	$x_n^3 ...$	y_n	$x_n \cdot y_n$	$x_n^2 \cdot y_n$
$\sum n$	$\sum_1^n x_i$	$\sum_1^n x_i^2$	$\sum_1^n x_i^3 ...$	$\sum_1^n y_i$	$\sum_1^n x_i \cdot y_i$	$\sum_1^n x_i^2 \cdot y_i$

(v) The simultaneous equations (1.18a) and the table can now be used to calculate the coefficients a_0, a_1, a_2, etc.

(vi) It is advisable, and informative, to establish the degree of accuracy, or "goodness of fit" of the correlation finally established. The "mean square error" or the RMS of the residuals is obtained from

$$\text{RMS} = \left\{ \sum_1^n [f(x_i) - \phi(x_i)]^2 \Big/ n \right\}^{1/2} \quad (1.19)$$

or expressed as a percentage

$$\text{RMS}(\%) = \frac{\text{RMS} \cdot 100}{f(x_{\text{mean}})} \quad (1.19)$$

where x_{mean} = arithmetic mean of x_i.

Note. If a "best straight line" is being sought, only coefficients a_0 and a_1 need to be established. For a quadratic curve a_0, a_1 and a_2 are needed and so on for higher order curves.

Example 1.8.1
Given table of x and y values:

x	0	1	2	3	4	5
y	0·82	3·46	5·92	8·57	11·07	13·64

(a) Find "best straight line".
(b) Check RMS of accuracy.
(For solution see p. 347.)

Example 1.8.2

Given table of x and y values:

x	0	1	2	3	4	5	6
y	-0.01	0.37	0.71	0.98	1.28	1.52	1.69

(a) Find best straight line.
(b) Find best quadratic curve.
(c) Calculate RMS for (a) and (b).
(d) Plot points and curves
(For solution see p. 348.)

1.9 FINITE DIFFERENCE EQUATIONS

1.9.1 Single variable. Consider an empirical curve with no known continuous function (or where such a function is complex and difficult to differentiate). We may wish to determine

  (i) The slope ($\equiv$ first derivative);
  (ii) The curvature ($\equiv$ second derivative);
  (iii) The rate of change of curvature ($\equiv$ third derivative), etc. at any particular point, (x_0, y_0) (see Fig. 1.6).

If we mark off the smallest convenient intervals on both sides of x_0 and read off corresponding y-values, we can evaluate approximate 1st, 2nd, 3rd, etc. finite differences as follows:

	1st Difference	2nd Difference	3rd Difference
y_3			
	$(y_3 - y_2)$		
y_2			
	$(y_2 - y_1)$		
y_1		$(y_2 - 2y_1 + y_0)$	
	$(y_1 - y_0)$		
y_0	$\tfrac{1}{2}(y_1 - y_{-1})$	$(y_1 - 2y_0 + y_{-1})$	$\tfrac{1}{2}(y_2 - 2y_1 + 2y_{-1} - y_{-2})$
	$(y_0 - y_{-1})$		
y_{-1}		$(y_0 - 2y_{-1} + y_{-2})$	
	$(y_{-1} - y_{-2})$		
y_{-2}			
	$(y_{-2} - y_{-3})$		
y_{-3}			

1. APPROXIMATE CORRELATIONS

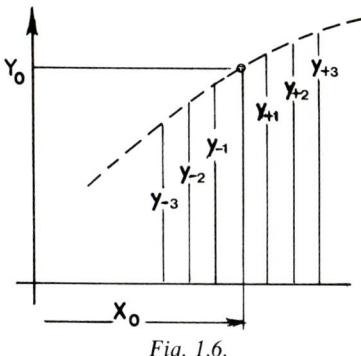

Fig. 1.6.

Referring to Fig. 1.6, we obtain

$$\frac{dy}{dx_0} \simeq \frac{1}{2h}(y_1 - y_{-1})$$

$$\frac{d^2 y}{dx_0^2} \simeq \frac{1}{h^2}(y_1 - 2y_0 + y_{-1})$$

$$\frac{d^3 y}{dx_0^3} \simeq \frac{1}{2h^3}(y_2 - 2y_1 + 2y_{-1} - y_{-2}) \tag{1.20}$$

etc.

1.9.2 Two variables. A function $z = f(x, y)$ represents a surface. Such a surface may be divided into a net or square mesh in the x–y plane as shown in Fig. 1.7. The z-values corresponding to each point in the mesh must be established. For any point (x_0, y_0), one can then estimate the approximate partial derivatives as follows:

$$\left(\frac{\partial z}{\partial x}\right)_0 \simeq \frac{1}{2h}(z_1 - z_3)$$

$$\left(\frac{\partial z}{\partial y}\right)_0 \simeq \frac{1}{2h}(z_2 - z_4)$$

$$\left(\frac{\partial^2 z}{\partial x^2}\right)_0 \simeq \frac{1}{h^2}(z_1 - 2z_0 + z_3)$$

$$\left(\frac{\partial^2 z}{\partial y^2}\right)_0 \simeq \frac{1}{h^2}(z_2 - 2z_0 + z_4) \tag{1.21}$$

etc.

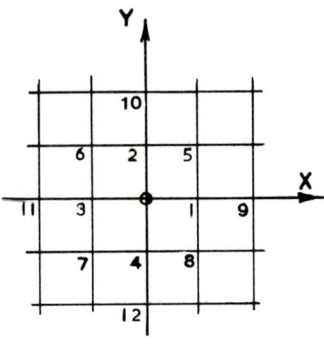

Fig. 1.7.

Example 1.9.1

Use finite differences to determine the deflections of a simply supported beam with a central load. Having found the deflections, determine
   (a) the slope as near to the ends as possible,
   (b) the curvature at the centre.

Divide the beam length into a number of equal parts to suit (say 6 parts).

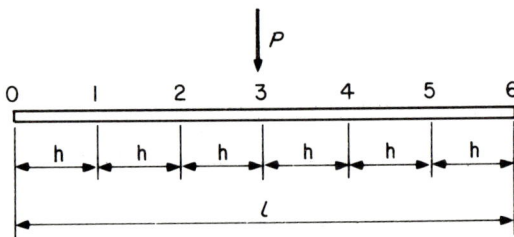

From basic beam theory, we have

$$\frac{d^2 y}{dx^2} = \frac{M}{EI} \qquad \text{(i)}$$

where

M = bending moment at x
y = deflection at x
E = Young's modulus of elasticity
I = 2nd moment of area of beam section

Hence

$$\begin{aligned} M_0 &= M_6 = 0 \\ M_1 &= M_5 = Pl/12 \\ M_2 &= M_4 = Pl/6 \\ M_3 &= \ldots Pl/4 \end{aligned} \qquad \text{(ii)}$$

Referring to equation (1.20), the 2nd approximate derivative can be expressed as a function of the deflections, viz. for point 3, $h = \frac{1}{6}l$.

$$h^2 \cdot \left(\frac{d^2 y}{dx^2}\right)_{(3)} \simeq (y_2 - 2y_3 + y_2) \simeq \frac{h^2 \cdot M_3}{EI} \quad \text{(iii)}$$

$$\simeq \frac{Pl^3}{144EI}$$

for points 2 and 4

$$h^2 \cdot \left(\frac{d^2 y}{dx^2}\right)_{(2 \text{ and } 4)} \simeq (y_3 - 2y_2 + y_1) \simeq \frac{h^2 \cdot M_2}{EI} \quad \text{(iv)}$$

$$\simeq \frac{Pl^3}{216EI}$$

for points 1 and 5

$$h \cdot \left(\frac{d^2 y}{dx^2}\right)_{(1 \text{ and } 5)} \simeq (y_2 - 2y_1 + y_0) \simeq \frac{h^2 \cdot M_1}{EI} \quad \text{(v)}$$

$$\simeq \frac{Pl^3}{432EI}$$

Solving the three simultaneous equations, and noting that $y_0 = y_6 = 0$ and $y_2 = y_4$ and $y_1 = y_5$, we get

$$y_3 = -0.022 \frac{Pl^3}{EI}$$

$$y_2 = -0.0185 \frac{Pl^3}{EI} \quad \text{(vi)}$$

$$y_1 = -0.0104 \frac{Pl^3}{EI}$$

It may be of interest to note the level of accuracy achieved by using 2 to 6 points as compared to the theoretical values.

No. of points	y_{max} by finite differences	Deviation from exact value (%)
2	-0.0313	50.5
4	-0.0234	12.5
6	-0.0220	5.78

y_{max}, theoretical $= -0.0208 \dfrac{Pl^3}{EI}$.

(a) To determine the slope at point **1**, we use equation (1.20). Thus,

$$\left(\frac{dy}{dx}\right)_{(1)} \simeq \frac{1}{2h}(y_2 - y_0) \simeq \tfrac{3}{1}(-0{\cdot}0185 - 0)\frac{Pl^3}{EI}$$

or

$$\left(\frac{dy}{dx}\right)_{(1)} \simeq -0{\cdot}0555\,\frac{Pl^2}{EI} \tag{vii}$$

(Note that the exact solution gives

$$\left(\frac{dy}{dx}\right)_{(1)} = -0{\cdot}0486\,\frac{Pl^2}{EI}$$

i.e. the approximate answer is 14% up.)

If we determine the slope at $l/12$ from the end, we get

$$\left(\frac{dy}{dx}\right)_{(\frac{1}{2})} \simeq \frac{1}{2h}(-0{\cdot}0104 - 0)\frac{Pl^3}{EI} \simeq -0{\cdot}062\,\frac{Pl^2}{EI}$$

and exact solution yields

$$\left(\frac{dy}{dx}\right)_{\frac{1}{2}(0-1)} = -0{\cdot}059\,\frac{Pl^2}{EI}$$

i.e. the approximate answer is 5·7% higher.

(b) To determine the curvature at point (3) (centre) we have

$$\left(\frac{d^2 y}{dx^2}\right)_{(3)} \simeq \frac{1}{h^2}(y_4 - 2y_3 + y_2)\frac{Pl^3}{EI}$$

$$\simeq \frac{1}{h^2}(2y_2 - 2y_3)\cdot\frac{Pl^3}{EI}$$

$$\simeq \frac{36}{l^2}[-2(0{\cdot}0185) + 2(0{\cdot}022)]\frac{Pl^3}{EI}$$

$$\simeq 0{\cdot}252\cdot\frac{Pl}{EI} \tag{viii}$$

The exact solution is

$$\left(\frac{d^2 y}{dx^2}\right)_{(3)} = 0{\cdot}25\cdot\frac{Pl}{EI}$$

Therefore, our approximate answer is +0·8% in error.

Example 1.9.2 Determine by "Finite Differences" the distribution of the shear-stress function, ϕ, for a square bar section in torsion.

1. APPROXIMATE CORRELATIONS

Notes. Relationships between the stress function ϕ and the shear stresses resulting from torsional deformation of bars of various non-circular sections have been developed by St Venant, Poisson and others. Assuming that warping takes place freely, they showed that for equilibrium,

$$\frac{\partial \tau_{yz}}{\partial y} + \frac{\partial \tau_{xz}}{\partial x} = 0 \tag{a}$$

and for deformation compatibility

$$\frac{\partial \tau_{xz}}{\partial y} - \frac{\partial \tau_{yz}}{\partial x} = -2G\theta \tag{b}$$

A stress function ϕ, defined by the following relationships, satisfies both equation (a) and (b), thus:

$$\tau_{xz} = \frac{\partial \phi}{\partial y} \quad \text{and} \quad \tau_{yz} \leqslant -\frac{\partial \phi}{\partial x} \tag{c}$$

Differentiating (c) we get

$$\text{from (a)} \quad \frac{\partial^2 \phi}{\partial x \, \partial y} - \frac{\partial^2 \phi}{\partial x \, \partial y} = 0 \ldots = \text{true} \tag{d}$$

$$\text{from (b)} \quad \frac{\partial^2 \phi}{\partial x^2} + \frac{\partial^2 \phi}{\partial y^2} = -2G\theta \tag{e}$$

Having established the relationship between the shear stress function ϕ and the angle of twist θ, we can now use the finite difference method to calculate ϕ values for any point in the section:
1. Draw the square section to scale.
2. Divide area into a grid of, say 16 and 36 small squares (see Figs 1.8 and 1.9). N.B. Due to symmetry points in one quadrant only need to be dealt with.

Case A

16 square grid, $h_A = \frac{1}{4}a$. From equation (1.21) we get for point (0)

$$(-2h_A^2 G\theta)_0 = [(\phi_1 - 2\phi_0 + \phi_3) + (\phi_4 - 2\phi_0 + \phi_2)],$$

and since

$$\phi_1 = \phi_2 = \phi_3 = \phi_4 \quad \text{(by symmetry)}$$

we have

$$(-2h_A^2 G\theta)_0 = [4\phi_1 - 4\phi_0]; \tag{Ai}$$

for point (1)

$$(-2h_A^2 G\theta)_1 = [(\phi_9 - 2\phi_1 + \phi_0) + (\phi_8 - 2\phi_1 + \phi_5)]$$

and since

$$\phi_9 = 0 \quad \text{and} \quad \phi_8 = \phi_5;$$

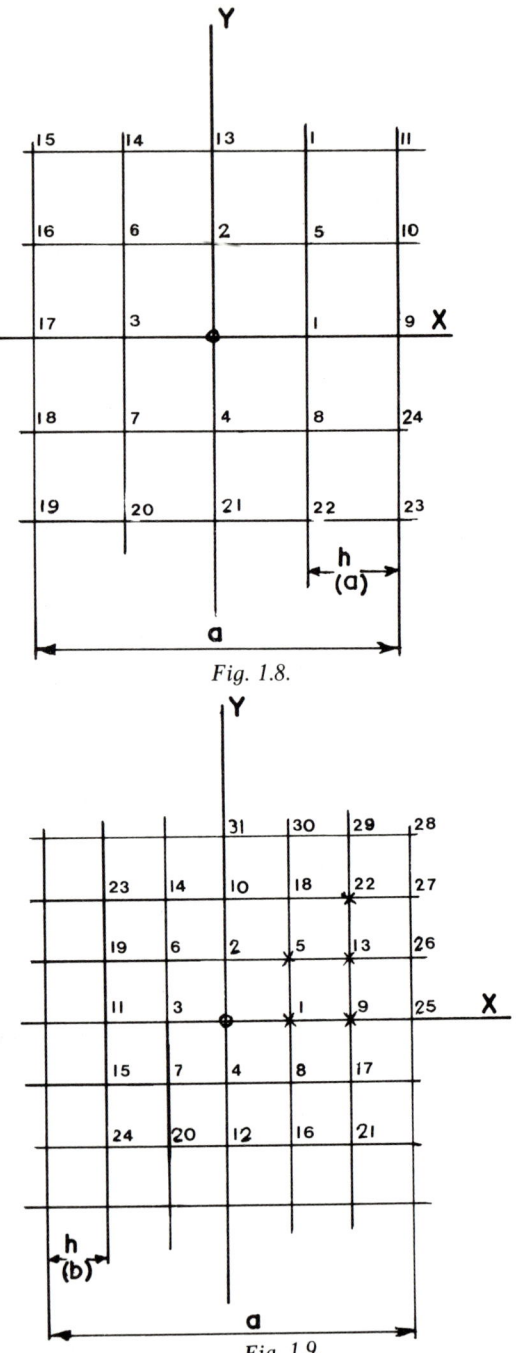

Fig. 1.8.

Fig. 1.9.

we have
$$(-2h_A^2 G\theta)_1 = [\phi_0 - 4\phi_1 + 2\phi_5]. \tag{Aii}$$
For point (5), we get
$$(-2h_A^2 G\theta)_5 = [2\phi_1 - 4\phi_5] \tag{Aiii}$$
The above three simultaneous equations may be solved and yield

$$\left. \begin{array}{l} \phi_{1_A} = 1.75 h_A^2 G\theta \\ \phi_{0_A} = 2.25 h_A^2 G\theta \\ \phi_{5_A} = 1.375 h_A^2 G\theta \\ \phi_{9_A} = 0 \end{array} \right\} \text{(for a 16 square matrix)} \tag{Aiv}$$

Case B
36 square matrix $h_B = a/6$. Following the same procedure as for Case A, we get six simultaneous equations which yield for ϕ_0
$$\phi_{0_B} = 5.1923 h_B^2 G\theta \tag{Bi}$$
Comparing ϕ_{0_A} with ϕ_{0_B} we have
$$\frac{\phi_{0_A}}{\phi_{0_B}} = \frac{2.25(h_A)^2}{5.1923(h_B)^2} = \frac{2.25}{5.1923}\left(\frac{6}{4}\right)^2 = \frac{2.25}{5.1925}\left(\frac{3}{2}\right)^2$$
$$\therefore \phi_{0_A} = 0.974\ \phi_B; \text{ or } \phi_{0_B} = 1.027\ \phi_{0_A}$$
i.e. the finer grid gives a ϕ_0 value 2.7% higher.

Table 1.2. Approximate computations.

Function	Relative error does not exceed when x varies between		
	0.1%	1%	10%
$\sin x \simeq x$	$\pm 0.077 = \pm 4.4°$	$\pm 0.245 = \pm 14.0°$	$\pm 0.786 = \pm 45.0°$
$\sin x \simeq x - \dfrac{x^3}{6}$	$\pm 0.580 = \pm 33.2°$	$\pm 1.005 = \pm 57.6°$	$\pm 1.632 = \pm 93.5°$
$\cos x \simeq 1$	$\pm 0.045 = \pm 2.6°$	$\pm 0.141 = \pm 8.1°$	$\pm 0.451 = \pm 25.8°$
$\cos x \simeq 1 - \dfrac{x^2}{2}$	$\pm 0.386 = \pm 22.1°$	$\pm 0.662 = \pm 37.9°$	$\pm 1.036 = \pm 59.3°$
$\tan x \simeq x$	$\pm 0.054 = \pm 3.1°$	$\pm 0.172 = \pm 9.8°$	$\pm 0.517 = \pm 29.6°$
$\tan x \simeq x + \dfrac{x^3}{3}$	$\pm 0.293 = \pm 16.8°$	$\pm 0.519 = \pm 29.7°$	$\pm 0.895 = \pm 51.3°$
$(a^2 + x)^{1/2} \simeq a + \dfrac{x}{2a}$	$\begin{cases} 0.093a^2 \\ -0.085a^2 \end{cases}$	$\begin{cases} +0.328a^2 \\ -0.247a^2 \end{cases}$	$\begin{cases} +1.545a^2 \\ -0.607a^2 \end{cases}$
$(a^2 + x)^{-1/2} \simeq \dfrac{1}{a} - \dfrac{x}{2a^3}$	$\begin{cases} +0.052a^2 \\ -0.051a^2 \end{cases}$	$\begin{cases} +0.166a^2 \\ -0.157a^2 \end{cases}$	$\begin{cases} +0.530a^2 \\ -0.448a^2 \end{cases}$
$\dfrac{1}{a+x} \simeq \dfrac{1}{a} - \dfrac{x}{a^2}$	$\pm 0.031a$	$\pm 0.099a$	$\pm 0.301a$
$e^x \simeq 1 + x$	± 0.045	$\begin{cases} +0.148 \\ -0.134 \end{cases}$	$\begin{cases} +0.502 \\ -0.375 \end{cases}$
$\ln(1+x) \simeq x$	± 0.002	± 0.020	$\begin{cases} +0.230 \\ -0.176 \end{cases}$

Table 1.3. Relationships between trigonometric and exponential functions.

$$\sin x = \frac{e^{xi} - e^{-xi}}{2i}$$

$$\cos x = \frac{e^{xi} + e^{-xi}}{2}$$

$$\sinh x = \frac{e^{x} - e^{-x}}{2}$$

$$\cosh x = \frac{e^{x} + e^{-x}}{2}$$

REFERENCES

Balfour, A. and McTernan, A. J. (1967). "Numerical Solutions to Equations." Heinemann Educational Books, London.

Bronstein, I. N. and Semendyayev, K. A. (1973). "A Guide Book to Mathematics." Springer Verlag, New York.

Bruckheimer, M., Gowar, N. W. and Scraton, R. A. (1968). "Mathematics for Technology—A New Approach." Chatto and Windus (Educational), London.

Morton, B. R. (1969). "Numerical Approximations." Routledge and Kegan Paul Ltd, London.

Chapter 2

Theory of Errors

The professional engineer baptized in the field of engineering practice is all too well aware that there are no exact data or "spot on" values. Every measurement, every numerical value given is valid only within certain limits or tolerances. It therefore follows that:
  (a) Computed results cannot be more accurate than the accuracy of the data used;
  (b) It is needless to carry out calculations to a greater degree of accuracy (i.e. number of decimals) than desirable or sensible with the data available;
  (c) The accuracy of functions derived from toleranced data should be determined.

2.1 DEFINITION OF ERRORS
(Bronstein and Semendyayev)

The *absolute error* may be defined as the difference between the correct and the approximate values, or the difference between the mean and the extreme values (in either direction relative to the mean). Thus if x = correct value and a = approximate value then the absolute error $\Delta a = (x - a)$ and

$$\Delta a \underset{\text{limit}}{\rightarrow} = (x - a) \underset{\text{limit}}{\rightarrow} \qquad (2.1)$$

The *relative error* is defined as the ratio of the absolute error and the approximate value, thus

$$\delta a = \frac{\Delta a}{a}$$

or

$$\delta a\% = \frac{\Delta a}{a} \cdot 100 \qquad (2.2)$$

2.2 ROUNDING OFF LAST FIGURES

It is useful to follow some common-sense rules and conventions when rounding off last decimal or digits in calculations:
  (i) Retain correct digits only in data or calculations.
  (ii) If the first discarded number is greater than 5, the preceding digit should be increased by 1.
  (iii) If the discarded figure equals 5, the last figure should be left, or made even.
  (iv) If the discarded figure is equal or less than 4, the preceding digit should be left unaltered.

These guide lines assist in averaging out cumulative errors when several numbers are involved in calculations.

2.3 OPERATIONS WITH APPROXIMATE NUMBERS

It is self-evident that the results of calculations with approximate numbers will also be approximate. What one wants to know is the degree of approximation in both. The following theorems relate the errors of the results to the errors of the data used.

Rule 1
The absolute error of a *sum* equals the sum of the absolute errors of the terms:

$$\Delta a_\Sigma = \sum_1^n \Delta a_i \tag{2.3}$$

e.g.

$$\Delta(a+b) = \Delta a + \Delta b \tag{2.4}$$

Rule 2

(a) The absolute error of a difference of *independent* terms equals the sum of the absolute errors of the terms, i.e.

$$\Delta(a-b) = \Delta a + \Delta b \text{ if (a) and (b) are independent terms} \tag{2.5}$$

(b) The absolute error of a difference of *correlated* terms equals the difference of the absolute errors of the terms, i.e.

$$\Delta(a-b) = \Delta a - \Delta b \text{ if (a) and (b) are correlated terms} \tag{2.6}$$

2. THEORY OF ERRORS

Proofs
For independent terms,
$$a = (x_a + \Delta a)$$
$$b = (x_b - \Delta b)$$

(Since terms are independent, the errors may be in opposite direction.)
$$\therefore (a-b) = (x_a + \Delta a) - (x_b - \Delta b)$$

or
$$(a-b) - (x_a - x_b) = \Delta a + \Delta b$$
$$\therefore \Delta(a-b) = \Delta a + \Delta b$$

For correlated terms,
$$a = (x_a + \Delta_a)$$
$$b = (x_b + \Delta_b)$$
$$\therefore (a-b) = (x_a + \Delta a) - (x_b + \Delta b)$$

or
$$(a-b) - (x_a - x_b) = \Delta a - \Delta b$$
$$\Delta(a-b) = (\Delta a - \Delta b)$$

Rule 3
The relative error of a sum lies between the greatest and least relative errors of the terms

$$\delta a_\Sigma = \frac{\Delta a_\Sigma}{a_\Sigma} = \frac{\Delta a_1 + \Delta a_2 + \Delta a_3 + \ldots}{a_1 + a_2 + a_3 + \ldots}$$
$$= \frac{n \cdot \Delta a_{mean}}{n \cdot a_{mean}} = \frac{\Delta a_{mean}}{a_{mean}} \qquad (2.7)$$

Rule 4
The relative error of a product equals the sum of the relative errors of the approximate factors
$$\delta(a \cdot b) = \delta a + \delta b$$

Proof
$$(x_a \cdot x_b) = (a + \Delta a)(b + \Delta b)$$
$$= (a \cdot b) + a\Delta b + b\Delta a + \Delta a \Delta b$$

$\therefore (x_a \cdot x_b - a \cdot b) \simeq a\Delta b + b\Delta a$ neglecting second order small terms, i.e.

$$\frac{\Delta(a \cdot b)}{a \cdot b} = \frac{\Delta b}{b} + \frac{\Delta a}{a}$$

i.e.

$$\delta(a \cdot b) = \delta a + \delta b \qquad (2.8)$$

Note. For correlated variables the errors are in the same direction and for independent ones they can be in either direction; the largest error arises when they are in the same direction as in proof above.

Rule 5

The relative error of a quotient is equal to
  (a) the sum of the relative errors of the numerator and denominator, for independent variables with errors of ($\pm$) values;
  (b) the difference of the relative errors of the numerator and denominator, for correlated variables with errors of ($\pm$) values.

Proofs

For correlated variables (Δa and Δb in same direction):

$$\frac{a}{b} = \frac{x_a - \Delta a}{x_b - \Delta b}$$

$$\therefore a(x_b - \Delta b) = b(x_a - \Delta a)$$

or

$$\frac{x_b}{b} - \frac{x_a}{a} = \frac{\Delta b}{b} - \frac{\Delta a}{a}$$

or

$$\left[\frac{x_a}{a} - \frac{x_b}{b}\right] = \left[\frac{\Delta a}{a} - \frac{\Delta b}{b}\right] \qquad (i)$$

Now,

$$\Delta\left(\frac{a}{b}\right) = \left[\frac{x_a}{x_b} - \frac{a}{b}\right]$$

$$\therefore \delta\left(\frac{a}{b}\right) = \left[\left(\frac{x_a}{x_b}\right) - \left(\frac{a}{b}\right)\right] / \left(\frac{a}{b}\right)$$

But

$$\left[\left(\frac{x_a}{x_b}\right) - \left(\frac{a}{b}\right)\right] / \left(\frac{a}{b}\right) = \left[\left(\frac{x_a}{a} - \frac{x_b}{b}\right)\left(\frac{a}{x_b}\right)\right] / \left(\frac{a}{b}\right) \simeq \left(\frac{x_a}{a}\right) - \left(\frac{x_b}{b}\right)$$

2. THEORY OF ERRORS

since
$$\left(\frac{a}{x_b}\right) \simeq \left(\frac{a}{b}\right)$$

Therefore
$$\delta\left(\frac{a}{b}\right) \simeq \left[\frac{x_a}{a} - \frac{x_b}{b}\right] \qquad \text{(ii)}$$

Thus from (i) and (ii), we get
$$\delta\left(\frac{a}{b}\right) \simeq (\delta a - \delta b) \quad \text{(for correlated variables)} \qquad (2.9)$$

(b) For *independent* variables, Δb can be in any direction relative to Δa. For the worst case, therefore,
$$\delta\left(\frac{a}{b}\right) = \delta a + \delta b \qquad (2.10)$$

Rule 6
The relative error of the nth power of an approximate number is n times the relative error of the base number, i.e.
$$\delta(a^n) = n \times \delta a$$

Proof
Let accurate value = x and approximate value = a. Then by definition,
$$\Delta(a^n) = (x^n) - (a^n)$$
or
$$\delta(a^n) = \frac{x^n - a^n}{a^n} = \left[\left(\frac{x}{a}\right)^n - 1\right] \qquad \text{(i)}$$

Also
$$x = a + \Delta a$$
$$\therefore x^n = (a + \Delta a)^n = a^n \cdot \left[1 + \frac{\Delta a}{a}\right]^n$$

Using Taylor's expansion, we get
$$x^n = a^n \left[1 + n \cdot \left(\frac{\Delta a}{a}\right) + \frac{n(n-1)}{2!} \cdot \left(\frac{\Delta a}{a}\right)^2 + \dots\right]$$

The first approximation for x^n then is
$$x^n \simeq a^n \left[1 + n\frac{\Delta a}{a}\right]$$

36 APPROXIMATE METHODS IN ENGINEERING DESIGN

or

$$\left[\left(\frac{x}{a}\right)^n - 1\right] \simeq n \cdot \left(\frac{\Delta a}{a}\right) \simeq n \cdot \delta a \quad \text{(ii)}$$

Comparing (i) and (ii), we have

$$\delta(a^n) \simeq n \cdot \delta a \quad (2.11)$$

2.4 ERROR OF A FUNCTION

Where functions other than the preceding forms arise, use is made of derivations and/or finite difference relationships. This is best illustrated by some typical examples.

Example

In a right-angled triangle the sides are $(a \pm \Delta a)$ and $(b \pm \Delta b)$. Find the accuracy of the angle ϕ, as shown:

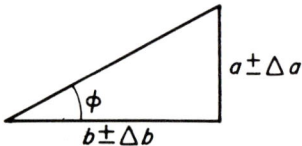

ϕ is a function of a and b, expressed as

$$\tan \phi = \left(\frac{a}{b}\right)$$

Using differentiation, we know that

$$d \tan \phi = (1 + \tan^2 \phi) \cdot d\phi \quad \text{(i)}$$

Alternatively,

$$d \tan \phi = d\left(\frac{a}{b}\right) = \frac{b \cdot da - a \cdot db}{b^2} \quad \text{(ii)}$$

Combining (i) and (ii) we get

$$d\phi = \frac{b \cdot da - a \cdot db}{b^2(1 + \tan^2 \phi)} = \frac{b \cdot da - a \cdot db}{b^2\left[1 + \left(\frac{a}{b}\right)^2\right]}$$

or

$$d\phi = \frac{b\cdot da - a\cdot db}{(a^2 + b^2)} \qquad \text{(iii)}$$

The corresponding finite difference equation is

$$\Delta\phi = \frac{b\cdot \Delta a \pm a\cdot \Delta b}{(a^2 + b^2)} \qquad \text{(iv)}$$

or expressed in terms of relative errors of the sides of the triangle

$$\Delta\phi = \frac{\delta a \pm \delta b}{\left(\dfrac{a}{b}\right) + \left(\dfrac{b}{a}\right)} \qquad \text{(v)}$$

Note. The ($\pm$) signs have been introduced in (iv) and (v) to allow for the possibility that the errors in a and b may be of opposite signs, in which case δa and δb can be additive and $\Delta\phi$ then is larger.

Example 2.1
Consider function $V = r^2 h$. If the errors of r and h are $\pm\Delta r$ and $\pm\Delta h$, respectively, what will be the error of V if r and h are independent variables?

Example 2.2
Given the relationship

$$z = \sqrt{\left(\frac{x}{1+y}\right)}$$

where x and y are independent variables, calculate the error in z if the errors of x and y are $\pm\Delta x$ and $\pm\Delta y$ respectively.

Example 2.3
$$y = x^3 - 2x^2 + x + 2$$
If x is an approximate number with error $\pm\Delta x$, what is the approximate value of y?

Example 2.4
Find the accuracy of the radius of a circle if the measurements of the ordinate and abscissa have known accuracies of $\pm\Delta y$ and $\pm\Delta x$ respectively.

Example 2.5
A circle has a diameter $D = 5\cdot 92$ with an absolute error of $\Delta D = \pm 0\cdot 005$. Find (a) the error of the circumference and (b) the error of the area.

Example 2.6
The dimensions of a triangle are: base = 10.35 ± 0.15 and height = 8.02 ± 0.05. Estimate the error in the area of the triangle.

Example 2.7
If $\tan \alpha = 0.818 \pm 0.002$, what is the tolerance (error) of α?
(For answers see p. 351.)

2.5 THE INVERSE PROBLEM

The inverse problem arises when a desired accuracy is stipulated for the result or for an assembly and the admissible errors of the data provided or the tolerances in the dimensions of the parts produced must be determined. Since more than one parameter or dimension is usually involved, a unique answer would require additional information about the data or the results.

We can illustrate the problem by a relatively simple example:

Example
One adjacent side of a right-angled triangle is about three times greater than the other. With what accuracy should the sides be measured so that the error of the angle determined by the tangent does not exceed $\pm 0°1'$?

We have the relationship
$$\tan \phi = \frac{a}{b}$$

From the example in Section 2.4, we have
$$\Delta \phi = \pm \frac{b \cdot \Delta a - a \cdot \Delta b}{a^2 + b^2}$$

Since $b \simeq 3a$
$$\Delta \phi = \pm \frac{a(3 \cdot \Delta a - \Delta b)}{10 a^2}$$

i.e.
$$\Delta \phi = \pm \frac{3 \cdot \Delta a - \Delta b}{10 a} \qquad (i)$$

There are three possible combinations of measurement errors of a and b.

Case 1

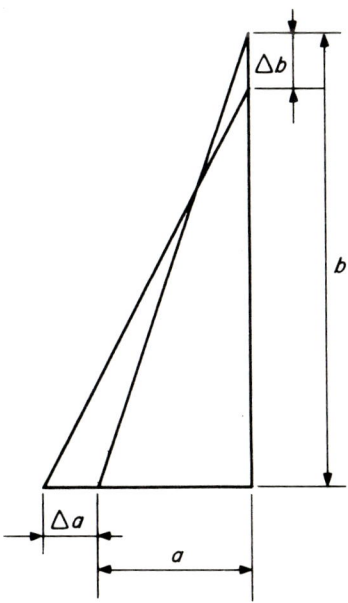

Errors in measurements are in opposite directions and of same magnitude for a and b

$$|\Delta a| = |\Delta b|$$

$$\therefore \Delta\phi = \pm \frac{-3\Delta a - \Delta b}{10a}$$

$$= \pm 0.40 \cdot \left(\frac{\Delta a}{a}\right) \text{ radians}$$

Now,

$$\Delta\phi = 0°1' = 0.000\,29 \text{ radians}$$

$$\therefore \delta a = 0.0007 = 0.07\% \text{ relative error}$$

i.e.

$$|\Delta a| = |\Delta b| = 0.0007 \cdot (a)$$

or

$$= \tfrac{1}{3}(0.0007) \cdot (b) = 0.000\,233 \cdot (b)$$

Case 2

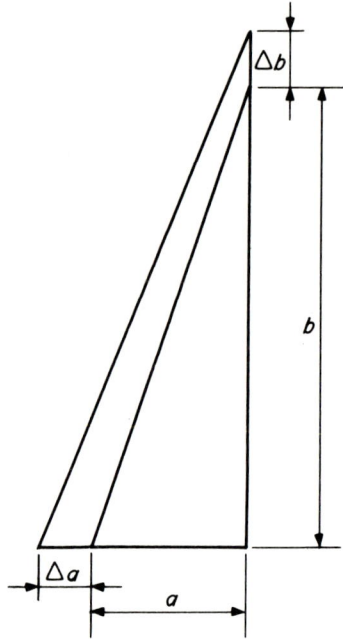

Errors of a and b are in the same direction and of equal magnitude, we have

$$\Delta\phi = \pm \frac{-3\Delta a - \Delta b}{10a}$$

$$= \pm \frac{2\Delta a}{10a}$$

$$\Delta\phi = \pm 0.20 \left(\frac{\Delta a}{a}\right)$$

$$\therefore \delta a = 5 \cdot \Delta\phi$$

or

$$\delta a = 5 \times 0.000\,29 = 0.145\% \text{ of } a$$

and

$$\Delta a = \Delta b = 0.001\,45 \cdot a$$

and

$$\delta b = 0.0483\% \text{ of } b$$

Case 3
Errors will vary in proportion to a and b respectively, i.e. relative errors will be constant.

2. THEORY OF ERRORS

Then for
(i) both errors in same direction,

$$\Delta\phi = \frac{3\Delta a - \Delta b}{10a} = \frac{3\Delta a - 3\Delta a}{10a} = 0$$

i.e. no effect on accuracy of angle Φ at all.
(ii) errors in opposite direction,

$$\Delta\phi = \frac{3\Delta a + 3\Delta a}{10a} = 0.60 \cdot \delta a$$

i.e.

$$0.000\,29 = 0.60 \cdot \delta a$$

$$\delta a = -\delta b = 0.048\%$$

Example 2.8
Determine the variation in stiffness of a cantilever leaf spring resulting from tolerances of its dimensions and the Young's modulus of the material. The following data concerning the leaf spring are available:

    length = 100 mm ± 2.0
    width = 20 mm ± 0.50
    thickness = 3 mm ± 0.04
    Young's modulus = $21 \cdot 10^4$ N mm$^{-2} \pm 3.33\%$

Example 2.9
Wallingford Spring Pressings Ltd produce Belleville disc springs and recommend the formulae below for calculations of suitable springs for functions required:

P = axial load (newtons)

P_{flat} = load to flatten disc

f = deflection (mm)

f/h = ratio deflection/free height

$s_{(x)}$ = stress at (x) in (N mm^{-2})

K_1, K_2, K_3 = factors depending on (D/d) and E ⎫
                                                                                                                     ⎬ (given in tables, p. 356)
C = factor depending on (h/t) and (f) ⎭

$$P_{flat} = K_1 \cdot 10^6 \cdot \left(\frac{h \cdot t^3}{D^2}\right) \text{ (newtons)} \qquad\qquad\qquad (i)$$

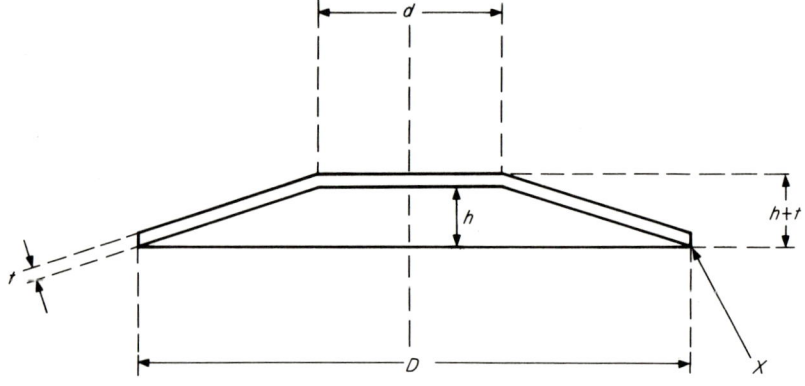

$P = C \cdot P_{\text{flat}}$ (ii)

$C = a\left[\left(1 - \dfrac{3a}{2} + \dfrac{a^2}{2}\right)\left(\dfrac{h}{t}\right)^2 + 1\right]$ (iii)

$a = f/h$ (iv)

$S_{(x)} = K_1 \cdot \dfrac{f}{D^2} \cdot 10^6 \cdot \left[(2K_3 - K_2)\left(h - \dfrac{f}{2}\right) + K_3 \cdot t\right] (N\ \text{mm}^{-2})$ (v)

Using correlations (i) and (ii) above, (v) can be written as

$S_{(x)} = \dfrac{P}{c} \cdot \dfrac{x}{t^2}\left[\dfrac{h}{t}\left(1 - \dfrac{x}{2}\right) \cdot (2K_3 - K_2) + K_3\right] (N\ \text{mm}^{-2})$ (vi)

where $x = (f/h)$

Select a suitable spring from the maker's catalogue with the deterministic dimensions given. Assign practical tolerance values bearing in mind methods of manufacture, and estimate their effect on P, P_{flat}, $S_{(x)}$ and the stiffness of the disc spring.

Example 2.10

Torque can be transmitted between a wheel and a shaft by relying only on an interference fit between them. Given the relationship for the radial pressure due to interference as

$$p_{\text{rad}} = \dfrac{E \cdot \gamma_{\text{rad}}(D^2 - d^2)}{d \cdot D^2}$$

for a wheel and shaft of similar materials, where

D = hub outer diameter

d = shaft diameter

2. THEORY OF ERRORS

γ_{rad} = radial interference between shaft and bore in the hub

p_{rad} = radial pressure

(a) Establish the variation in transmittable torque as a function of tolerances in D, d and γ_{rad}, assuming E and coefficient of friction to be constant. (b) Using the relationship(s) from (a), calculate the transmitted torque variation for

$$D = 98 \pm 0.120 \text{ mm}$$
$$d = 50 \text{ mm}$$

and ISO interference fit,

$$\text{H7 for hole}$$
$$\text{S7 for shaft}$$

Example 2.11

The semi-empirical correlation for forced convection heat transfer between a fluid flowing inside a tube and the tube wall is

$$Q = 0.027 \cdot \left[\frac{\mu_{bf}}{\mu_s}\right]^{0.14} \cdot \text{Re}^{0.80} \cdot \text{Pr}^{0.33} \cdot \frac{k}{d} \cdot (t_w - t_{bf}) \text{ in BTU ft}^{-2} \text{ h}^{-1}$$

where

μ_{bf} = viscosity of bulk fluid
μ_s = viscosity of fluid at wall temperature
Re = Reynolds number
Pr = Prandtl number
k = conductivity of fluid
d = inside diameter of tube
t_w = wall temperature
t_{bf} = bulk-fluid temperature

It is acceptable to assume that μ_{bf}, μ_s, k and C_p (specific heat) vary insignificantly with small variations in temperature.

Estimate variations in Q allowing for tolerances or errors in
(1) tube and fluid bulk temperatures (t_w and t_{bf})
(2) tube bore diameter (d)
(3) fluid density (ρ)
(4) fluid velocity (v)

(For solutions to Examples 2.8 to 2.11, see pp. 354–362.)

2.6 INFLUENCE ON DESIGN

The design of a system of components and parts cannot be undertaken satisfactorily without a careful analysis of the interaction and influence of all relevant elements such as materials, shapes, tolerances, surface finishes, etc. This needs to be done for new assemblies and also if and when they have been affected by deformation and wear (Prokinov, 1974). Hansen (1970) emphasizes this by saying that "it is important that adequate provision be made, at an early stage in the design of complex mechanisms, for the adjustment of key components to take into account the small variations in dimensions which are inevitable in production, and also to enable compensation to be made for the effect of wear".

To illustrate the influence of tolerances, user convenience and cost, we borrow an example from Hansen's (1968) book on systematic design.

A telemeter is required to measure a maximum of 100 m distance with an accuracy of 0·1 m. The instrument should be convenient for mobile use, and cost should be kept to a minimum. Limiting tolerances of all parts must be established and the methods of achieving them indicated.

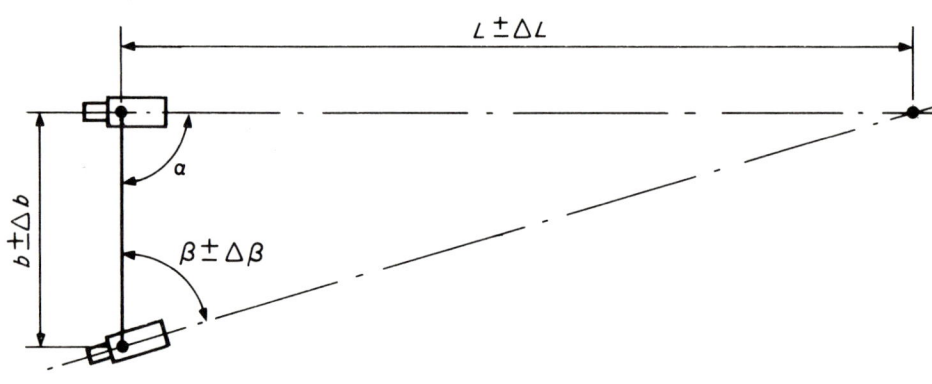

We have

$L = 100$ m $\quad \Delta L = \pm 0·1$ m
$b = ?$ $\quad \Delta b = $ minimum practical
$\alpha = 90°$ $\quad \Delta \alpha = 0$, set up on assembly
$\beta = ?$ $\quad \Delta b$, depends on instrument used

The relationship between L, b and β is

$$L = b \cdot \tan \beta \qquad (i)$$

Differentiating, $dL = (db \cdot \tan \beta) + (b \cdot \tan^2 \beta \cdot d\beta)$, or expressed as a finite

2. THEORY OF ERRORS

difference equation,

$$\Delta L = \Delta b \cdot \tan \beta + b \cdot \tan^2 \beta \cdot \Delta \beta \qquad \text{(ii)}$$

Since β will be close to 90°, $\tan \beta \gg 1$,

$$\therefore \Delta L \simeq \Delta b \cdot \frac{L}{b} + b \cdot \left(\frac{L^2}{b}\right) \cdot \Delta \beta$$

or

$$\Delta L \simeq \frac{L}{b}[\Delta b + L \cdot \Delta \beta],$$

i.e.

$$\Delta b \simeq \frac{b}{L} \cdot \Delta L - L \cdot \Delta b \qquad \text{(iii)}$$

Selecting possible values for b, we can plot a family of lines for Δb v. $\Delta \beta$ as shown in Fig. 2.1.

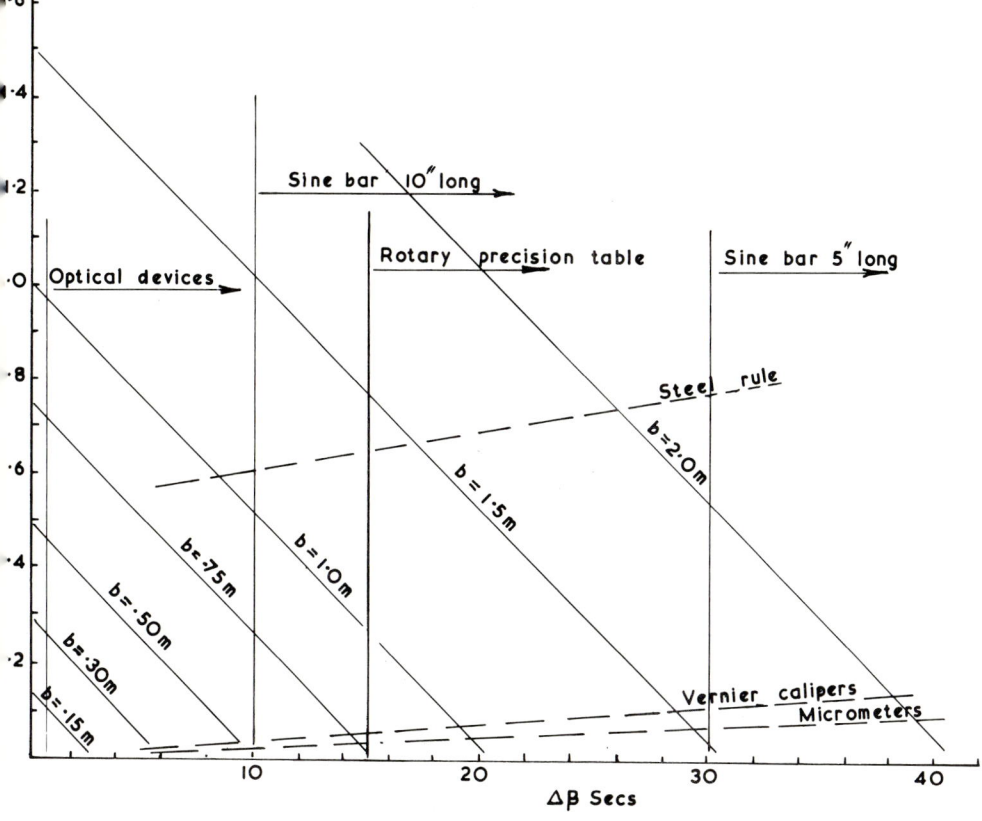

Fig. 2.1. Design of telemeter.

46   APPROXIMATE METHODS IN ENGINEERING DESIGN

To enable the designer to make a sensible decision as to construction and size of the Telemeter, information must be gathered on existing instrumentation, their accuracy and cost. Table 2.1 represents typical data.

Table 2.1. Possible instrumentation for telemeter; accuracy and cost

Instrument	Accuracy	1976 cost (£)
A. LINEAR MEASUREMENT		
1. Steel rules		
2·0 m long	0·75 mm	50·0
1·5 m long	0·66 mm	40·0
0·30 m long	0·50 mm	2·0
2. Vernier calipers		
2·0 m	0·15 mm	375
1·0 m	0·08 mm	200
0·30 m	0·03 mm	70
0·15 m	0·03 mm	30
3. Micrometer		
1·0 m	0·05 mm	250
0·30 m	0·015 mm	90
B. ANGULAR MEASUREMENT		
1. Simple protractor	30'	5
2. Vernier protractor		
Plain	4'	35
Large diameter and eye piece	1'	50
3. Sine bars		
5 in long	30"	40 (plus cost of slip gauges)
10 in long	10"	70
4. Rotary table with micrometer worm	15–20"	100
5. Optical devices	<1"	3000 and over

In Fig. 2.1 lines corresponding to accuracy ranges of (1) linear measuring devices and (2) angular measuring devices have been added. To choose a combination of (1) and (2) that will yield the accuracy of L specified, the value of b has to be such that a point on a b-line must lie above curves (1) and to the right of a line (2). For example:

  (i) A steel rule may be used in combination with a rotary precision table if the length of b is 1·4 m or greater;

2. THEORY OF ERRORS

(ii) b may be reduced to 0.8 m if vernier calipers are used for (1), retaining the rotary table for (2).

Clearly, the smallest size telemeter can be designed by using high-precision optical devices, but it can be seen from Table 2.1 that these are very expensive. A compromise between size and cost is needed.

Sine bars are ruled out because of the inconvenience of having to use slip-gauges. Protractors require b to be longer than 2 m, which is impractical. We are therefore left with the rotary table as the cheaper alternative to optical devices. Table 2.2 lists possible combinations and their estimated costs.

Table 2.2. Cost of feasible combinations for a telemeter.

Linear devices	Length (m)	Rotary device	Combined cost £ (estimated)
Micrometer	0.8	Rotary table	350
Vernier	1.0	Rotary table	300
Steel rule	1.5	Rotary table	140

Whilst the steel rule is the cheapest, it is also the longest. If one considers the ancillary parts which will be required, the apparent cost advantage of the steel rule construction will be reduced. The choice seems, therefore, to be between a vernier or a micrometer construction using an 0.8 m long slide for a rotary precision table on which is mounted the viewing telescope. The design has to be taken to a more advanced stage before a final cost decision can be taken.

A further example from Hansen (1968) is shown in Table 2.3. This concerns the function of an electric relay and tabulates the possible deviations due to material properties, manufacture, and the effects of environment and age-deterioration on the performance of the relay.

Table 2.3. Determination of "virtual deviations" of an electric relay, considering system, function and parameters.[a]

Function	Parameter	Relationship	Virtual deviation	External influences
CLOSING OF RELAY Voltage Coil (Flow)	U R, w	$i = U/R$; ΔR $\theta = i \cdot w$; Δw		

Table 2.3.

Function	Parameter	Relationship	Virtual deviation	External influences
Core Armature Knife-edge Air-gap	μ_C, i_C μ_A, l_A μ_K, l_K μ_0, l_0 } μ_λ l_λ		$\Delta\mu_\lambda$ Δl_λ Δl_0	Ageing
Magnetic flux		$\phi = \dfrac{\theta}{\sum \dfrac{\mu_\lambda \cdot A_\lambda}{l_\lambda}}$	Stray flux	
Force of attraction	A_0	$F = \dfrac{\phi^2}{\mu_0 \cdot A_0}$	ΔA_0 $l_0 \not\to 0$	Vibrations $\to \Delta l_0$
Armature lever	a, b		Friction	
Force of closure		$F_S = F \cdot \dfrac{a}{b}$	ΔF ... Influenced by location and movements: gravitational field and mass/inertia	
Contact Spring A	Cross-section length material		Δh Δb Δl	Softening and relaxation
Deflection of A		$f_A = \dfrac{4 \cdot F_S \cdot l^3}{E \cdot b \cdot h^3}$		
Contact A Air-gap Contact B	h_A h_L h_B		Form and pressure depend on height and width of A and B	
Deflection of B		$f_B = f(f_A, h_L ...)$	Δh_L	
Contact press		F_C		
Contact resistance		$R_C = f(F_C)$		Vibrations change R_C
Opening of Relay				
Breaking circuit permanent magnet		$U \to 0$	$F \not\to 0$	

*After Hansen (1968).

It is becoming clear that we must not only be aware of performance deviations but we must also be concerned with operational reliability. This will be considered in Chapter 7.

REFERENCES

British Standard 4500 (1970). Data sheet 4500A.
Bronstein, I. N. and Semendyayev, K. A. (1973). "A Guide Book to Mathematics." Springer Verlag, New York.
Hansen, F. (1968). "Konstruktionsystematik." VEB Verlag Technik, Berlin.
Hansen, F. (1970). "Adjustment of Precision Mechanisms." Iliffe Books Ltd, London.
Pronikov, A. S. (1974). "Dependability and Durability of Engineering Products" (Trans. T. T. Furman). Butterworths, London.
Wallingford Spring Pressings Ltd. "Design and Selection of Spring Washers."

Chapter 3

Statistical Errors—Empirical Correlations

In Chapter 2 we considered the total errors of individual parameters and developed methods of assessing their effect on the overall functional performance. The resultant error was that derived from the worst possible interaction of all individual "sure-fit" errors. In practical situations in an assembly or system all individual items or variables are seldom at their worst values simultaneously. The larger the number of variables or dimensions, the lower the probability that a worst "sure-fit" situation will occur.

3.1 STATISTICAL DESIGN FACTORS

Gladman (1968) studied variations in tolerances of machined components which tended generally to show a random (Gaussian) distribution. For assemblies of manufactured components he suggested that the "probable maximum variation in tolerances" may be estimated from the relationship

$$P = \pm \frac{1 \cdot 8 \cdot (S)}{\sqrt{n}} \qquad (3.1)$$

where

P = probable maximum variation in overall tolerance
(S) = maximum possible variation of tolerance sum from "sure-fit" considerations
n = number of dependent variables, or dimensions, in total assembly, or chain

3.2 TOLERANCE AND STANDARD DEVIATION

Gladman (1968) quantified the relationship between tolerances and standard deviations (s.d.) of machined components and assemblies by the use of a

3. STATISTICAL ERRORS

"design factor", thus

$$\sigma_j = \tfrac{1}{2}(\rho_j \cdot x_j)$$
$$\sigma_{Ej} = \tfrac{1}{2}(\rho_{Ej} \cdot x_{Ej})$$
(3.2)

where

σ_j = s.d. of individual dimensions, or single component
ρ_j = design factor of single component or dimension
x_j = mean value of single component or dimension

and

σ_{Ej} = s.d. of sum dimensions or assembly
ρ_{Ej} = design factor of sum dimensions or assembly
x_{Ej} = mean value of sum dimensions or assembly

Also that

$$(\sigma_{Ej})^2 = \sum_1^n (b_i^2 \cdot \sigma_i^2) \quad \text{(valid for } \sigma_i \leqslant 0{\cdot}2 \cdot x_i)$$
(3.3)

where

$$b_i = \frac{\partial(f(x_i))}{\partial x_i}$$
(3.4)

i.e. the partial derivative of the function with respect to x_i. Gladman (1968) recommends the following values for the design factors:

$\rho_j = 0{\cdot}45$, for machined dimensions in medium to large quantity production, where tolerance levels are $\geqslant$ IT6 (see Table 3.1)

and

$\rho_j = 0{\cdot}85$, for dimension with tolerance levels $\leqslant$ IT6, in small quantity production

He considered it safe to assume a *normal*, i.e. Gaussian, distribution of tolerance variations. For "sum" dimensions and assemblies he proposed

$\rho_{Ej} = 0{\cdot}33$, for $0{\cdot}27\%$ out of tolerance band
$\rho_{Ej} = 0{\cdot}388$, for $1{\cdot}0\%$ out of tolerance band

Examples
Apply Gladman's design factors to the examples in preceding sections and comment on differences between results obtained.

Table 3.1. ISO, IT6 tolerance grade (after Gladman, 1968)

Basic diameter over and inclusive of	IT6
0·04 → 0·12	0·25
0·12 → 0·24	0·30
0·24 → 0·40	0·40
0·40 → 0·71	0·40
0·71 → 1·19	0·50
1·19 → 1·97	0·60
1·97 → 3·15	0·70
3·15 → 4·73	0·90
4·73 → 7·09	1·00
7·09 → 9·85	1·20
9·85 → 12·41	1·20
12·41 → 15·75	1·40
15·75 → 19·69	1·60

3.3 RELIABILITY OF FATIGUE CALCULATIONS

Values of fatigue strength or endurance limit of materials given in technical publications generally represent results from tests on standard specimens. The endurance strength of a component made from such a material is then obtained from

$$f_e \, \text{comp} = f_e \cdot \left(\frac{A \cdot B \cdot C \cdot D}{K} \right) \tag{3.5}$$

where

f_e = material endurance limit of standard test specimen at a specified number of cycles of reversed bending stress
A = factor for stress cycles different from reversed bending
B = factor for size different from the standard 0·3 inch diameter test specimen
C = factor for surface finishes different from "polished" test specimens
D = factors for surface conditions such as interference fits, heat treatment, residual stresses, etc.
K = stress concentration factor, i.e. the increase in local stress levels relative to the norminal value

Values for A, B, C, D, and K are to be found in most strength of materials reference books.

The f_e deterministic values published usually represent a mean value from a largish number of tests. Stullen et al. (1956) indicated that these values represented a 50% reliability level with a standard deviation of about 8%. They recommended the following correlation for different reliability levels

$$f'_e = K_c . f_e \qquad (3.6)$$

where

$$K_c = 1 - 0.08 . D \qquad (3.7)$$
$$D = \text{Deviation Factor, as from Table 3.2.}$$

Table 3.2. Deviation Factor (after Stullen *et al.*, 1956).

Reliability level (survival rate) (%)	D factor
90	1·30
95	1·60
99	2·30
99·9	3·1
99·99	3·7

Thus, for example, to obtain 99·9% reliability against fatigue failure, using f_e from standard data, we have to reduce the stress amplitudes to

$$f'_e = 0.752 f_e \quad (\text{for } R = 99.9\%)$$

3.4 DYNAMIC LIFE/LOAD CAPACITY OF ROLLING ELEMENT BEARINGS

Manufacturers of rolling element bearings define the nominal or rated life of a bearing as the number of revolutions which will be reached or exceeded by 90% of a fairly large sample. This bearing life is designated $L = 1$ in Fig. 3.1 and Table 3.3, after Palmgren (1959). Figure 3.1 shows the usual life distribution curve which may be approximated by the relationship

$$\log \frac{1}{S} = (L_s)^a . \log\left(\frac{1}{0.9}\right), \qquad (3.8)$$

If the life for $S = 0.90$ is equal to 1×10^6 r.p.m., a = dispersion exponent, which for commonly used bearing steels is in the range 1·1 to 1·5, usually taken as

$$a = \tfrac{10}{9} \text{ for ball bearings}$$

and

$$a = \tfrac{9}{8} \text{ for roller bearings}$$

Table 3.3. Life of bearings (after Palmgren, 1959).

L (millions of revolutions)	$\dfrac{C}{P}$ Ball bearings	$\dfrac{C}{P}$ Roller bearings	L (millions of revolutions)	$\dfrac{C}{P}$ Ball bearings	$\dfrac{C}{P}$ Roller bearings	L (millions of revolutions)	$\dfrac{C}{P}$ Ball bearings	$\dfrac{C}{P}$ Roller bearings
0.5	0.793	0.812	100	4.64	3.98	1 500	11.4	8.97
0.75	0.909	0.917	120	4.93	4.20	1 600	11.7	9.15
1	1	1	140	5.19	4.40	1 700	11.9	9.31
1.5	1.14	1.13	160	5.43	4.58	1 800	12.2	9.48
2	1.26	1.23	180	5.65	4.75	1 900	12.4	9.68
3	1.44	1.39	200	5.85	4.90	2 000	12.6	9.78
4	1.59	1.52	250	6.30	5.24	2 200	13	10.1
5	1.71	1.62	300	6.69	5.54	2 400	13.4	10.3
6	1.82	1.71	350	7.05	5.80	2 600	13.8	10.6
8	2	1.87	400	7.37	6.03	2 800	14.1	10.8
10	2.15	2	450	7.66	6.25	3 000	14.4	11
12	2.29	2.11	500	7.94	6.45	3 500	15.2	11.5
14	2.41	2.21	550	8.19	6.64	4 000	15.9	12
16	2.52	2.30	600	8.43	6.81	4 500	16.5	12.5
18	2.62	2.38	650	8.66	6.98	5 000	17.1	12.9
20	2.71	2.46	700	8.88	7.14	5 500	17.7	13.2
25	2.92	2.63	750	9.09	7.29	6 000	18.2	13.6
30	3.11	2.77	800	9.29	7.43	7 000	19.1	14.2
35	3.27	2.91	850	9.47	7.56	8 000	20	14.8
40	3.42	3.02	900	9.65	7.70	9 000	20.8	15.4
45	3.56	3.13	950	9.83	7.82	10 000	21.5	15.8
50	3.68	3.23	1000	10	7.94	12 500	23.2	16.9
60	3.91	3.42	1100	10.3	8.17	15 000	24.7	17.9
70	4.12	3.58	1200	10.6	8.39	17 500	26	18.7
80	4.31	3.73	1300	10.9	8.59	20 000	27.1	19.5
90	4.48	3.86	1400	11.2	8.79	25 000	29.2	20.9

3. STATISTICAL ERRORS

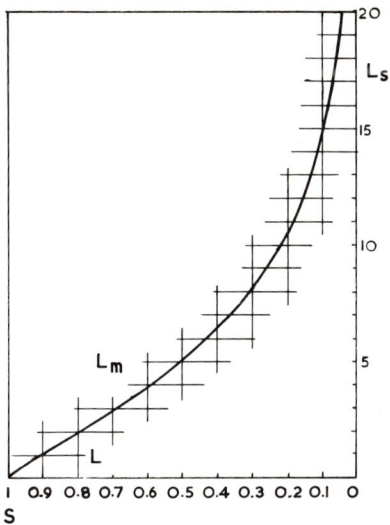

Fig. 3.1. The carrying capacity of ball and roller bearings.

then

L = life reached or exceeded by S proportion of all bearings tested.

Note: $L_s = 5 \times 10^6$ r.p.m. for $S = 50\%$ of bearings, cfr. $L_s = 1 \times 10^6$ r.p.m. for $S = 90\%$ of bearings.

Manufacturers usually recommend a number of operational empirical factors for temperature, type of loading, inner or outer race rotating, lubrication, etc., in order to establish an equivalent "Basic Dynamic Capacity" $= C$, corresponding to 1×10^6 revolutions and 90% reliability. The actual life of a bearing is then given by the relationship

$$L = \left(\frac{C}{P}\right)^n \tag{3.9}$$

where

C = Basic dynamic load capacity for 10^6 revolutions
P = equivalent applied load
n = index depending on type of bearing used. Usually,
$n = 3$ for ball bearings
$n = \frac{10}{3}$ for roller bearings

It is clear from the relationship of equation (3.8) and Fig. 3.1 that the life calculation, or load capacity, found by using equation (3.9) is related to a statistical 90% reliability level. If a different reliability level is required,
  (a) at the same load, the number of revolutions (life) will have to be reduced or increased, as the case may be;

(b) at the same number of revolutions (life), the load will have to be reduced or increased, as the case may be.

3.5 EFFECT OF MISALIGNMENTS ON THE LIFE OF ROLLING ELEMENT BEARINGS

3.5.1 Rated life. The International Standards Organization (ISO) definition of *rated life* is the number of revolutions that 90% of all bearings will complete or exceed before first evidence of fatigue failure develops. It is valid only for perfectly aligned bearings, i.e. where the axes of the inner and outer races are "perfectly parallel".

Misalignment can be caused by errors in manufacture and/or assembly and also by deflections from external loads. (See Figs 3.2 and 3.3.) Rigid, i.e. non-self-aligning bearings act as elastic fixture points which experience built-in bending moments with consequent internal eccentric load distributions and reduced bearing life.

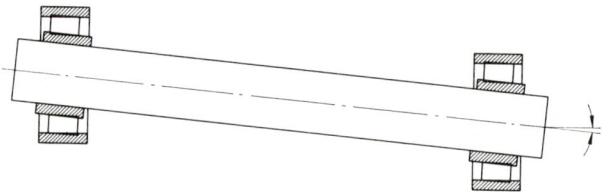

Fig. 3.2. Misalignment due to radial offset.

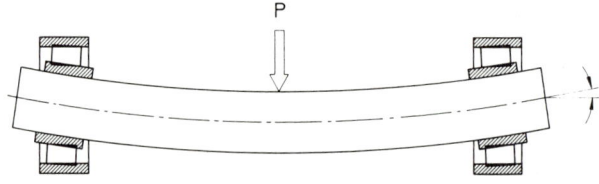

Fig. 3.3. Misalignment due to shaft deflection.

3.5.2 Roller bearings. The idealized load distributions for perfectly aligned plain and crowned rollers are shown in Figs 3.4 and 3.5. Tilting the inner race relative to the outer race causes eccentric pressure distribution on single- and on double-row rollers (see Figs 3.6 and 3.7). An exact three-dimensional stress analysis is complex and, for design purposes, impractical.

Brändlein (1971–2) used a simplified model and neglected the variation in shear stress distribution in the cross-section along the rollers and also any end

3. STATISTICAL ERRORS

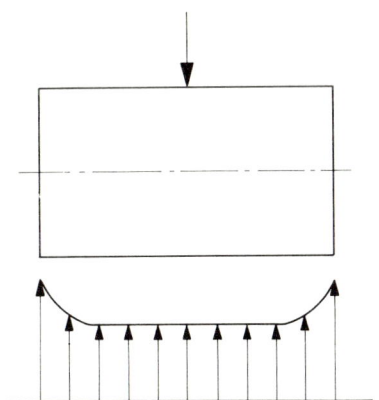

Fig. 3.4. Load distribution along a straight roller.

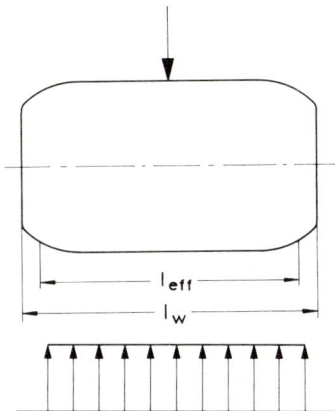

Fig. 3.5. Load distribution along a partially crowned roller.

effects. He obtained correlations for (P_R/P_{req}) v. the eccentric loading $e/(l_{eff}/2)$. Figure 3.8 shows the effects of misalignment (i.e. eccentric loading on a roller) on load and life capacities:

P_R = actual radial load capacity of a single roller
P_{req} = effective radial load capacity of a single roller
L_0 = standard ISO life
L_{verk} = actual life

It can be seen that the actual life of a roller reduces very rapidly with increased misalignment; it equals $(\frac{1}{3} . L_0)$ when the pressure at one end reaches zero.

For practical design purposes it is useful to specify a maximum safe misalignment angle, θ_{perm}. This is selected to correspond to a life reduction of

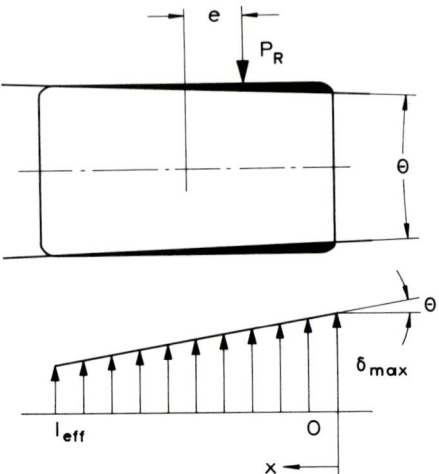

Fig. 3.6. Deformation of a misaligned eccentrically loaded roller.

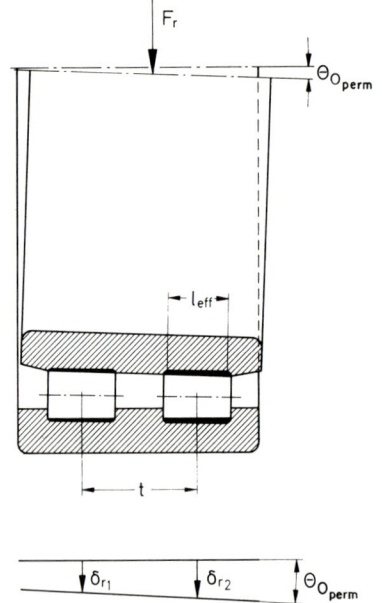

Fig. 3.7. Misaligned double row cylindrical roller roller bearing.

20%, i.e. $L_{\text{verk}} = 0.80 L_0$. An expression for θ_{perm} is given by Brändlein (1971–2) for clearance free roller type bearings, viz.

$$\theta_{\text{perm}} = B \cdot \left[\frac{F_r}{C_0}\right]^{0.925} \text{ angular minutes} \qquad (3.10)$$

3. STATISTICAL ERRORS

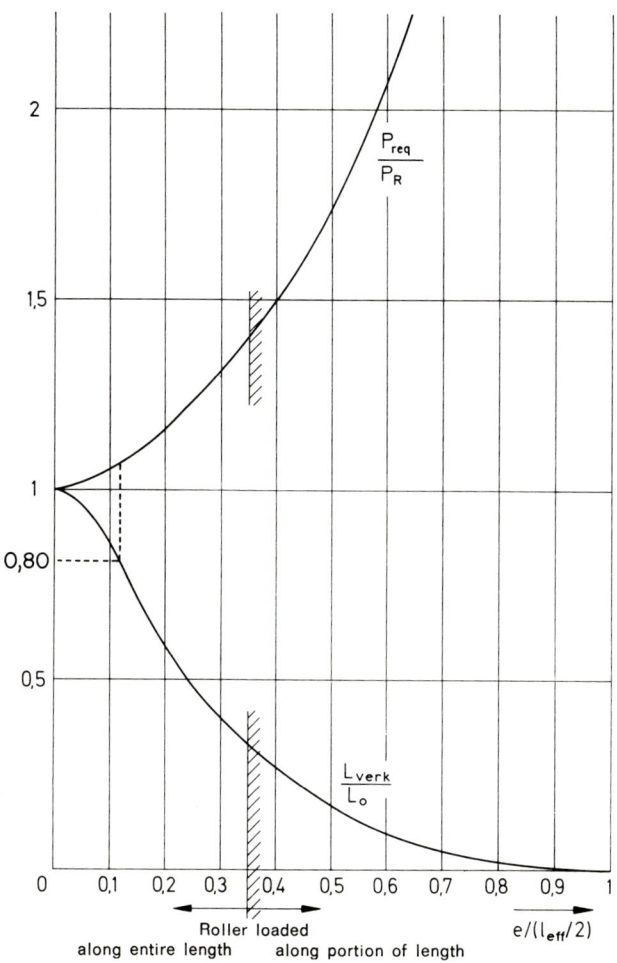

Fig. 3.8. Effect of misalignment on equivalent roller load and life.

Values for θ_{perm} v. B and (F_r/C_0) are given in Table 3.4. For various types of roller bearings; F_r represents the actual radial load applied and C_0 is the static load capacity. It will be noted that θ_{perm} does not exceed 6·6 angular minutes for general mechanical engineering applications. For installations demanding ultra-high accuracy—zero clearance and often pre-loaded—the permissible misalignment is much smaller. (Ref. FAG. publ. No. 41 116 EA.)

3.5.4 Ball bearings.
Two types of ball bearings were considered by Brändlein (1971–2):
  (i) deep groove ball bearings, and

Table 3.4. Values of constant B and permissible misalignment.

Bearing type and series	B	$F_r/C_0 = 0.1$		$F_r/C_0 = 0.2$		$F_r/C_0 = 0.3$	
		$\theta_{0\,\text{perm}}$ (seconds of arc)	Permissible misalignment angle between inner and outer ring (microns/100 mm)	$\theta_{0\,\text{perm}}$ (seconds of arc)	Permissible misalignment angle between inner and outer ring (microns/100 mm)	$\theta_{0\,\text{perm}}$ (seconds of arc)	Permissible misalignment angle between inner and outer ring (microns/100 mm)
Cylindrical roller bearings single row							
Type NU							
Series 10, 2, 3, 4, 2E, 3E	20	2·4	70	4·5	130	6·6	190
Other types							
Series 10, 2, 3, 4, 2E, 3E	13	1·5	45	2·9	85	4·3	125
Series 22, 23, 22E, 23E	9	1·1	30	2·0	80	2·9	85
Series 49	4	0·5	15	0·9	25	1·3	40
Tapered roller bearings							
Series 302, 303, 313	11	1·3	40	2·5	70	3·6	100
Series 320X, 322, 323	7	0·8	25	1·6	45	2·3	85
Cylindrical roller bearings double row							
Series NN 30K	4	0·5	15	0·9	25	1·3	40
Series NNU 49K	3	0·4	10	0·7	20	1·0	30

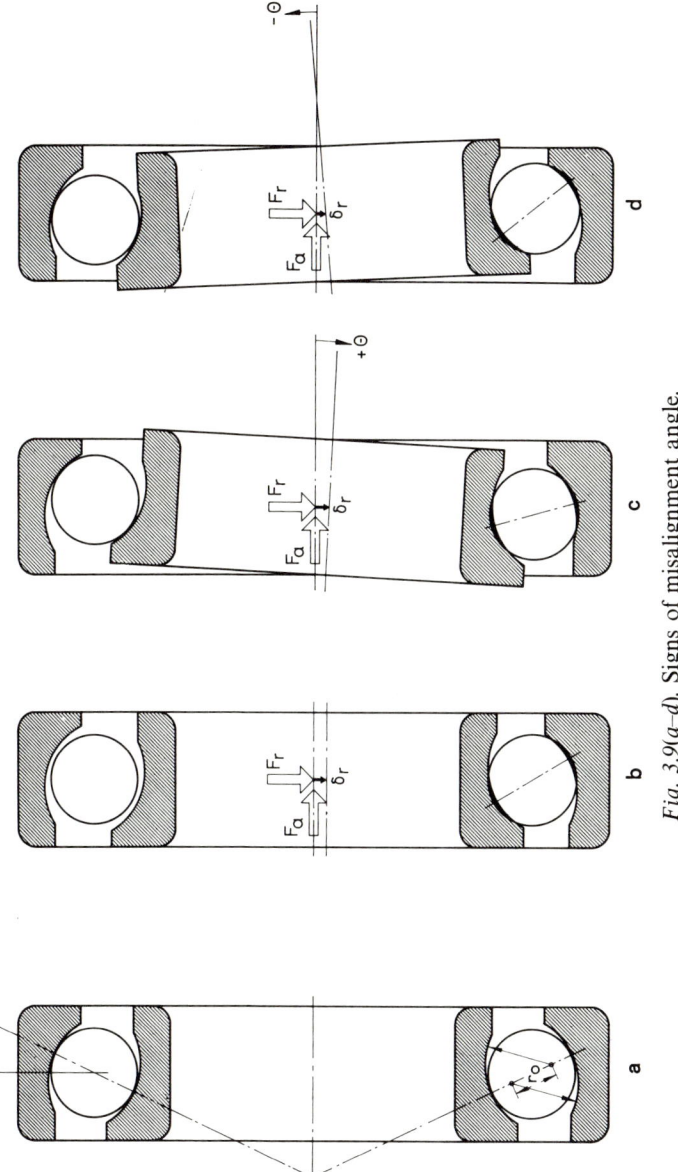

Fig. 3.9(a–d). Signs of misalignment angle.

(ii) angular contact bearings.

Figure 3.9 shows three types of misalignment possible:
(1a) an unloaded aligned bearing;
(1b) a radially displaced bearing;
(1c) positive misalignment relative to direction of radial load;
(1b) negative misalignment relative to radial load.

Whereas positive misalignment tends to decrease the effective (and resultant) axial load on the lower radial-load-carrying balls, negative misalignment increases the effective axial (and resultant) load on these balls.

Figure 3.10 shows the increase in effective load and the change in distribution in angular contact bearings at various misalignment angles. The distribution of loading on the balls is not seriously affected for angles of $\theta < 9'$, but the distribution rapidly worsens for $\theta > 9'$. At large angles of θ the contact ellipse between ball and raceway extends beyond the edge of the race.

As in the case of roller bearings, Brändlein (1971–2) defined a critical misalignment angle θ_{perm} as the value corresponding to a reduction of 20% in the life of the bearing under nominally the same resultant load. Beyond θ_{perm} the reduction in life is rapid and bearings become unsafe for practical engineering applications.

The procedure recommended by the FAG Company is to establish the θ_{perm} value for the bearing intended to be used and then calculate the maximum θ the working installation may experience. If $\theta \leqslant \theta_{perm}$, the design would be safe and the life of the bearing could be expected to be at least 80% of the L_0 value for $\theta = 0$.

3.5.3 Procedure for calculating θ_{perm} (80% life)

Step 1. For bearing selected calculate

$$K = (\text{Constant}) \cdot C_0 \qquad (3.11)$$

where

C_0 = static load capacity
Constant = 50 for deep groove ball bearings, series 60, 62, 63, 64 and 160 $\qquad$ (3.12a)
Constant = 66 for angular contact bearings, series 72B and 73B $\qquad$ (3.12b)

Step 2. Calculate

$$\rho_{F_r} = \frac{F_r}{K} \qquad (3.13)$$

where

F_r = actual radial applied load
ρ_{F_r} = abscissa in Fig. 3.11

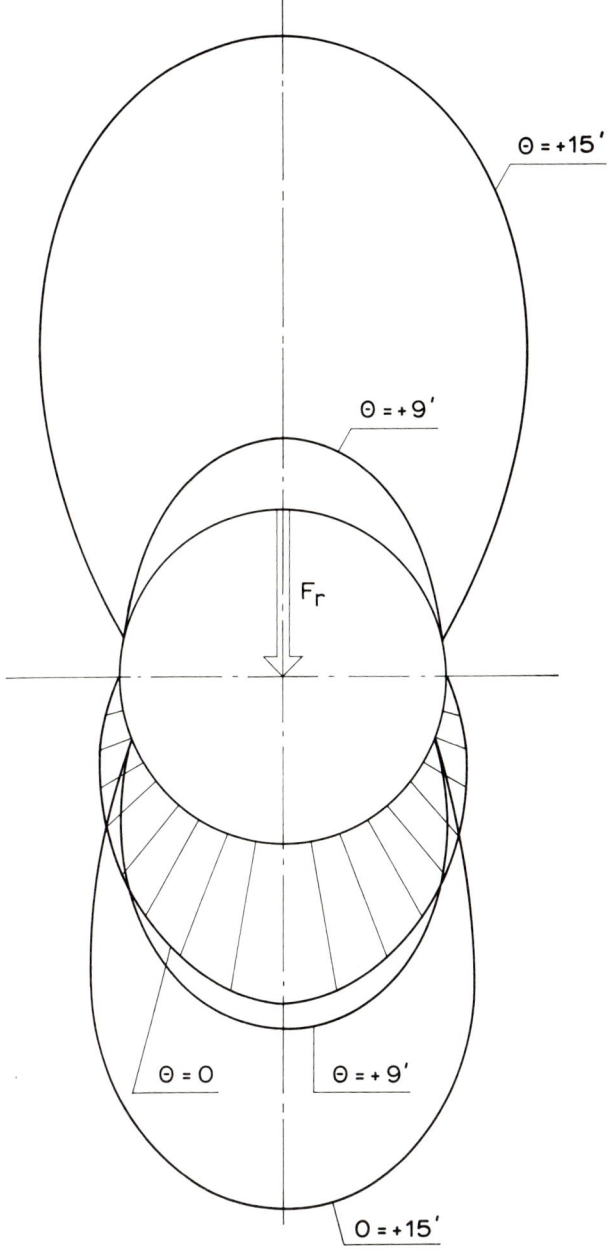

Fig. 3.10. Load distribution in an angular contact ball bearing 7212B with $\alpha_0 = 40°$, $F_r = 500$ kg and various misalignment angles.

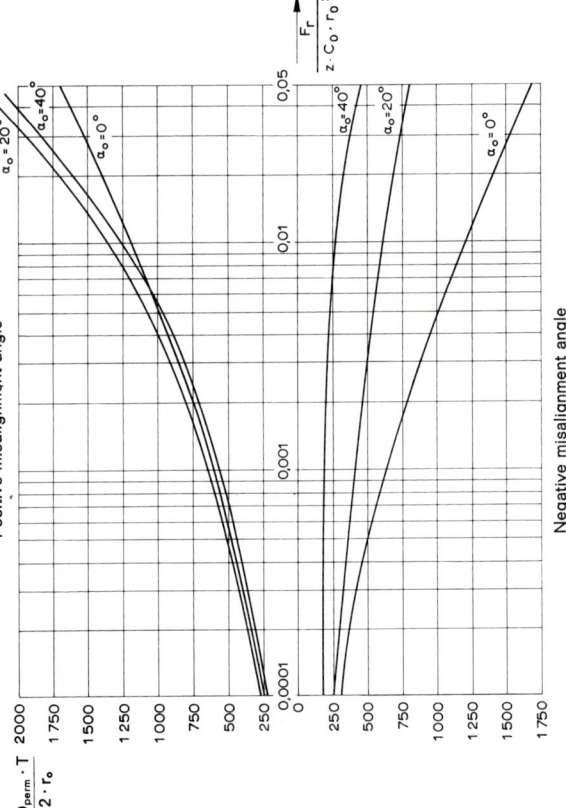

Fig. 3.11. Permissible misalignment for 20% life reduction.

3. STATISTICAL ERRORS

Step 3. From Fig. 3.11 find $\rho_{\theta\text{perm}}$ on the ordinate, using the appropriate transmission angle α_0 for the bearing and ρ_{F_r} from Step 2:

$\alpha_0 = 0$ for deep groove ball bearings
$\alpha_0 = $ values given in, or derived from data in manufacturer's catalogues for angular contact bearings

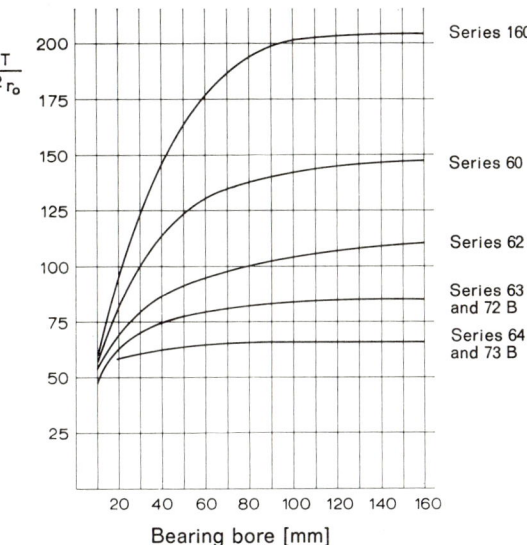

Fig. 3.12. The ratio $T/2r_0$ for deep groove ball bearings and angular contact bearings.

Step 4. From Fig. 3.12 find $(T/2r_0)$, and thence calculate

$$\theta_{\text{perm}} = \frac{\rho_{\theta\text{perm}}}{(T/2 \cdot r_0)} \qquad (3.14)$$

where $T = $ pitch diameter of balls or rollers in the bearing. A combined nomogram for deep-groove ball bearings is given in Fig. 3.13.

3.5.5 Effect of radial clearance.
Brändlein (1971–2) claims that radial clearance has an "insignificant effect on θ_{perm}". This may be true for new and perfectly fitted bearings, but the example below shows that the effect can be significant:

Consider deep groove ball bearing FAG 6208:

bore, $d = 40$ mm
static load capacity, $C_0 = 1600$ kp
radial applied load, $F_r = 160$ kp
∴ $F_r/C_0 = 0.1$

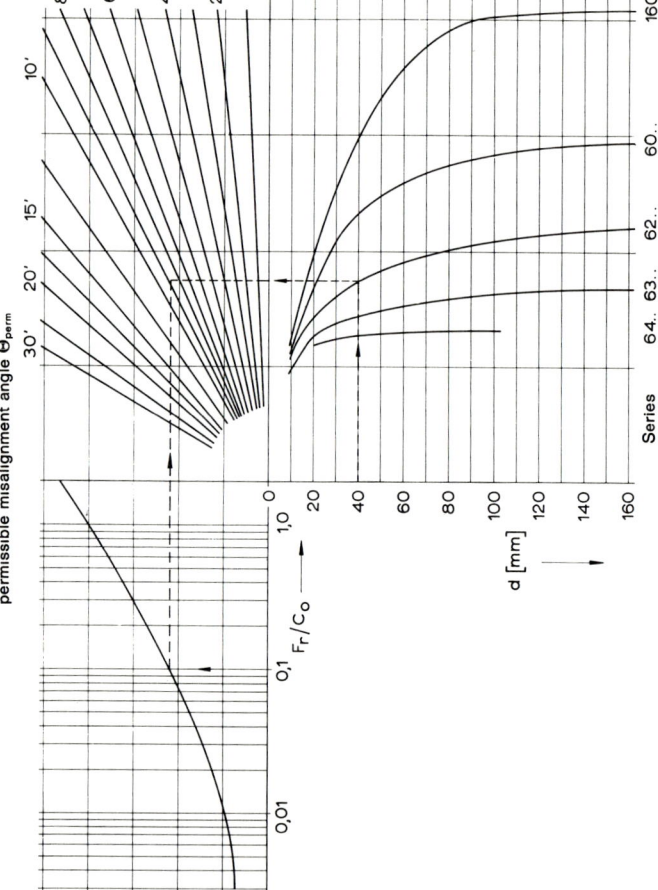

Fig. 3.13. Nomogram for the calculation of the permissible misalignment angle for deep groove ball bearings. *Example*: deep groove ball bearing 6208, bore $d = 40$ mm, static load rating $C_0 = 1600$ kg, radial load $F_r = 160$ kg, $F_r/C_0 = 0{\cdot}1$, permissible misalignment angle $\Theta_{perm} \approx 8{\cdot}5'$.

3. STATISTICAL ERRORS

For deep groove ball bearing $\alpha_0 = 0$.

(a) If a standard "zero" clearance bearing is used, we obtain from the combined nomogram (Fig. 3.13),

$$\theta_{perm} \simeq 8.5 \text{ angular minutes for } 80\% \text{ life.}$$

(b) If due to installation or wear an *in situ* radial clearance, $e = 0.0254$ mm ($\equiv 0.001$ inches) exists, what would θ_{perm} be?

From Fig. 3.12 we get

$$\left(\frac{T}{2 \cdot r_0}\right) = 87.5 \quad \text{for } d = 40 \text{ mm, and series 62 bearings} \tag{i}$$

$$T = \frac{D+d}{2}, \quad \text{approximately}$$

$$\therefore T \simeq \tfrac{120}{2} \simeq 60 \text{ mm} = \text{pitch diameter of balls} \tag{ii}$$

$$\therefore r_0 \simeq \tfrac{30}{87.5} \simeq 0.3428 \tag{iii}$$

and, therefore $e/(r_0) = 0.074$. \hfill (iv)

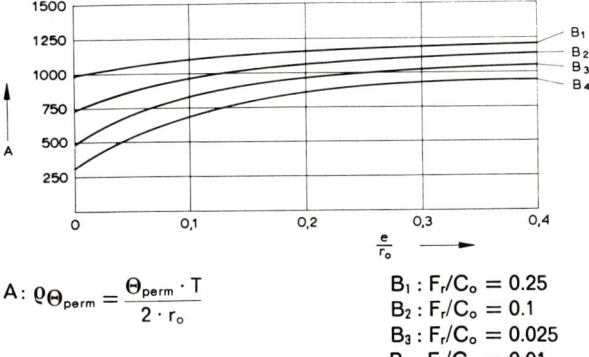

A: $\varrho\Theta_{perm} = \dfrac{\Theta_{perm} \cdot T}{2 \cdot r_0}$

$B_1 : F_r/C_0 = 0.25$
$B_2 : F_r/C_0 = 0.1$
$B_3 : F_r/C_0 = 0.025$
$B_4 : F_r/C_0 = 0.01$

Fig. 3.14. The effect of radial clearance on the permissible misalignment angle Θ_{perm} of deep groove ball bearings.

Now, referring to Fig. 3.14, we get

$$\text{for } \left(\frac{e}{r_0}\right) = 0.074$$

and

$$B_2 = 0.10; \quad \left(= \frac{F_r}{C_0}\right),$$

$$A = \rho_{\theta_{perm}} = 875 \tag{v}$$

Thus,
$$\theta_{perm} = \tfrac{875}{87\cdot5} \simeq 10 \text{ angular minutes} \tag{vi}$$

We see, therefore, that a radial clearance of 0·0254 mm for this bearing increases the permissible misalignment angle from 8·5′ to 10′. If operationally such radial clearance is acceptable, this could make the difference between using a simple deep groove ball bearing instead of having to install a more expensive self-aligning bearing.

REFERENCES

Brändlein, J. (1971–2). The effect of misalignment on rolling element bearings. *Ball and Roller Bearing Engineering* **X**, 32–37.

Gladman, C. A. (1968). Techniques for applying probability to the tolerancing of machined components, *C.I.R.P.* Vol. XVIII.

Palmgren, A. (1959). "Ball and roller bearing engineering." SKF Industries Inc., U.S.A.

Stulen, F. B., Cummings, H. N. and Schulte, W. C. (1956). "Conference on Fatigue of Metals." Institute of Mechanical Engineers.

Chapter 4

Types of Distributions

In analysing data it is helpful to collect identical values together and to show the frequency with which each individual value is observed in the total number of observations. The frequency distribution reveals how the frequencies of the values are spread or distributed over the full range of values.

A random variable is called *discrete* if it can take a *finite* or a number of *unique* values.

A variable is called *continuous* if it can take any real value in a given range.

Examples of discrete variables:
   Number of children born in any given period of time;
   Number of people attending football matches;
   Annual death rates or accidents;
   Number of failures of machine parts;
   Number of heads and tails in throwing of a coin.

Examples of continuous variables:
   Dimensions and weights of machined parts;
   Temperature measurements on a process plant;
   Strength of a material;
   Hardness levels;
   Rainfall per annum;
   Velocity of a vehicle;
   Power of animal or engine.

When a continuous variable can be represented by very small differences between distinct observed values, they can be represented by "grouped" frequencies to facilitate calculations.

4.1 THE BINOMIAL DISTRIBUTION

The binomial distribution is related to the expansion $(R \pm F)^n$ and allows the prediction of success (R) or failure (F) of discrete events in a number (n) trials.

This distribution is of practical use in research and industrial inspection. Its application is best illustrated by the following two examples:

Example 4.1 Two items are taken from a large batch containing 10% defectives. What are the probabilities that
  (a) both items are defective?
  (b) one is defective and one is satisfactory?
  (c) both items are satisfactory (= neither is defective)?

The binomial distribution for *2 items* becomes

$$F_n = (R+F)^2 = R^2 + 2RF + F^2$$

i.e.

$(R)^2$ = probability of both satisfactory ($\equiv$ no defectives)
$(2RF)$ = probability of one satisfactory and one defective
$(F)^2$ = probability of both defective

Now,

$$F = 0{\cdot}1 \quad \text{and} \quad R = 0{\cdot}9$$

$$(F+R = 1) \quad \text{for whole batch}$$

∴ (a) = $F^2 = 0{\cdot}01 = 1\%$ chance of both being defective.
  (b) = $2RF = 2 \times 0{\cdot}1 \times 0{\cdot}9 = 0{\cdot}18 = 18\%$ change of one defective and one satisfactory.
  (c) = $R^2 = 0{\cdot}81 = 81\%$ chance of both being satisfactory.
  Total = 100%.

Example 4.2 Throw one dice to obtain either No. 1 or No. 6, both being successes. What is the probability of success with 3 throws?

R = probability of success for any *one* throw = $\frac{2}{6} = \frac{1}{3}$
F = probability failure for any *one* throw = $\frac{4}{6} = \frac{2}{3}$

For three throws, the binomial series becomes

$$R_n = (R+F)^3 = R^3 + 3R^2 F + 3RF^2 + F^3$$

where

R^3 = probability of 3 successes
$3R^2 F$ = probability of 2 successes and 1 failure
$3RF^2$ = probability of 1 success and 2 failures
F^3 = probability of 0 success and 3 failures

4. TYPES OF DISTRIBUTIONS

Thus for 3 throws, Nos 1 and 6 both = successes on dice:

$$R_n = 3 \text{ successes, } p = (\tfrac{1}{3})^3 = \tfrac{1}{27}$$
$$R_n = 2 \text{ successes, } p = 3(\tfrac{1}{3})^2 \cdot (\tfrac{2}{3}) = \tfrac{6}{27}$$
$$R_n = 1 \text{ success, } p = 3(\tfrac{1}{3}) \cdot (\tfrac{2}{3})^2 = \tfrac{12}{27}$$
$$R_n = 0 \text{ success, } p = (\tfrac{2}{3})^3 = \tfrac{8}{27}$$

probabilities of obtaining Nos 1 or 6

$$\text{Total} = \tfrac{27}{27} = 1$$

(N.B. Use Pascal's triangle for coefficients.)

n								
1				1				
2			1	2	1			
3			1	3	3	1		
4		1	4	6	4	1		
5	1	5	10		10	5	1	
6	1	6	15	20	15	6	1	

The binomial distribution clearly becomes cumbersome for a large number of events or trials, especially if only isolated or rare occurrences are being considered.

4.2 POISSON'S DISTRIBUTION

Poisson's distribution is useful when studying rare or isolated occurrences over a long period of time, or for large number of trials. If an average frequency of occurrence is known, or can be established, the Poisson's distribution enables one to predict the probability of such an occurrence during a limited period of time, or a smaller number of trials.

As before, let

$$R = \text{probability of success}$$
$$F = \text{probability of failure}$$

We know, by definition that $(R+F) = 1$. Thus also $(R+F)^n = 1$. If $n =$ very large, the binomial distribution is impractical to apply.

Poisson made use of the infinite series:

$$e^z = \left(1 + z + \frac{z^2}{2!} + \frac{z^3}{3!} + \frac{z^4}{4!} + \dots + \frac{z^n}{n!}\right) \quad (4.1)$$

and since $e^z \cdot e^{-z} = 1$, he substituted these for R and F and defined

$z \equiv$ the expected *average* number of occurrences over a specified period of time, number of trials or quantity of continuum.

Thus, given an *average* frequency of occurrence z,

the probability of observing an event
$$\begin{cases} 0 \text{ times} = 1 \cdot e^{-z} \\ 1 \text{ time} = z \cdot e^{-z} \\ 2 \text{ times} = \dfrac{z^2}{2!} \cdot e^{-z} \\ 3 \text{ times} = \dfrac{z^3}{3!} \cdot e^{-z} \\ 4 \text{ times} = \dfrac{z^4}{4!} \cdot e^{-z} \\ n \text{ times} = \dfrac{z^n}{n!} \cdot e^{-z} \end{cases}$$

Example 4.3

No. of goals scored	0	1	2	3	4	5	6	7	
No. of times *teams* scored so many goals in a match	95	158	108	63	40	9	5	2	Σ = 480 matches
Total no. of goals	0	158	216	189	160	45	30	14	Σ = 812

Thus 480 recorded matches resulted in a total of 812 goals, i.e.

$$\text{average goals per match} = \tfrac{812}{480} = 1 \cdot 70 = z$$

Using Poisson's distribution to predict the probability (i.e. frequency) of scoring 0, 1, 2, 3, 4, 5, 6 or 7 goals in any one match gives

$$e^{-z} = e^{-1 \cdot 7} = 0 \cdot 183$$

and

$$P_n = 0 \cdot 183 \left[1 + 1 \cdot 7 + \frac{(1 \cdot 7)^2}{2!} + \frac{(1 \cdot 7)^3}{3!} + \ldots + \frac{(1 \cdot 7)^7}{7!} \right]$$

No. of goals	0	1	2	3	4	5	6	7
P_n	0·183	0·311	0·264	0·150	0·0637	0·0217	0·00613	0·0015
$P_n \times 480$ matches	88	149	126	72	31	10·4	3	1
Actual frequency	95	158	108	63	40	9	5	2

Fair agreement!

4. TYPES OF DISTRIBUTIONS

Example 4.4 An electric servo-motor is known to have a probability of failure of 0·005 for any one routine operation. Compute the approximate probability of (a) at least one failure, and (b) at least two failures in 1000 operations of the system.

We have

$z_1 = 0{\cdot}005$ for a single operation

$\therefore z_{1000} = 5$ for 1000 operations (i.e. the average number of expected failures over 1000 operations)

Now, Poisson's distribution is

$$p_f = e^{-5}\left[1 + 5 + \frac{5^2}{2!} + \frac{5^3}{3!} + \ldots + \frac{5^n}{n!}\right]$$

where

$p_{f_0} = e^{-5}$, i.e. probability of *no* failure in 1000 operations

$p_{f_1} = e^{-5} \times 5$, i.e. probability of *one* failure in 1000 operations

$p_{f_n} = e^{-5} \times \dfrac{5^n}{n!}$, i.e. probability of n failures in 1000 operations

Thus, for (a),

$$p_{f_{n \geq 1}} = (1 - p_{f_0}) = 1 - e^{-5} = 1 - 0{\cdot}0067 = 0{\cdot}993$$

i.e. the probability of *at least one* failure in 1000 operations = 99·3%.

For (b),

$$p_{f_{n \geq 2}} = 1 - (0{\cdot}0067 + 0{\cdot}0335) = 0{\cdot}9598$$

i.e. there is a 95·98% chance that *two or more* failures will occur during 1000 operations of the servo-motor.

Example 4.5 On engineering drawings the frequency of error in dimensions (e.g. decimal points, digits, tolerances, etc.) have been observed to be 0·10% average. How many dimensions may a drawing contain if the probability of finding *one or more* dimensional errors is to be less than 0·05? (For answer see p. 363.)

4.3. CONTINUOUS PROBABILITY DISTRIBUTIONS

A continuous probability distribution function must embrace a large population, even if only an abstract one. (Histograms are for discrete smaller numbers only.)

Definition of terms

$F(t)$ = unreliability = probability of failure relative to time (or number of cycles)
$R(t)$ = reliability = probability of survival
Thus, by definition,

$$F(t) + R(t) = 1 \text{ for total population} \tag{4.2}$$

$f(t)$ = probability distribution function,
i.e. = probability density function

Thus

$$F(t) = \int_0^t f(t) \, dt \tag{4.3}$$

and

$$f(t) = \frac{dF(t)}{dt} \tag{4.4}$$

$\lambda(t)$ = local failure rate

$$= \frac{\text{number of failures per unit time at any given time of life}}{\text{number of items exposed to failure at the same instant of time, i.e. number still surviving}} \tag{4.5}$$

Thus

$$\lambda(t) = \int_t^{t+1} f(t) \, dt \bigg/ R(t)$$

Since $dt = 1$ unit of time,

or
$$\left.\begin{array}{l} \lambda(t) = \dfrac{f(t)}{R(t)} = \dfrac{f(t)}{1 - F(t)} \\[2mm] \lambda(t) = \dfrac{dF(t)}{dt} \cdot \dfrac{1}{R(t)} \end{array}\right\} \tag{4.6}$$

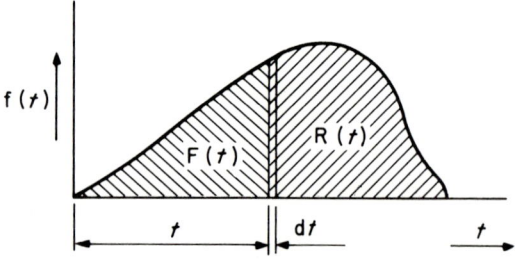

Area under curve = 1. Integrating (4.6) from 0 to t, we get

$$\int_0^t \lambda(t) \, dt = \int_0^t \frac{dF(t)}{R(t)}$$

or

$$\int_0^t \lambda(t) \, dt = \int_0^t \frac{d[1-R(t)]}{R(t)} = -\int_0^t \frac{dR(t)}{R(t)}$$

$$= [-\log_e R(t)]$$

$$\therefore R(t) = \exp\left[-\int_0^t \lambda(t) \, dt\right] \tag{4.7}$$

Thus from (4.6)

$$f(t) = \lambda(t) \cdot \exp\left[-\int_0^t \lambda(t) \, dt\right] \tag{4.8}$$

4.4 PATTERN OF FAILURE RATES OF MECHANICAL EQUIPMENT

A large number of equipment and machinery exhibit three more or less distinct phases of wear or deterioration:

Early failures. Due to material faults, manufacturing faults, running-in tests, etc. These rates of failures are highest at the beginning and generally reduce towards a steady rate of failure.

Constant or steady failure rates. Generally random failures caused by accidents, natural calamities, human errors, etc. Examples are birds striking aircraft, storms, nails in tyres, dirt in fuel, defective service, unexpected overloads, etc.

Increasing failure rates. Due to fatigue, wear-out corrosion, radiation effects, etc.

The life pattern of such equipment is usually represented by a typical "bathtub" curve, as shown in Fig. 4.1. Each of these phases may be described by a suitable probability distribution.

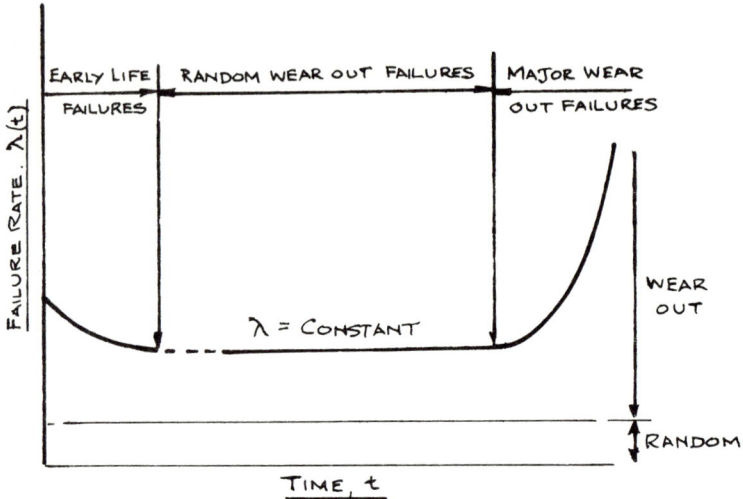

Fig. 4.1. Complete life failure pattern for complex maintained machinery. (After Carter, 1972, by permission of Macmillan and Co. Ltd.)

4.5 WEIBULL PROBABILITY DISTRIBUTIONS

Equation (4.7) read

$$R(t) = \exp\left[-\int_0^t \lambda(t)\,dt\right] \qquad (4.7)$$

To represent a variety of failure rates it would be useful to establish an expression for $\lambda(t)$ which permitted variability so as to fit a range of probable failures. A simple form could be:

$$\lambda(t) = a(t)^b \qquad (4.9)$$

Equation (4.7) then becomes

$$R(t) = \exp\left[-\frac{a \cdot t^{b+1}}{b+1}\right] = \exp\left[-\left(\frac{t}{A}\right)^{b+1}\right] \qquad (4.10)$$

where

$$A = \left[\frac{b+1}{a}\right]^{[1/(b+1)]}$$

A model of this type was proposed by Weibull in 1915. Thus,

$$R(t) = \exp\left[-\left(\frac{t-\gamma}{\eta}\right)^\beta\right] \qquad (4.11)$$

where,

t = time to failure
γ = time at which $F(t) = 0$, i.e. $R(t) = 1$
    = datum parameter
η = characteristic life = scale parameter
β = shape parameter

Equation (4.11) yields the following relationships:

$$F(t) = 1 - \exp\left[-\left(\frac{t-\gamma}{\eta}\right)^\beta\right] \qquad (4.12a)$$

$$f(t) = \frac{dF(t)}{dt} = \frac{\beta \cdot (t-\gamma)^{\beta-1}}{\eta^\beta} \cdot \exp\left[-\left(\frac{t-\gamma}{\eta}\right)^\beta\right] \qquad (4.12b)$$

$$\lambda(t) = \frac{f(t)}{R(t)} = \beta \cdot \frac{(t-\gamma)^{\beta-1}}{\eta^\beta} \qquad (4.12c)$$

We have noted that three parameters control the Weibull distribution correlation, viz. β, γ and η. Figures 4.2 and 4.3 show characteristic curves for $f(t)$ and $\lambda(t)$ plotted for various values of β, taking $\gamma = 0$ at $t = 0$, and $\eta = 1$.

It will be noted from (4.12a) that

$$R(t) = \exp\left[-\left(\frac{t-\gamma}{\eta}\right)^\beta\right]$$

If $\eta = (t - \gamma)$ we get

$$R(t) = \exp[(-1)^\beta] = e^{-1} = 0.368$$

or

$$F(t) = 63.2\%$$

The scaling constant η can thus be seen to represent the time from $\gamma = 0$ (the starting point) to when 63.2% of the population can be expected to have failed. It is independent of the value of β, i.e. independent of the shape of the distribution curves. From these plots we note that:

(i) When $\beta < 1$, $\lambda(t)$ decreases with time; the distribution represents a decreasing failure rate, i.e. a running-in period.
(ii) When $\beta = 1$, $\lambda(t)(1/\eta)$ = constant.
(iii) When $1 < \beta < 2$, the $f(t)$ distribution is skewed and the failure rate, $\lambda(t)$ increases with time, i.e. a wear-out period.
(iv) When $\beta > 2$, the $f(t)$ distribution tends to asymptote at $t = 0$ and $t = \infty$ and is more symmetrical as β increases.

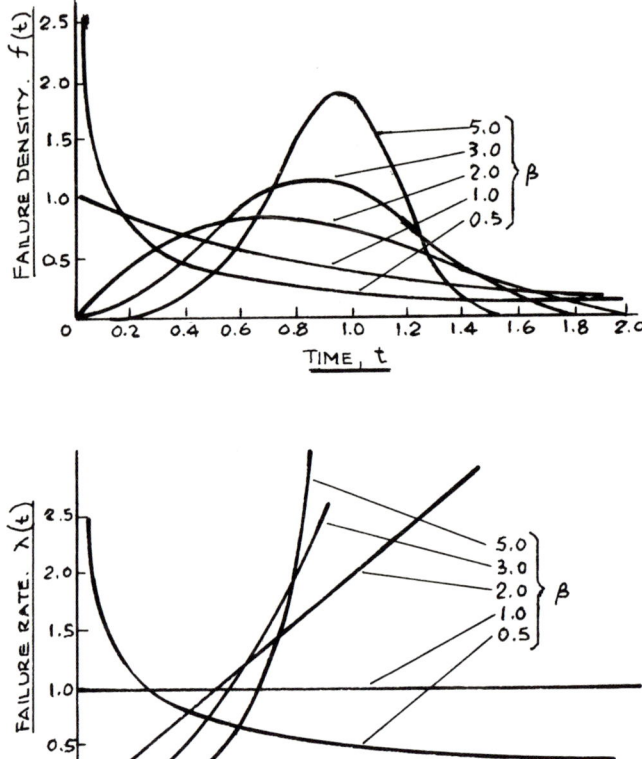

Fig. 4.2. Effect of β on failure density and failure rate for $t_0 = 0$ and $\eta = 1{\cdot}0$ in the Weibull distribution. (After Carter, 1972, by permission of Macmillan and Co. Ltd.)

(v) When $\beta = 3{\cdot}2$, Weibull's distribution is a good approximation to a *normal* Gaussian distribution, i.e. it represents a perfectly random distribution of failures.

Weibull's distribution is clearly very versatile and adaptable to various distribution (failure) patterns, but it tends to be rather complex to use. For engineering design we shall concentrate on the *normal* distribution, albeit with some caution and reservation when appropriate.

Example Establish a Weibull distribution for the expected life of rolling element bearings. (Refer to Section 3.4, equation (3.8).) According to Palmgren

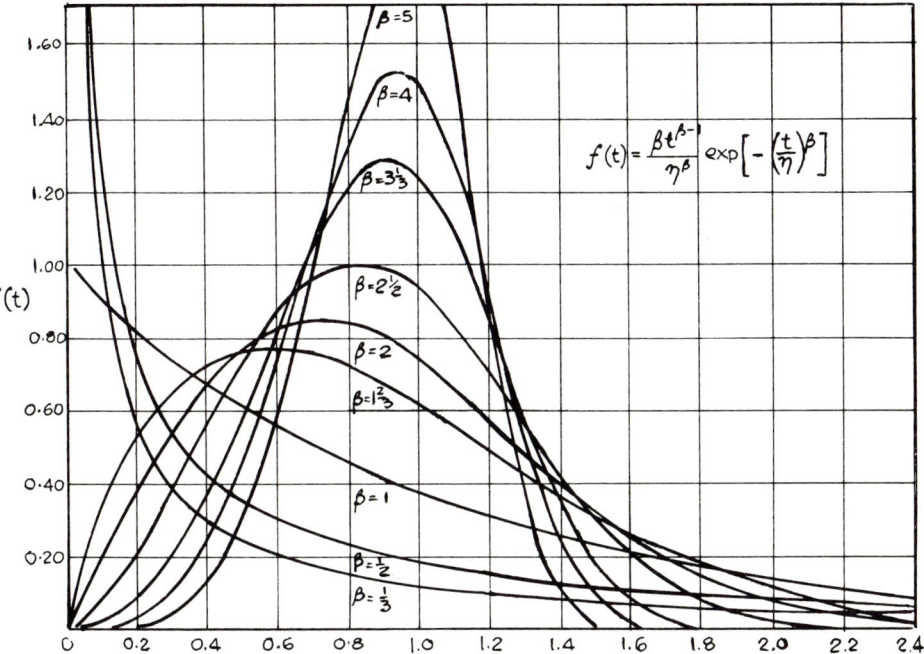

Fig. 4.3. The Weibull probability density function for various values of β when $\eta = 1$, $y = 0$. (After Bompas-Smith, 1973, by permission of McGraw-Hill Book Co.)

(1959), Figure 3.4 may be represented by the expression:

$$\log\left(\frac{1}{S}\right) = (L_s)^a \log\left(\frac{1}{0\cdot 90}\right) \qquad \text{(i)}$$

where

$S \equiv R(t) =$ probability of survival after time (t)

$L_s \equiv t =$ time in number of revolutions

$a =$ dispersion index

We can re-write (i) as

$$\frac{1}{R(t)} = \frac{1}{0\cdot 90} \cdot \exp(t^a)$$

or

$$R(t) = 0\cdot 90 \cdot \exp(-t^a) \qquad \text{(ii)}$$

By definition,

$$F(t) = 1 - R(t) = 1 - 0\cdot 90 \cdot \exp(-t^a) \qquad \text{(iii)}$$

and

$$f(t) = -\frac{dR(t)}{dt} = 0.90 \cdot a \cdot t^{(1-a)} \cdot \exp(-t^a) \quad \text{(iv)}$$

thus

$$\lambda(t) = \frac{f(t)}{R(t)} = a \cdot t^{(1-a)} \quad \text{(v)}$$

This compares with a Weibull distribution where $\gamma = 0$, $\eta = 1$ and $\beta = a$. In our case, $a = 1 \cdot 10 – 1 \cdot 50$, according to Palmgren. The failure density curves for rolling element bearings will lie somewhere between values of $\beta = 1$ and $\beta = 2$ (see Fig. 4.3).

4.6 THE NORMAL (GAUSSIAN) DISTRIBUTION

This distribution is used extensively in quality control, manufacture and many other areas where the effect(s) of many variables tend to combine.

The Central Limit Theorem states that "The *sum* of a number of independent statistical distributions of any form is approximately normally distributed when the number is large"

or

"The distribution of the mean of samples of about equal size taken at *random* from *independent* distributions approaches the *normal* distribution as the sample size increases, regardless of the form of the independent individual distributions."

Example. Fatigue strength of a component depends upon:

  Material properties
  Size of component
  Shape of component
  Surface finish
  Crystal-structure
  Micro-flaws
  Residual stresses
  Applied stresses (loads)
  Environmental conditions (heat, corrosion, etc.)

All of which are variable and independent of each other.
The normal distribution is a function of two main parameters:

(1) the Mean (arithmetic)
(2) the Standard Deviation (= measure of "spread")

4. TYPES OF DISTRIBUTIONS

(1) **The arithmetic mean**
 (a) For discrete numbers, or events (tables or histograms),

$$\bar{x} = \frac{1}{n}\sum_{1}^{n}(x_i) \tag{4.13}$$

 (b) For continuous function (distribution),

$$\bar{x}_{n\to\infty} = \frac{\int_0^{x_{max}} f(x)\cdot x\cdot dx}{\int_0^{x_{max}} f(x)\cdot dx} \tag{4.14}$$

Thus, $\bar{x} \equiv$ the centre of gravity of the distribution area under $f(x)$ relative to the $f(x)$ axis, and

$$f(x) \equiv \text{the frequency or density of } events$$

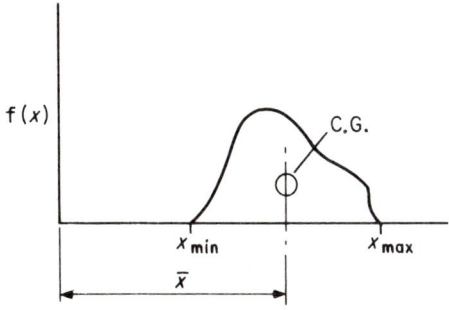

(2) **The standard deviation**
 (a) For discrete numbers or events (tables or histograms):

$$S = \left[\sum_{1}^{n}(x-\bar{x})^2 \Big/ n\right]^{1/2} \tag{4.15}$$

and

$$S^2 = \text{variance} = \left[\sum_{1}^{n}(x-\bar{x})^2 \Big/ n\right] \tag{4.16}$$

If one expands equation (4.16), it can be written as follows:

$$S^2 = \left[\left(\sum_{1}^{n} x^2 \Big/ n\right) - 2\cdot\bar{x}\cdot\left(\sum_{1}^{n} x \Big/ n\right) + \left(\sum_{1}^{n}(\bar{x})^2 \Big/ n\right)\right]$$

or
$$S^2 = \left(\sum_1^n x^2/n\right) - (\bar{x})^2$$

or
$$S^2 = \left(\sum_1^n x^2/n\right) - \left(\sum_1^n x/n\right)^2$$

i.e.
$$S = \left[\left(\sum_1^n x^2/n\right) - \left(\sum_1^n x/n\right)^2\right]^{1/2} \tag{4.17}$$

We note that in this form S can be derived without the need to calculate $\bar{x}$ (as in equation (4.16)).

(b) For continuous function

The standard deviation or variance can be expressed as follows:

$$\sigma^2 = \frac{\int f(x).(x-\bar{x})^2.dx}{\int f(x).dx} \tag{4.18}$$

This may be expanded similarly to the above. We get,

$$\sigma^2 = \frac{\int f(x).x^2.dx - 2\int f(x).x.\bar{x}.dx + \int f(x).\bar{x}^2.dx}{\int f(x).dx}$$

$$= \frac{\int f(x).x^2 dx}{\int f(x).dx} - 2\bar{x}.\frac{\int f(x).x.dx}{\int f(x).dx} + \frac{\int f(x).dx(\bar{x})^2}{\int f(x) dx}$$

$$= \frac{\int f(x).x^2.dx}{\int f(x).dx} - 2(\bar{x})^2 + (\bar{x})^2$$

i.e.
$$\sigma^2 = \frac{\int f(x).x^2.dx}{\int f(x).dx} - (\bar{x})^2 \tag{4.19}$$

It will be observed that the standard deviation σ represents the radius of gyration of the distribution area about the ordinate [$f(t)$] through the mean, $\bar{x}$. The variance is obtained, as can be seen, by dividing the second moment of area about the mean, $\bar{x}$, by the area of the distribution curve above the x-axis.

4. TYPES OF DISTRIBUTIONS

Referring to Section 1, equation (1.19) it will be seen that the standard deviation is also identical to the root-mean-square value of errors. It is useful to work out a few examples for simple distributions.

Example 4.6 Given a rectangular distribution, compute the mean and the standard deviation:

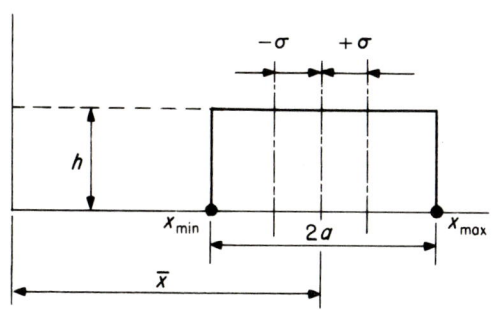

$$\bar{x} = (x_{min} + a)$$

$$\sigma^2 = \frac{I_{\bar{x}}}{\text{area}}$$

$$I_{\bar{x}} = \int_{-a}^{+a} f(x) \cdot dx \cdot (x - \bar{x})^2,$$

or take

$$\bar{x} = 0; \quad I_{\bar{x}} = \int_{-a}^{+a} f(x) \cdot x^2 \cdot dx$$

Then

$$I_{\bar{x}} = h \cdot \frac{(2 \cdot a)^3}{12} = \tfrac{2}{3} \cdot a^3 \cdot h$$

$$\text{Area} = h \cdot 2(a) = 2ah$$

$$\sigma^2 = \frac{I_{\bar{x}}}{A} = \tfrac{1}{3}a^2$$

or

$$\sigma = \sqrt{(\tfrac{1}{3})} \cdot a = 0{\cdot}5774 \cdot a$$

Example 4.7 Determine the standard deviation for a triangular distribution as shown:

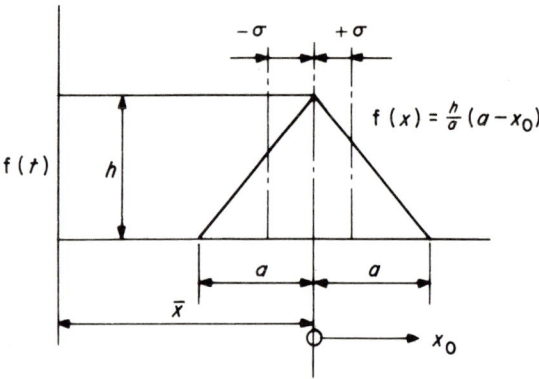

Distribution is symmetrical about the mean $= \bar{x}$. Consider half of distribution:

$$A_{1/2} = \tfrac{1}{2}(a \cdot h)$$

$$\tfrac{1}{2}(I_{\bar{x}}) = \int_0^{+a} \frac{h}{a}\cdot(a-x_0)\cdot x_0^2 \cdot dx_0 = \frac{h}{a}\left[\frac{ax_0^3}{3} - \frac{x_0^4}{4}\right]$$

or

$$\tfrac{1}{2}\cdot I_{\bar{x}} = \left[\frac{a^4}{3} - \frac{a^4}{4}\right]\cdot\frac{h}{a} = \frac{h}{12\cdot a}\cdot(a^4) = \frac{h\cdot a^3}{12}$$

$$\therefore \sigma^2 = \frac{h\cdot a^3}{12}\cdot\left(\frac{2}{a\cdot h}\right) = \frac{a^2}{6}$$

or

$$\sigma = 0{\cdot}408 \cdot a$$

Example 4.8 Compute the standard deviation for a parabolic distribution as shown.

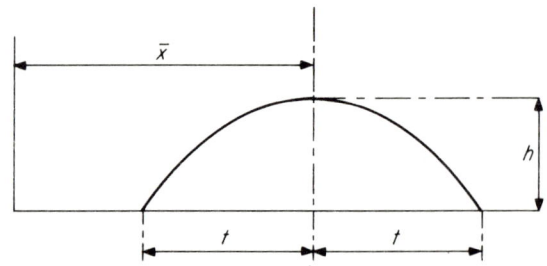

4. TYPES OF DISTRIBUTIONS

Example 4.9 Establish the mean ($\bar{x}$) and the standard deviation (σ) for the skewed distribution given.

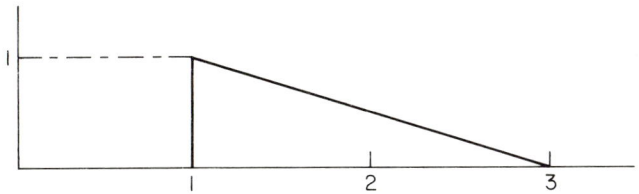

Example 4.10 For the distribution shown below, calculate the mean ($\bar{x}$) and the standard deviation

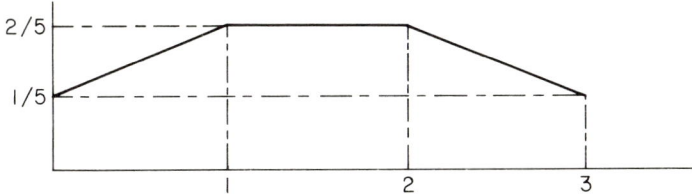

(For answers to Examples 4.8, 4.9 and 4.10 see pp. 363–365.) In the above examples, used for calculating the mean ($\bar{x}$) and the variance (S), we did not have true probability distributions. A true probability distribution, as indicated earlier, is such that the area under the distribution (density) curve is equal to unity; i.e.

$$\int_{t_1}^{t_2} f(t) \cdot dt = 1$$

The mathematical probability density function chosen by Gauss was

$$y = \frac{1}{\sigma\sqrt{2\pi}} \cdot \exp\left[-\frac{(x-\bar{x})^2}{2 \cdot \sigma^2}\right] \quad (4.20)$$

Denoting $(x-\bar{x})/\sigma = z$ representing the number of standard deviations within the interval x to $\bar{x}$, the scale on the time-axis (x-axis) makes $\sigma = 1$. Thus equation (4.20) becomes

$$y = \frac{1}{\sigma\sqrt{2\pi}} \cdot \exp\left(\frac{-z^2}{2}\right), \quad (4.21)$$

or expressing y as a function of z and noting that $\sigma = 1$, we write

$$f(z) = \frac{1}{\sqrt{2\pi}} \cdot \exp\left(\frac{-z^2}{2}\right) \quad (4.22)$$

Integrating between $z = -\infty$ to $+\infty$ (Bronstein and Semendyayev, 1973) we get

$$\int_{-\infty}^{+\infty} f(z)\,dz = \int_{-\infty}^{+\infty} \frac{1}{\sqrt{2\pi}} \cdot \exp\left(\frac{-z^2}{2}\right) = 1$$

From our definition of terms, equations (4.2) to (4.5), the Gaussian functions become

$$F(z) = \frac{1}{\sqrt{2\pi}} \cdot \int_{-\infty}^{z} \exp\left(\frac{-z^2}{2}\right) \cdot dz \qquad (4.23)$$

$$R(z) = \frac{1}{\sqrt{2\pi}} \cdot \int_{z}^{+\infty} \exp\left(\frac{-z^2}{2}\right) \cdot dz \qquad (4.24)$$

$$\lambda(z) = \frac{\exp\left(\dfrac{-z^2}{2}\right)}{\displaystyle\int_{z}^{+\infty} \exp\left(\dfrac{-z^2}{2}\right) \cdot dz} \qquad (4.25)$$

and

$$f(z) = \frac{1}{\sqrt{2\pi}} \cdot \exp\left(\frac{-z^2}{2}\right) \qquad (4.26)$$

Remembering that $z = [(x - \bar{x})/\sigma]$, making $\sigma = 1$ along the z axis (abscissa), we note the effect of the magnitude of σ on the shape of the distribution curves (see Fig. 4.4). It is also clear from the above definitions that the area under the $f(z)$ curve

(a) from $-\infty$ to $z = F(z)$ = probability of failure
(b) from z to $+\infty = R(z)$ = probability of survival

For most practical applications the effective area (e.g. sure-fit limits) is conventionally regarded as lying between

$$z = (-3\sigma) \text{ to } (+3\sigma); \quad \text{area} = 99\cdot73\%$$

and

$$z = (-4\sigma) \text{ to } (+4\sigma); \quad \text{area} = 99\cdot994\%$$

Fifty per cent of the whole population (or all events) will lie between

$$z = (-0\cdot6745 \cdot \sigma) \text{ and } (+0\cdot6745 \cdot \sigma)$$

Values of $f(z)$, $F(z)$, $R(z)$ and $\lambda(z)$ have been tabulated in standard probability tables and often represented in graphical form (see Fig. 4.5 and Table 4.1).

4. TYPES OF DISTRIBUTIONS

$$f(t) = \frac{1}{\sqrt{(2\pi)}} \cdot e^{-t^2/2}$$

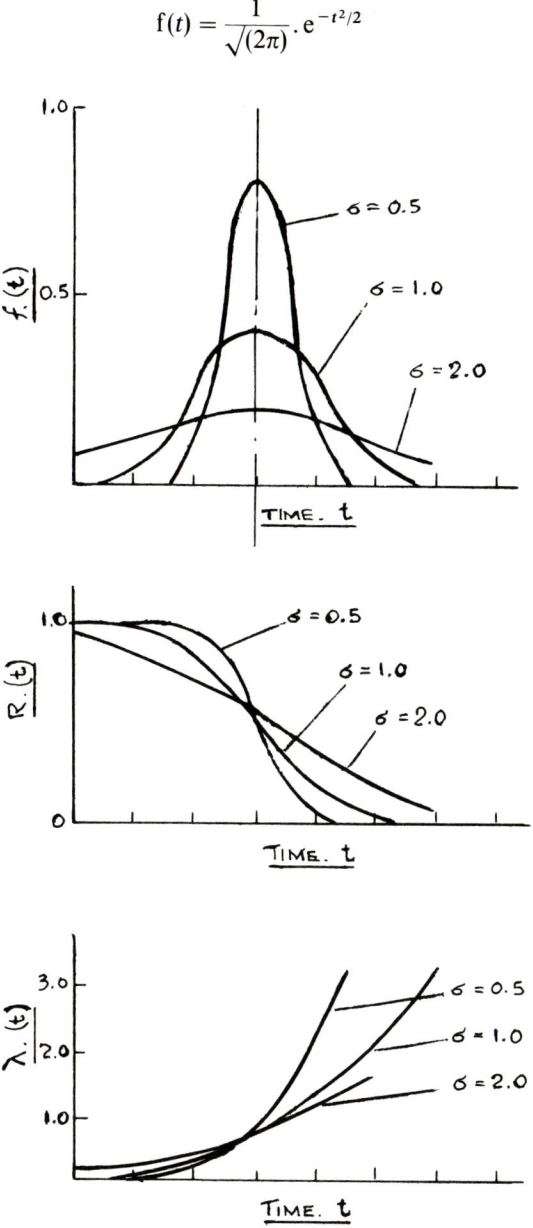

Fig. 4.4. Variation of $f(t)$, $R(t)$ and $\lambda(t)$ for the normal distribution of the failure v. time t and standard deviation σ. (After Carter, 1972, by permission of Macmillan and Co. Ltd.)

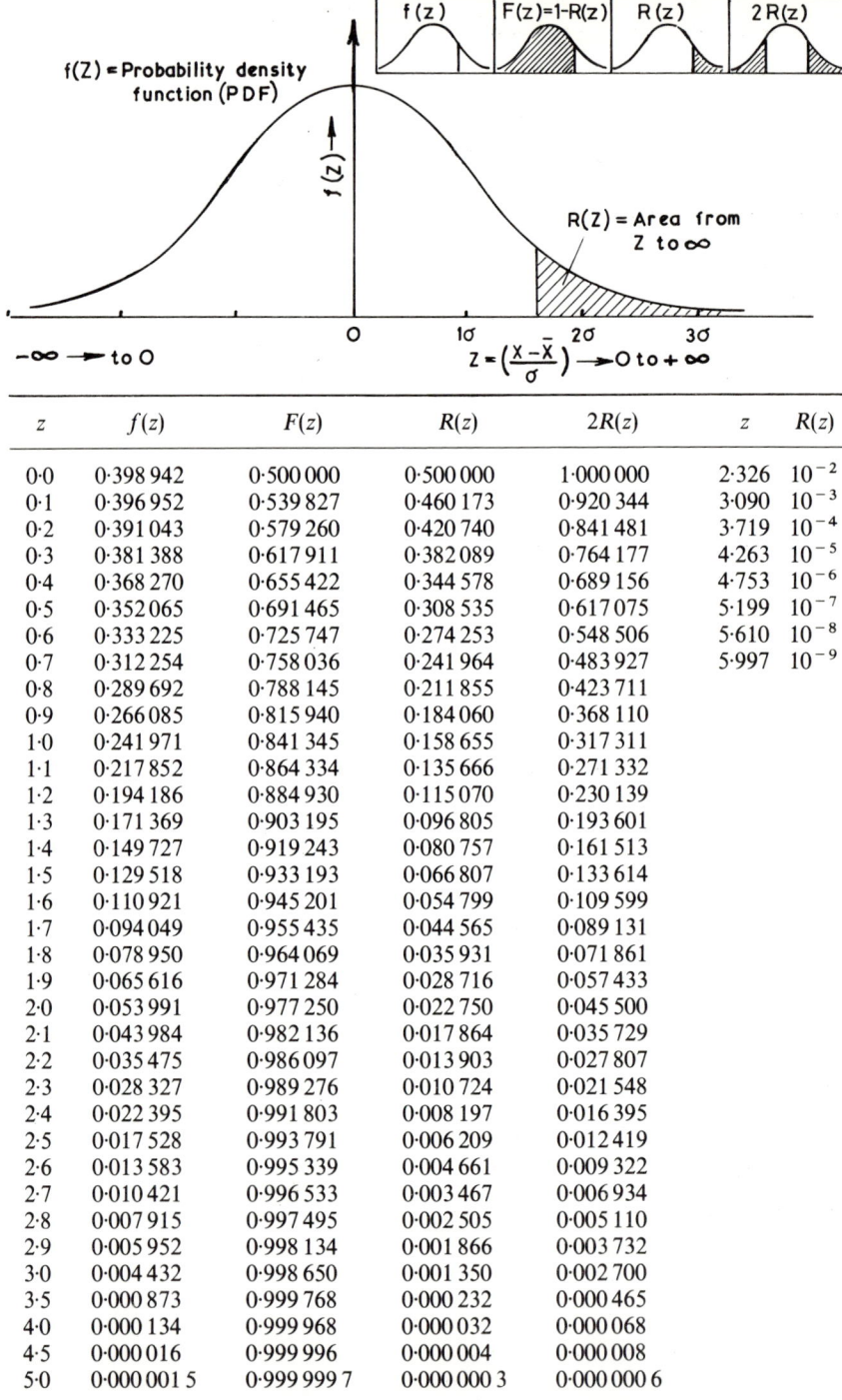

z	f(z)	F(z)	R(z)	2R(z)	z	R(z)
0·0	0·398 942	0·500 000	0·500 000	1·000 000	2·326	10^{-2}
0·1	0·396 952	0·539 827	0·460 173	0·920 344	3·090	10^{-3}
0·2	0·391 043	0·579 260	0·420 740	0·841 481	3·719	10^{-4}
0·3	0·381 388	0·617 911	0·382 089	0·764 177	4·263	10^{-5}
0·4	0·368 270	0·655 422	0·344 578	0·689 156	4·753	10^{-6}
0·5	0·352 065	0·691 465	0·308 535	0·617 075	5·199	10^{-7}
0·6	0·333 225	0·725 747	0·274 253	0·548 506	5·610	10^{-8}
0·7	0·312 254	0·758 036	0·241 964	0·483 927	5·997	10^{-9}
0·8	0·289 692	0·788 145	0·211 855	0·423 711		
0·9	0·266 085	0·815 940	0·184 060	0·368 110		
1·0	0·241 971	0·841 345	0·158 655	0·317 311		
1·1	0·217 852	0·864 334	0·135 666	0·271 332		
1·2	0·194 186	0·884 930	0·115 070	0·230 139		
1·3	0·171 369	0·903 195	0·096 805	0·193 601		
1·4	0·149 727	0·919 243	0·080 757	0·161 513		
1·5	0·129 518	0·933 193	0·066 807	0·133 614		
1·6	0·110 921	0·945 201	0·054 799	0·109 599		
1·7	0·094 049	0·955 435	0·044 565	0·089 131		
1·8	0·078 950	0·964 069	0·035 931	0·071 861		
1·9	0·065 616	0·971 284	0·028 716	0·057 433		
2·0	0·053 991	0·977 250	0·022 750	0·045 500		
2·1	0·043 984	0·982 136	0·017 864	0·035 729		
2·2	0·035 475	0·986 097	0·013 903	0·027 807		
2·3	0·028 327	0·989 276	0·010 724	0·021 548		
2·4	0·022 395	0·991 803	0·008 197	0·016 395		
2·5	0·017 528	0·993 791	0·006 209	0·012 419		
2·6	0·013 583	0·995 339	0·004 661	0·009 322		
2·7	0·010 421	0·996 533	0·003 467	0·006 934		
2·8	0·007 915	0·997 495	0·002 505	0·005 110		
2·9	0·005 952	0·998 134	0·001 866	0·003 732		
3·0	0·004 432	0·998 650	0·001 350	0·002 700		
3·5	0·000 873	0·999 768	0·000 232	0·000 465		
4·0	0·000 134	0·999 968	0·000 032	0·000 068		
4·5	0·000 016	0·999 996	0·000 004	0·000 008		
5·0	0·000 001 5	0·999 999 7	0·000 000 3	0·000 000 6		

Fig. 4.5. Standard normal probability distribution.

Table 4.1(a). Values of the distribution function of the normal distribution[a]

$$\Phi(u) = \Phi\left(\frac{t-\mu}{\sigma}\right) = \frac{1}{\sqrt{2\pi}} \int_{-x}^{u} \exp\left(-\frac{x^2}{2}\right) dx, \quad \Phi(-u) = 1 - \Phi(u).$$

u	0·00	0·01	0·02	0·03	0·04	0·05	0·06	0·07	0·08	0·09
0·0	0·5000	0·5040	0·5080	0·5120	0·5160	0·5199	0·5239	0·5279	0·5319	0·5359
0·1	0·5398	0·5438	0·5478	0·5517	0·5557	0·5596	0·5636	0·5675	0·5714	0·5753
0·2	0·5793	0·5832	0·5871	0·5910	0·5948	0·5987	0·6026	0·6064	0·6103	0·6141
0·3	0·6179	0·6217	0·6255	0·6293	0·6331	0·6368	0·6406	0·6443	0·6480	0·6517
0·4	0·6554	0·6591	0·6628	0·6664	0·6700	0·6736	0·6772	0·6808	0·6844	0·6879
0·5	0·6915	0·6950	0·6985	0·7019	0·7054	0·7088	0·7123	0·7157	0·7190	0·7224
0·6	0·7257	0·7291	0·7324	0·7357	0·7389	0·7422	0·7454	0·7486	0·7517	0·7549
0·7	0·7580	0·7611	0·7642	0·7673	0·7704	0·7734	0·7764	0·7794	0·7823	0·7852
0·8	0·7881	0·7910	0·7939	0·7967	0·7995	0·8023	0·8051	0·8079	0·8106	0·8133
0·9	0·8159	0·8186	0·8212	0·8238	0·8264	0·8289	0·8315	0·8340	0·8365	0·8389
1·0	0·8413	0·8438	0·8461	0·8485	0·8508	0·8531	0·8554	0·8577	0·8599	0·8621
1·1	0·8643	0·8665	0·8686	0·8708	0·8729	0·8749	0·8770	0·8790	0·8810	0·8830
1·2	0·8849	0·8869	0·8888	0·8907	0·8925	0·8944	0·8962	0·8980	0·8997	0·9015
1·3	0·9032	0·9049	0·9066	0·9082	0·9099	0·9115	0·9131	0·9147	0·9162	0·9177
1·4	0·9192	0·9207	0·9222	0·9236	0·9251	0·9265	0·9279	0·9292	0·9306	0·9319
1·5	0·9332	0·9345	0·9357	0·9370	0·9382	0·9394	0·9406	0·9418	0·9429	0·9441
1·6	0·9452	0·9463	0·9474	0·9484	0·9495	0·9505	0·9515	0·9525	0·9535	0·9545
1·7	0·9554	0·9564	0·9573	0·9582	0·9591	0·9599	0·9608	0·9616	0·9625	0·9633
1·8	0·9641	0·9649	0·9656	0·9664	0·9671	0·9678	0·9686	0·9693	0·9699	0·9706
1·9	0·9713	0·9719	0·9726	0·9732	0·9738	0·9744	0·9750	0·9756	0·9761	0·9767
2·0	0·9773	0·9778	0·9783	0·9788	0·9793	0·9798	0·9803	0·9808	0·9812	0·9817
2·1	0·9821	0·9826	0·9830	0·9834	0·9838	0·9842	0·9846	0·9850	0·9854	0·9857
2·2	0·9881	0·9864	0·9868	0·9871	0·9875	0·9878	0·9881	0·9884	0·9887	0·9890
2·3	0·9893	0·9896	0·9898	0·9901	0·9904	0·9906	0·9909	0·9911	0·9913	0·9916
2·4	0·9918	0·9920	0·9922	0·9925	0·9927	0·9929	0·9931	0·9932	0·9934	0·9936
2·5	0·9938	0·9940	0·9941	0·9943	0·9945	0·9946	0·9948	0·9949	0·9951	0·9952
2·6	0·9953	0·9955	0·9956	0·9957	0·9959	0·9960	0·9961	0·9962	0·9963	0·9964
2·7	0·9965	0·9966	0·9967	0·9968	0·9969	0·9970	0·9971	0·9972	0·9973	0·9974
2·8	0·9974	0·9975	0·9976	0·9977	0·9977	0·9978	0·9979	0·9979	0·9980	0·9981
2·9	0·9981	0·9982	0·9983	0·9983	0·9984	0·9984	0·9985	0·9985	0·9986	0·9986
3·0	0·9987	0·9987	0·9987	0·9988	0·9988	0·9989	0·9989	0·9989	0·9990	0·9990

[a] H. F. Zalud and A. H. Zaludova, personal communication.

Table 4.1(b). Selected values[a] of
$$\Phi(u) = \frac{1}{\sqrt{2\pi}} \int_{-\infty}^{u} \exp\left(-\frac{x^2}{2}\right) dx.$$

u	$\Phi(u)$	u	$\Phi(u)$
−3·29	0·0005	0·00	0·50
−3·09	0·001	0·25	0·60
−2·58	0·005	0·52	0·70
−2·33	0·01	0·84	0·80
−1·96	0·025	1·28	0·90
−1·64	0·05	1·64	0·95
−1·28	0·10	1·96	0·975
−0·84	0·20	2·33	0·99
−0·52	0·30	2·58	0·995
−0·25	0·40	3·09	0·999
−0·00	0·50	3·29	0·9995

[a] H. F. Zalud and A. H. Zaludova, personal communication.

4.7 GROUPED FREQUENCIES

When large numbers of readings or data are involved it is convenient to group these into finite intervals or cells. The height of each cell is then a measure of the number of readings or data within the interval. The diagrams thus produced are called histograms.

Table 4.2

Class mid mark in an exam (x)	Frequency (f)	(f. x)
59·5	1	59·5
69·5	2	139·0
79·5	9	715·5
89·5	22	1969·0
99·5	33	3283·5
109·5	22	2409·0
119.5	8	956·0
129·5	2	259·0
139·5	1	139·5
	$\Sigma f = 100$	$\Sigma(f.x) = 9930$

From these totals,
$$\bar{x} = \frac{\Sigma(f.x)}{\Sigma(f)} = \frac{9930}{100} = 99\cdot30$$

4. TYPES OF DISTRIBUTIONS

Table 4.2 gives data which are presented in histogram form in Fig. 4.6. To establish a best-fit normal distribution for this grouped frequency, we must evaluate its arithmetic mean and standard deviation. We proceed as follows:

1. Assume an approximate mean mid-mark, say $x_0 = 99.5$.

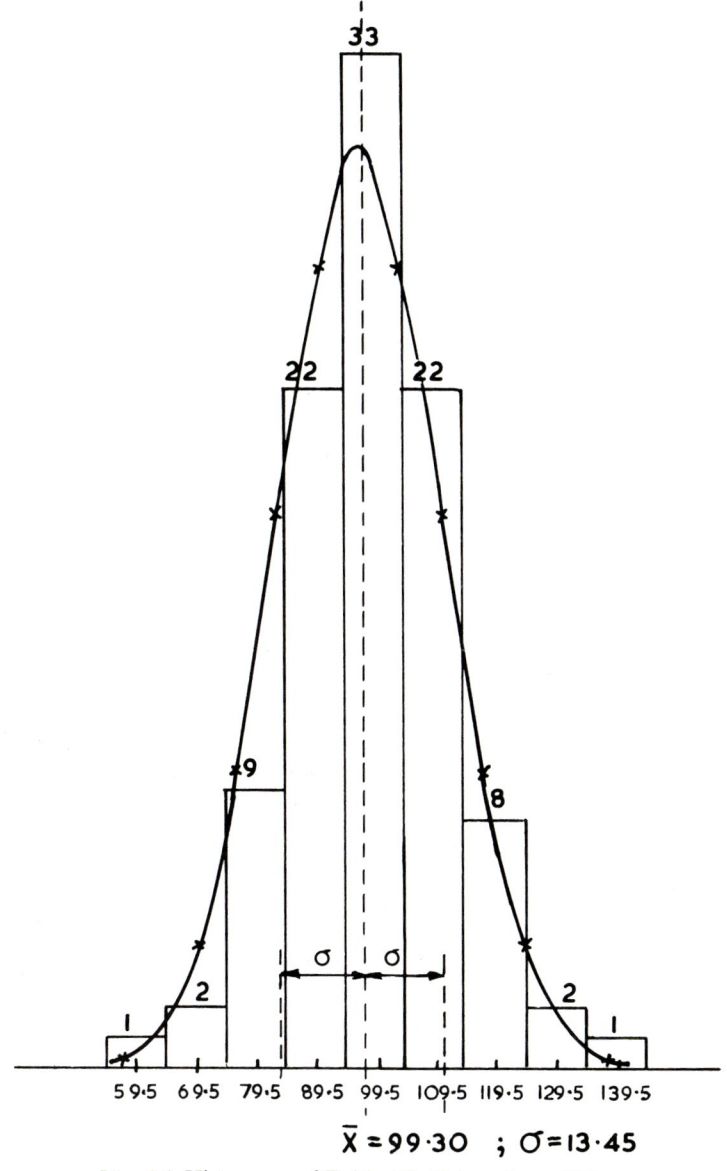

Fig. 4.6. Histogram of Table 4.2. (Interval $c = 10$.)

2. The number of intervals of any class mark from the x_0 value may be written as

$$t = \frac{x - x_0}{c} \quad \text{or} \quad x = x_0 + c \cdot t \qquad (i)$$

Multiplying both sides by the frequency (f) we get

$$f \cdot x = f \cdot x_0 + f \cdot c \cdot t$$

3. Table 4.2 can then be re-calculated to read

$$\frac{\sum f \cdot x}{\sum f} = \frac{\sum f \cdot x_0}{\sum f} + \frac{\sum f \cdot c \cdot t}{\sum f}$$

or,

$$\left. \begin{array}{c} \bar{x} = x_0 + c \cdot \bar{t} \\ \\ \text{since} \quad \dfrac{c \cdot \sum f \cdot t}{\sum f} = c \cdot \bar{t} \end{array} \right\} \qquad (ii)$$

4. The re-calculated parameters are shown in Table 4.3.

Table 4.3

	Class mid-mark x	t	t^2	Frequency f	$(f \cdot t)$	$f \cdot t^2$
	59·5	−4	16	1	−4	16
	69·5	−3	9	2	−6	18
	79·5	−2	4	9	−18	36
	89·5	−1	1	22	−22	22
$x_0 =$	99·5	0	0	33	0	0
	109·5	+1	1	22	+22	22
	119·5	+2	4	8	+16	32
	129·5	+3	9	2	+6	18
	139·5	+4	16	1	+4	16
	$C = 10$	—		$\sum f = 100$	$\sum ft = -2$	180

We get from equation (ii)

$$\bar{x} = 99 \cdot 5 + 10 \left(\frac{-2}{100} \right)$$

or

$$\bar{x} = 99 \cdot 5 - 0 \cdot 2 = 99 \cdot 3 \qquad (iii)$$

(which corresponds to the value calculated at the bottom of Table 4.2).

4. TYPES OF DISTRIBUTIONS

5. To determine the standard deviation, we may start with the expression given in equation (4.17), viz.

$$\sigma^2 = \left[\frac{\Sigma(x^2)}{n} - \left(\frac{\Sigma x}{n}\right)^2\right]$$

Alternatively one can define the variance as for a continuous function (see equation (4.18), thus

$$\sigma^2 = \frac{\Sigma[f(x-\bar{x})^2]}{\Sigma f}, \tag{4.27}$$

or

$$\sigma^2 = \left[\frac{\Sigma(f.x^2)}{\Sigma(f)} - (\bar{x})^2\right] \tag{4.28}$$

Equation (4.28) may be re-written in the following way:

$$\sigma^2 = \left[\frac{\Sigma(f.x^2)}{\Sigma(f)} - (\bar{x})^2\right] + [2x_0.\bar{x} - 2x_0.\bar{x}] + [x_0^2 - x_0^2]$$

or

$$\sigma^2 = \left[\frac{\Sigma(f.x^2)}{\Sigma(f)} - 2x_0\frac{\Sigma(f.x)}{\Sigma(f)} + \frac{x_0^2\Sigma(f)}{\Sigma(f)}\right] - (\bar{x} - x_0)^2$$

or

$$\sigma^2 = \frac{\Sigma f.(x-x_0)^2}{\Sigma(f)} - (\bar{x} - x_0)^2$$

and further

$$\sigma^2 = c^2\left[\frac{\Sigma f.(x-x_0/c)^2}{\Sigma(f)} - \left(\frac{\bar{x}-x_0}{c}\right)^2\right] \tag{4.29}$$

But we have denoted

$$t = \left(\frac{x - x_0}{c}\right)$$

therefore equation (4.29) becomes

$$\sigma^2 = c^2\left[\frac{\Sigma(f.t^2)}{\Sigma(f)} - (\bar{t})^2\right] = c^2\left[\frac{\Sigma(f.t^2)}{\Sigma(f)} - \frac{\Sigma(f.t)^2}{\Sigma(f)}\right] \tag{4.30}$$

which has a form similar to equation (4.28).

Using values from Table 4.3, we obtain

$$\sigma = 10\left[\frac{180}{100} - \left(\frac{-2}{100}\right)^2\right]^{1/2}$$

or

$$\sigma = 13{\cdot}415$$

From equation (4.21), we have

$$f(z) = \frac{1}{\sigma\sqrt{2\pi}} \cdot \exp\left(\frac{-z^2}{2}\right)$$

In our histogram the total area is equal to the total frequency multiplied by the cell interval, i.e.

$$\text{area} = 100 \times 10 = 1000 \text{ units},$$

and $\sigma = 13{\cdot}415$. The ordinates for the corresponding normal curve would, therefore, be

$$y = \frac{1000}{13{\cdot}415} \cdot \left[\frac{1}{\sqrt{2\pi}} \cdot \exp\left(\frac{-z^2}{2}\right)\right]$$

or

$$y = 74{\cdot}54 \left[\frac{1}{\sqrt{2\pi}} \cdot \exp\left(\frac{-z^2}{2}\right)\right]$$

The factor in the bracket for various values of z are given in the table on Fig. 4.5. Thus:

z	0	0·5	1·0	1·5	2·0	2·5	3	3·5
$f(z)$	0·399	0·352	0·242	0·130	0·054	0·018	0·0044	0·0008
y	29·74	26·24	18·0	9·7	4·01	1·34	0·33	0·060

The corresponding "normal" curve is shown superimposed in Fig. 4.6.

Example 4.11 The length of 80 items of mass-produced components were found to have a dimensional range of 42·23 mm to 42·67 mm. The block diagram in Fig. 4.7 gives the values of sizes and frequency of their occurrence. The distribution of measured lengths have been divided into 10 equal intervals of 0·05 mm.

(a) Estimate the mean and standard deviation of the histogram distribution.
(b) Given the lower limit = 42·20 and the upper limit = 42·60, what is the probability that the length of a component will exceed the upper limit?
(c) Trace the corresponding "normal" distribution curve on the histogram.

(For solution see p. 366.)

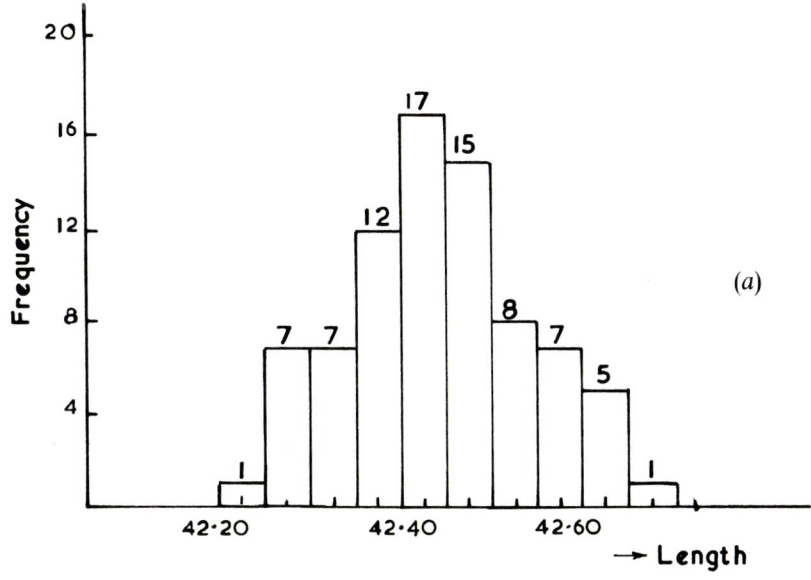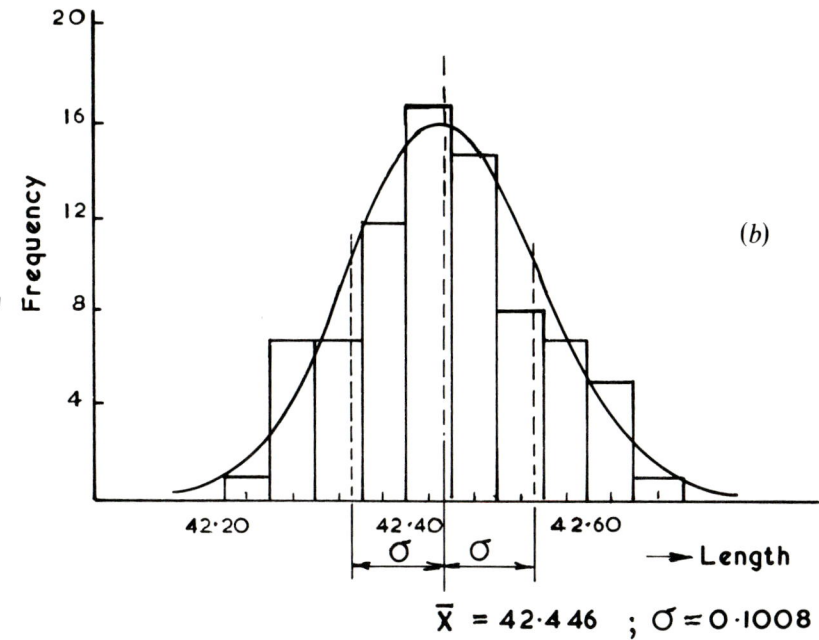

Fig. 4.7(a) and (b).

4.8 DISTRIBUTION OF ERRORS

Experimental and field results invariably involve some errors. These may be classified as:

(a) Systematic errors, due to definite causes. Such errors may be removed or allowed for.
(b) Random errors, due to a large number of causes. Such errors cannot be completely excluded; only some mean value correction may be introduced.

According to Bronstein and Semendyayev (1973), experimental data indicate that the following characteristics have been found to apply to the distribution of such random errors:

(1) The density function f(ε) is an even function, i.e.

$$f(-\varepsilon) = f(+\varepsilon)$$

The errors with different signs relative to a mean value are equally likely to occur.

(2) The density function f(ε) is decreasing for $x > 0$; i.e. large errors are less frequent than small ones.

(3) In most cases the best approximation to reality is the normal distribution:

$$f(\varepsilon) = \frac{1}{\sigma\sqrt{2\pi}} \cdot \exp\left[-\frac{\varepsilon^2}{2 \cdot \sigma^2}\right]$$

the function stretching from $-\infty$ to $+\infty$, and σ = the standard deviation, or the root mean square of the error ε. From equations (4.17) and (4.19), we see that

$$\sigma^2 = \overline{(\varepsilon^2)} - (\bar{\varepsilon})^2$$

or

$$\sigma^2 = \int_{-\infty}^{+\infty} \varepsilon^2 \cdot f(\varepsilon) \cdot d\varepsilon$$

where

$\overline{(\varepsilon^2)}$ = mean of the squares of all errors
$(\bar{\varepsilon})^2$ = the square of the arithmetic mean of the errors

(4) The simple mean of the absolute value of all errors is

$$\eta = \frac{1}{n}\sum_{1}^{n}|\varepsilon_i| \qquad (4.31a)$$

or

$$\eta = |\bar{\varepsilon}| = 2 \int_0^\infty |\varepsilon| \cdot f(\varepsilon) \cdot d\varepsilon, \quad \text{which has a finite value} \qquad (4.31b)$$

(5) The probable error r is defined as a quantity such that the probability of the error being less or greater than r in the absolute is equal to $\frac{1}{2}$; i.e.

$$\int_{-r}^{+r} f(\varepsilon) \cdot d\varepsilon = \tfrac{1}{2} \qquad (4.32)$$

From standard probability tables, we obtain

$$r = 0\cdot 6749 \cdot \sigma$$

(6) The measure of accuracy is defined as (Bronstein and Semendyayev, 1973)

$$h = \frac{1}{\sqrt{2 \cdot \sigma}}; \quad \text{accuracy} \propto \frac{1}{\text{spread}}, \qquad (4.33a)$$

or

$$h = \frac{0\cdot 707}{\sigma}; \quad h \propto \frac{1}{\sigma} \qquad (4.33b)$$

(7) The various parameters noted above are related as follows

$$\eta = \frac{1}{\sqrt{\pi \cdot h}} = \sqrt{\frac{2}{\pi}} \cdot \sigma \qquad (4.34a)$$

$$\sigma = \frac{1}{\sqrt{2 \cdot h}} = \sqrt{\frac{\pi}{2}} \cdot \eta \qquad (4.34b)$$

$$h = \frac{1}{\sqrt{\pi \cdot \eta}} = \frac{1}{\sqrt{2 \cdot \sigma}} \qquad (4.34c)$$

Defining $\rho = r \cdot h$, we also have

$$\sigma = \frac{r}{\sqrt{2} \cdot \rho}; \quad r = \rho \cdot \sqrt{2} \cdot \sigma = \rho \cdot \sqrt{\pi} \cdot \eta$$

The following numerical values are useful to have at hand when calculating these parameters:

$$\frac{1}{\sqrt{2}} = 0\cdot 707; \quad \frac{1}{\sqrt{\pi}} = 0\cdot 5642; \quad \sqrt{\frac{2}{\pi}} = 0\cdot 7979$$

$$\rho = 0\cdot 4769; \quad \rho\sqrt{2} = 0\cdot 6749; \quad \rho \cdot \sqrt{\pi} = 0\cdot 8454$$

4.9 EFFECTS OF SAMPLE SIZE

(a) If a limited number of readings, i.e. representative sample, has been taken of a quantity x, each with the same degree of accuracy, and assuming the errors to be normally distributed, then the mean of the readings would be

$$\hat{\mu} \simeq \frac{1}{n}\sum_1^n (x_1 + x_2 + \ldots + x_n) \qquad (4.35)$$

or

$$\hat{\mu} \simeq \frac{1}{n}\sum_1^n (x_i)$$

This estimate of $\hat{\mu}$ will correspond fairly well with the true value x. Similarly the first estimate of the sample variance would be

$$s^2 \simeq \frac{1}{n}\sum_1^n (x_i - \hat{\mu})^2 \simeq \frac{1}{n}\sum_1^n (\varepsilon_i)^2 \qquad (4.36)$$

However, the accuracy of the sum variance will depend largely upon the number (n) of the readings taken. Statistical analysis has shown that a more representative estimate of the variance of the whole population is obtained by using "Bessel's correction", i.e.

$$\sigma^2 = \frac{n}{n-1} \cdot s^2 \qquad (4.37a)$$

or

$$\sigma^2 = \frac{1}{n-1}\sum_1^n (\varepsilon_i)^2 \qquad (4.37b)$$

The simple arithmetic mean in this case then becomes

$$\eta = \frac{1}{\sqrt{n.(n-1)}} \cdot \sum_1^n |\varepsilon_i| \qquad (4.38a)$$

or

$$\eta \simeq \frac{1}{(n-\frac{1}{2})} \cdot \sum_1^n |\varepsilon_i| \quad \text{(from Taylor's Series)} \qquad (4.38b)$$

(b) If the readings of x_i are not made with the same accuracy, but with different variances, the estimates of $\hat{\mu}$ and s^2 have to be modified. In this case one has to use "weighted" readings, viz.

$$\hat{\mu} = w_1 . x_1 + w_2 . x_2 + \ldots + w_n . x_n$$

or

$$\hat{\mu} = \sum_1^n (w_i . x_i) \qquad (4.39)$$

4. TYPES OF DISTRIBUTIONS

In order to ensure that $\hat{\mu}$ is as nearly equal to the true mean μ, representing the actual value of x, the weighting values of w_i have to be chosen so as to satisfy two conditions:

(i) $\sum_{1}^{n} w_i = 1$ (4.40)

(ii) the resultant variance of $\hat{\mu}$ should be as small as possible. We have

$$s^2 = \text{var}(w_1 . x_1 + w_2 x_2 + w_3 x_3 + \ldots + w_n x_n)$$

or

$$s^2 = [w_1^2 \sigma_1^2 + w_2^2 \sigma_2^2 \ldots + w_n^2 . \sigma_n^2]$$

or

$$s^2 = \sum_{1}^{n} (w_i^2 . \sigma_i^2) \qquad (4.41)$$

To make $\text{var}(\hat{\mu})$ a minimum, we make the partial derivatives of $\text{var}(\hat{\mu})$ relative to each w_i equal to zero, i.e.

$$\frac{\partial(s^2)}{\partial w_1} = \frac{\partial(s^2)}{\partial w_2} = \frac{\partial(s^2)}{\partial w_n} = 0$$

We aim to establish a relationship between the weighting (w_i) and the variance (σ_i^2) for each reading. This is best demonstrated with a simple example. Consider three readings, x_1, x_2 and x_3, which have variances σ_1^2, σ_2^2 and σ_3^2 respectively. We have

$$s^2 = [w_1^2 \sigma_1^2 + w_2^2 \sigma_2^2 + w_3^2 \sigma_3^2]$$
$$= [w_1^2 \sigma_1^2 + w_2^2 \sigma_2^2 + (1 - w_1 - w_2)^2 . \sigma_3^2]$$

To minimize (s^2), i.e. to make $\hat{\mu}$ as accurate as possible, we make

$$\frac{\partial(s^2)}{\partial w_1} = \frac{\partial(s^2)}{\partial w_2} = \frac{\partial(s^2)}{\partial w_3} = 0$$

This yields

$$2w_1 . \sigma_1^2 - 2(1 - w_1 - w_2) \sigma_3^2 = 0 \qquad \text{(i)}$$

and

$$2w_2 . \sigma_2^2 - 2(1 - w_1 - w_2) . \sigma_3^2 = 0 \qquad \text{(ii)}$$

which combined give

$$w_2 = w_1 . \left(\frac{\sigma_1}{\sigma_2}\right)^2 \qquad \text{(iii)}$$

Inserting (iii) into (i) and (ii), we get

$$w_1 = \frac{\sigma_2^2 \cdot \sigma_3^2}{\sigma_1^2 \cdot \sigma_2^2 + \sigma_1^2 \cdot \sigma_3^2 + \sigma_2^2 \cdot \sigma_3^2}$$

and

$$w_2 = \frac{\sigma_1^2 \cdot \sigma_3^2}{\sigma_1^2 \cdot \sigma_2^2 + \sigma_1^2 \cdot \sigma_3^2 + \sigma_2^2 \cdot \sigma_3^2}$$

We may thus define the weighting w_i as

$$\left. \begin{array}{c} w_i = \dfrac{k}{\sigma_i^2} \\[1em] \text{where} \\[1em] k = \dfrac{\sigma_1^2 \cdot \sigma_2^2 \cdot \sigma_3^2}{\sigma_1^2 \cdot \sigma_2^2 + \sigma_1^2 \cdot \sigma_3^2 + \sigma_2^2 \cdot \sigma_3^2} \end{array} \right\} \quad (4.42)$$

Considering equations (4.40) and (4.42), we note that

$$\sum_1^n w_i = \sum_1^n \left(\frac{k}{\sigma_i^2}\right) = 1,$$

or

$$k = 1 \Big/ \left[\sum_1^n \left(\frac{1}{\sigma_i^2}\right)\right] \qquad (4.43)$$

And from (4.41) and (4.42), we get

$$s^2 = \sum_1^n \left(\frac{k}{\sigma_i^2}\right)^2 \cdot \sigma_i^2 = \sum \left(\frac{k^2}{\sigma_i^2}\right)$$

Using equation (4.43) for k, this becomes

$$s^2 = \frac{\sum(1/\sigma_i^2)}{(\sum(1/\sigma_i^2))^2} = \frac{1}{\sum(1/\sigma_i^2)}$$

i.e.

$$s^2 = k \qquad (4.44)$$

The derivation of the best mean value ($\hat{\mu}$) and its variance (s^2) of a number of sample readings (x_i) with variances (σ_i^2), is well demonstrated by the following example:

Readings (x_i)	Variances (σ_i^2)	Standard deviations (σ_i)
4·5	0·2	0·447
4·3	0·4	0·632
4·7	0·3	0·548
4·6	0·1	0·316

Estimate the best mean value and the variance of these readings. From equation (4.43), we get

$$k = \frac{1}{\sum(1/\sigma_i^2)} = \left(\frac{1}{0.2} + \frac{1}{0.4} + \frac{1}{0.3} + \frac{0}{0.1}\right)^{-1}$$

or

$$k = (0.048)$$

And from equations (4.39) and (4.42)

$$\hat{\mu} = k \cdot \sum\left(\frac{x_i}{\sigma_i^2}\right) = k \cdot \left[\frac{4.5}{0.2} + \frac{4.3}{0.4} + \frac{4.7}{0.3} + \frac{4.6}{0.1}\right]$$

or

$$\hat{\mu} = 0.048[94.92] = 4.556$$

and

$$s^2 = 0.048 \quad (\text{equation (4.44)})$$

REFERENCES

Bompas-Smith, J. H. (1973). "Mechanical Survival." McGraw-Hill, New York.
Bronstein, I. N. and Semendyayev, K. A. (1973). "A Guide Book to Mathematics." Springer Verlag, New York.
Carter, A. D. S. (1972). "Mechanical Reliability." Macmillan, London.
Moroney, M. J. (1964). "Facts from Figures." Penguin, Harmondsworth.
Palmgren, A. (1959). "Ball and Roller Bearing Engineering". SKF Industries Inc., U.S.A.

Chapter 5

The Algebra of Random Numbers

5.1 DEFINITIONS

(a) A *deterministic variable* is a single discrete number, i.e. a single value number. Examples of discrete numbers are:
 4 wheels on a car
 25 teeth on a gear
 60 minutes in one hour
 π = ratio of perimeter to diameter of a circle = 3·141 59 ...
 e = 2·718 28 ...
It is assumed that the reader is familiar with the algebra and calculus of deterministic numbers.

(b) A *random variable* is a set of real numbers corresponding to the outcome of a series of experiments, tests, data measurements, etc. Examples of random variates are:
 Tensile strength, as determined from a number of tests, each test result being slightly different.
 Dimensions of manufactured components, which vary between specified limits.
 Time as read from a watch. The accuracy of reading will depend on how slow or fast the watch goes and what its initial setting was.
 Examination performance of students, which will vary for each student and as between students.
Random variates are described by,
 (i) a *mean value*, $\bar{x}$ or μ
 (ii) the *standard variation*, s or σ
(N.B. These and other statistical terms have been defined in Chapter 4.)

5.2 BASIC STATISTICAL THEORY

We need a special algebra to operate with random variates (or numbers). Haugen (1968) examines a number of basic algebraic operations, summarized

5. THE ALGEBRA OF RANDOM NUMBERS

in Table 5.1. The derivations are based on basic statistical theory and on a "partial derivative method" which is applicable for values of $\sigma < 0.20\mu$ (see Section 5.3). A few proofs of algebraic operations with random variates are illustrated on p. 106.

5.3 PARTIAL DERIVATIVE METHOD

To determine σ_ϕ let $\phi = f(x, y)$ where x and y are normally distributed and independent random variates. Then

$$\phi_i = f(x_i, y_i) \tag{5.1}$$

$$\bar{\phi} = f(\bar{x}, \bar{y}) \tag{5.2}$$

Denoting $|x_i - \bar{x}| = \delta x_i$ and $|y_i - \bar{y}| = \delta y_i$ being small differences compared to the mean values of x and y, we can write (5.1) as

$$\phi_i = f[(\bar{x} + \delta x_i), (\bar{y} + \delta y_i)] \tag{5.3}$$

Using Taylor's expansion (see Chapter 1), we can write (5.3) as follows:

$$\phi_i = f(\bar{x}, \bar{y}) + \frac{\partial \phi}{\partial x} \cdot \delta x_i + \frac{\partial \phi}{\partial y} \cdot \delta y_i + \dots \quad \text{higher orders}$$

Thus,

$$\phi_i \simeq f(\bar{x}, \bar{y}) + \frac{\partial \phi}{\partial x} \cdot \delta x_i + \frac{\partial \phi}{\partial y} \cdot \delta y_i \quad \text{if } \delta x_i \text{ and } \delta y_i \text{ are small} \tag{5.4}$$

From equation (5.2) we have

$$f(\bar{x}, \bar{y}) = \bar{\phi} = \frac{1}{n}\sum_1^n (\phi_i)$$

and equation (5.4) then becomes

$$(\phi_i - \bar{\phi}) = \delta\phi_i = \frac{\partial \phi}{\partial x} \cdot \delta x_i + \frac{\partial \phi}{\partial y} \cdot \delta y_i \tag{5.5}$$

Now, by definition,

$$\sigma_\phi^2 = \frac{1}{n}\sum_1^n (\phi_i - \bar{\phi})^2 = \frac{1}{n}\sum_1^n (\delta\phi_i)^2 \tag{5.6}$$

So, squaring (5.5), we get

$$\sigma_\phi^2 = \frac{1}{n}\left[\left(\frac{\partial \phi}{\partial x}\right)^2 \left\{\sum_1^n (\delta x_i)^2\right\} + \left(\frac{\partial \phi}{\partial y}\right)^2 \left\{\sum_1^n (\delta y_i)^2\right\} + 2\left(\frac{\partial \phi}{\partial x}\right)\cdot\left(\frac{\partial \phi}{\partial y}\right)\cdot\left\{\sum_1^n (\delta x_i \cdot \delta y_i)\right\}\right] \tag{5.7}$$

Table 5.1. Basic algebra for normally distributed random variables
(after Haugen, 1968, reprinted by permission of John Wiley and Sons Inc.)

Function	Correlated variables $(-1 \leqslant r \leqslant +1)$	Independent variables $(r = 0)$	Method of derivation
$z = a$		$\bar{z} = a$ (a = constant) $\sigma_z = 0$	
$z = a \cdot x$		$\bar{z} = a\bar{x}$ (a = constant) $\sigma_z = a \cdot \sigma_x$	
$z = x + a$		$\bar{z} = \bar{x} + a$ (a = constant) $\sigma_z = \sigma_x$	
$z = x \pm y$	$\bar{z} = \bar{x} \pm \bar{y}$ $\sigma_z = (\sigma_x^2 + \sigma_y^2 \pm 2r\sigma_x \cdot \sigma_y)^{1/2}$	$\bar{z} = \bar{x} \pm \bar{y}$ $\sigma_z = (\sigma_x^2 + \sigma_y^2)^{1/2}$	Statistical
$z = x \cdot y$	$\bar{z} = \bar{x} \cdot \bar{y} + r \cdot \sigma_x \cdot \sigma_y$ $\sigma_z = [(\bar{x}^2 \cdot \sigma_y^2 + \bar{y}^2 \cdot \sigma_x^2 + \sigma_x^2 \cdot \sigma_y^2)(1 + r^2)]^{1/2}$ or $\sigma_z \approx [\bar{x}^2 \cdot \sigma_y^2 + \bar{y}^2 \sigma_x^2 + 2r \cdot \bar{x} \cdot \bar{y} \cdot \sigma_x \cdot \sigma_y]^{1/2}$	$\bar{z} = \bar{x} \cdot \bar{y}$ $\sigma_z = [\bar{x}^2 \cdot \sigma_y^2 + \bar{y}^2 \cdot \sigma_x^2 + \sigma_x^2 \cdot \sigma_y^2]^{1/2}$ or $\sigma_z \approx [\bar{x}^2 \cdot \sigma_y^2 + \bar{y}^2 \sigma_x^2]^{1/2}$	Approximate: partial derivative
$z = x^2$	$\bar{z} = \bar{x}^2 + \sigma_x^2$ $\sigma_z = [4\bar{x}^2 \cdot \sigma_x^2 + 2\sigma_x^4]^{1/2}$ or $\sigma_z \approx [4\bar{x}^2 \sigma_x^2]^{1/2}$		

Table 5.1 (cont)

Function	Correlated variables ($-1 \leqslant r \leqslant +1$)	Independent variables ($r = 0$)	Method of derivation
$z = x/y$	$\bar{z} \simeq \bar{x}/\bar{y} + \sigma_y \cdot \bar{x}/\bar{y}^2(\sigma_y/\bar{y} - r \cdot \sigma_x/\bar{x})$	$\bar{z} = \bar{x}/\bar{y}$	
		$\sigma_z = 1/\bar{y} \cdot \left[\dfrac{\bar{x}^2 \cdot \sigma_y^2 + \bar{y}^2 \cdot \sigma_x^2}{\bar{y}^2 + \sigma_y^2} \right]^{1/2}$	Statistical
	$\sigma_z \simeq \bar{x}/\bar{y} \cdot [\sigma_x^2/\bar{x}^2 + \sigma_y^2/\bar{y}^2 - 2 \cdot r \cdot \sigma_x/\bar{x} \cdot \sigma_y/\bar{y}]^{1/2}$	or $\sigma_z \approx 1/\bar{y}^2 \cdot [\bar{x}^2 \cdot \sigma_y^2 + \bar{y}^2 \cdot \sigma_x^2]^{1/2}$	Approximate: partial derivative
$z = 1/x$		$\bar{z} \approx 1/\bar{x}$	
		$\sigma_z \approx \sigma_x/\bar{x}^2$	
$z = \sqrt{x}$	$\bar{z} \simeq [\tfrac{1}{2}(\bar{x} + \sqrt{\bar{x}^2 - \sigma_x^2})]^{1/2}$		
	$\sigma_z \approx \tfrac{1}{2} \cdot \sigma_x \cdot 1/\bar{z}$		
$z = ax^2 + bx + c$		$\bar{z} = a(\bar{x}^2 + \sigma_x^2) + b \cdot \bar{x} + c$	Statistical
		$\sigma_z^2 = \sigma_x^2(2a \cdot \bar{x} + b)^2 + 2a^2 \sigma_x^4$	
		or $\sigma_z \approx \sigma_x^2(2a \cdot \bar{x} + b)^2$	Approximate: partial derivative

Owing to the symmetry of a normal distribution the last term in the bracket is zero, and equation (5.7) reduces to

$$\sigma_\phi^2 = \frac{1}{n}\left[\left(\frac{\partial \phi}{\partial x}\right)^2\left\{\sum_1^n (\delta x_i)^2\right\} + \left(\frac{\partial \phi}{\partial y}\right)^2\left\{\sum_1^n (\delta y_i)^2\right\}\right]$$

But by definition

$$\frac{1}{n}\sum_1^n (\delta x_i)^2 = \sigma_x^2 \quad \text{and} \quad \frac{1}{n}\sum_1^n (\delta y_i)^2 = \sigma_y^2$$

therefore

$$\sigma_\phi^2 = \left[\left(\frac{\partial \phi}{\partial x}\right)^2 \cdot \sigma_x^2 + \left(\frac{\partial \phi}{\partial y}\right)^2 \cdot \sigma_y^2\right]$$

or, in general,

$$\sigma_\phi^2 = \left[\sum_1^n \left(\frac{\partial \phi}{\partial x_i}\right)^2 \cdot \sigma_{x_i}^2\right] \tag{5.8}$$

To ensure that the standard deviation σ_ϕ and the partial derivatives of the function are constant, the latter are taken on $x = \bar{x}, y = \bar{y}, m = \bar{m}$, etc. Thus the general partial derivative equation is

$$\sigma_\phi^2 = \left[\sum_1^n \left(\frac{\partial \phi}{\partial \bar{x}_i}\right)^2 \cdot \sigma_{x_i}^2\right] \tag{5.9}$$

where $i = 1, 2, 3, \ldots$ = number of variables in the function.

N.B. We observed in equation (5.4) that higher orders of δx_i and δy_i were neglected, as a first approximation. It has been verified that this approximation differs insignificantly from the correct value if $\sigma_i \leq 0.2 \cdot \bar{x}$; i.e. the standard deviation of the variables relative to the mean value is $\leq 20\%$.

5.4 EXAMPLES OF ALGEBRAIC DERIVATIONS

Case 1 Addition and subtraction of *independent* random variates. Let $(z, \sigma_z) = (x, \sigma_x) \pm (y, \sigma_y)$. It is clear from the illustration $\bar{z} = \bar{x} \pm \bar{y}$, and, generally, $z_i = x_i \pm y_i$.

Now by definition for standard deviation,

$$\sigma_z^2 = \frac{1}{n}\sum (z_i - \bar{z})^2$$

i.e.

$$= \frac{1}{n}\sum [(x_i \pm y_i) - (\bar{x} \pm \bar{y})]^2$$

5. THE ALGEBRA OF RANDOM NUMBERS

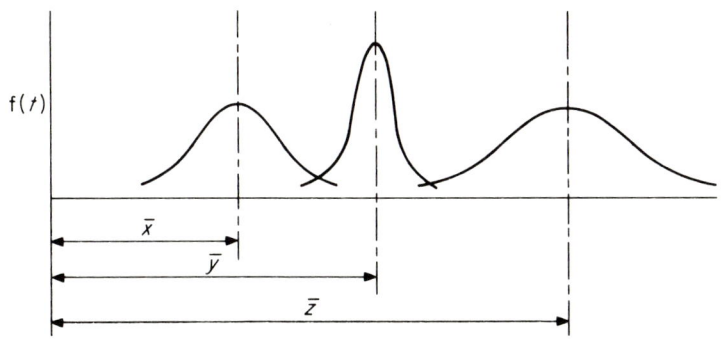

$$= \frac{1}{n} \sum [(x_i - \bar{x}) \pm (y_i - \bar{y})]^2$$

$$= \frac{1}{n} \sum [(x_i - \bar{x})^2 + (y_i - \bar{y})^2 \pm 2(x_i - \bar{x})(y_i - \bar{y})]$$

Owing to the symmetry of a normal distribution, the third term in the bracket is zero:

$$\therefore \sigma_z^2 = \frac{1}{n} \sum [(x_i - \bar{x})^2 + (y_i - \bar{y})^2]$$

or

$$\sigma_z^2 = \sigma_x^2 + \sigma_y^2 \quad (= \text{variance})$$

or

$$\sigma_z = [\sigma_x^2 + \sigma_y^2]^{1/2} \quad (= \text{standard deviation})$$

Case 2 $z = x^2$, i.e. $z = x \cdot x$. We note that x^2 is a function directly correlated to x; i.e. $r = 1$. Now by definition,

$$\sigma_x^2 = \frac{1}{n} \sum_1^n (x_i - \bar{x})^2$$

or

$$n \cdot \sigma_x^2 = \sum_1^n (x_i)^2 + \sum_1^n (\bar{x})^2 - 2 \sum_1^n (x_i \cdot \bar{x})$$

$$= \sum_1^n (x_i)^2 + n(\bar{x})^2 - 2\bar{x} \cdot n \cdot \bar{x}$$

$$\therefore n \cdot \sigma_x^2 = \sum_1^n (x_i)^2 - n(\bar{x})^2$$

or

$$\sigma_x^2 = \frac{1}{n}\sum_1^n (x_i)^2 - (\bar{x})^2 \qquad \text{(i)}$$

Since,

$$z = x^2 \quad \text{and} \quad \bar{z} = \frac{1}{n}\sum_1^n (x_i^2) \qquad \text{(ii)}$$

then combining (ii) and (i), we have

$$\sigma_x^2 = \bar{z} - (\bar{x})^2$$

or

$$\bar{z} = (\bar{x})^2 + (\sigma_x)^2 \quad \text{(See Table 5.1)} \qquad \text{(iii)}$$

Now to determine σ_z, we use the expression for $z = (x \cdot y)$ given in Table 5.1 for fully correlated variables, i.e. $r = 1$.

$$\sigma_{(x \cdot y)} = \bar{x} \cdot \bar{y} \left[\left\{ \frac{\sigma_x^2}{(\bar{x})^2} + \frac{\sigma_y^2}{(\bar{y})^2} + \frac{\sigma_x^2 \cdot \sigma_y^2}{(\bar{x})^2 (\bar{y})^2} \right\} \cdot (1 + r^2) \right]^{1/2} \qquad \text{(iv)}$$

for $x = f(y)$ and $r = 1$, i.e. $x^2 = f(x)$ and $r = 1$. Thus,

$$\sigma_{(x^2)} = (\bar{x})^2 \cdot \left[\left\{ 2 \cdot \frac{\sigma_x^2}{(\bar{x})^2} + \frac{\sigma_x^2}{(\bar{x})^4} \right\} \cdot (1 + 1) \right]^{1/2}$$

$$\therefore \sigma_{(x^2)} = \{4 \cdot (\bar{x})^2 \cdot \sigma_x^2 + 2 \cdot \sigma_x^4\}^{1/2} \qquad \text{(v)}$$

If $\sigma_x \leqslant 0 \cdot 20 \cdot (\bar{x})$, the fourth order term of σ_x may be neglected, i.e.

$$\sigma_{(x^2)} \simeq 2\bar{x} \cdot \sigma_x \qquad \text{(vi)}$$

Using the partial derivative method, we obtain,

$$\bar{z} = (\bar{x})^2 + \sigma_x^2 \quad \text{from (iii)}$$

$$\therefore \frac{\partial z}{\partial \bar{x}} = 2(\bar{x})$$

Thus,

$$\sigma_z^2 = \sum_1^n (2\bar{x})^2 \cdot \sigma_x^2$$

$$\therefore \sigma_{(x^2)} \simeq 2\bar{x} \cdot \sigma_x$$

which is the same as equation (vi) above.

Case 3 $z = \sqrt{x}$

(Note: The derivation and final result in Haugen (1968) is in error.) We write from Case 2 above

$$\bar{x} = (\bar{z})^2 + \sigma_z^2 \qquad \text{(vii)}$$

5. THE ALGEBRA OF RANDOM NUMBERS

and
$$\sigma_x = 2\sigma_z \cdot [(\bar{z})^2 + \tfrac{1}{2}\sigma_z^2]^{1/2} \qquad \text{(viii)}$$

Now, if $\sigma_z \ll \bar{z}$, then $\tfrac{1}{2}\sigma_z^2 \ll (\bar{z})^2$

$$\therefore \sigma_x \simeq 2\bar{z} \cdot \sigma_z \qquad \text{(ix)}$$

From (vii), we get
$$\bar{x} = (\bar{z})^2 + \left(\frac{\sigma_x}{2\bar{z}}\right)^2$$

i.e.
$$4 \cdot (\bar{z})^2 \cdot \bar{x} - 4(\bar{z})^4 - \sigma_x^2 = 0$$

or
$$(\bar{z})^4 - (\bar{z})^2 \cdot \bar{x} + \tfrac{1}{4}\sigma_x^2 = 0$$

which solved for $(\bar{z})^2$, yields
$$(\bar{z})^2 = \frac{\bar{x}}{2} \pm \sqrt{\frac{(\bar{x})^2}{4} - \frac{\sigma_x^2}{4}}$$

We know that $\bar{x}$ can still be $+$ve, even if $\sigma_x \to 0$, $\therefore$ only the $+$ve sign in front of the square root is valid, i.e.
$$(\bar{z})^2 = \frac{\bar{x}}{2} + \sqrt{\frac{(\bar{x})^2}{4} - \frac{\sigma_x^2}{4}}$$

or
$$\bar{z} = [\tfrac{1}{2}(\bar{x} + \sqrt{(\bar{x})^2 - \sigma_x^2})]^{1/2} \qquad \text{(x)}$$

Inserting $\sigma_z \simeq \sigma_x / 2\bar{z}$ from (ix), we get
$$\sigma_z \simeq \frac{\sigma_x}{2[\tfrac{1}{2}(\bar{x} + \sqrt{\bar{x}^2 - \sigma_x^2})]^{1/2}}$$

i.e.
$$\sigma_z \simeq \frac{\sigma_x}{[2(\bar{x} + \sqrt{\bar{x}^2 - \sigma_x^2})]^{1/2}} \qquad \text{(xi)}$$

which for $\sigma_x^2 \ll \bar{x}$ yields
$$\sigma_z \simeq \frac{\sigma_x}{2\sqrt{\bar{x}}}$$

which corresponds to the result obtained when using the partial derivative method.

Case 4 $Z = x^3$

Let $y = x^2$; $\therefore Z = y \cdot x$, where y and x are fully correlated. From Case 2

$$\bar{y} = \bar{x}^2 + \sigma_x^2$$

and

$$\sigma_y = (4\bar{x}^2 \cdot \sigma_x^2 + 2\sigma_x^4)$$

or

$$\sigma_y \simeq 2\bar{x} \cdot \sigma_x \quad \text{for } \sigma_x \ll \bar{x} \qquad \text{(xii)}$$

To calculate σ_z we use the expression for a "product" given in Table 5.1,

$$\sigma_z = \{\bar{x}^2 \cdot \sigma_y^2 + \bar{y}^2 \cdot \sigma_x^2 + 2r \cdot \bar{y} \cdot \bar{x} \cdot \sigma_y \cdot \sigma_x\}^{1/2}$$

Inserting the appropriate values, assuming $\sigma_x \ll \bar{x}$ and $r = 1$ (fully correlated variables), we get

$$\sigma_z = \left[\begin{array}{c} \bar{x}^2 \cdot (2\bar{x} \cdot \sigma_x)^2 + (\bar{x} + \sigma_x^2)^2 \cdot \sigma_x^2 \\ + 2 \cdot (\bar{x}^2 + \sigma_x^2) \cdot \bar{x} \cdot (2\bar{x} \cdot \sigma_x) \cdot \sigma_x \end{array} \right]^{1/2}$$

This yields

$$\sigma_z = [9\bar{x}^4 \cdot \sigma_x^2 + 6\bar{x}^2 \cdot \sigma_x^4 + \sigma_x^6]^{1/2} \qquad \text{(xiii)}$$

which can be written as

$$\sigma_z = 3 \cdot \sigma_x \cdot [(\bar{x}^2 + \tfrac{1}{3}\sigma_x^2)^2]^{1/2}$$

or

$$\left. \begin{array}{l} \sigma_z = 3\sigma_x \cdot [\bar{x}^2 + \tfrac{1}{3}\sigma_x^2] \\ \sigma_z = 3\bar{x}^2 \cdot \sigma_x + \sigma_x^3 \end{array} \right\} \qquad \text{(xiv)}$$

or

To obtain $\bar{z}$, we use the product rule in Table 5.1; thus,

$$\bar{z} = \bar{y} \cdot \bar{x} + r \cdot \sigma_y \cdot \sigma_x$$

i.e.

$$\bar{z} = (\bar{x}^2 + \sigma_x^2) \cdot \bar{x} + 1 \cdot (2\bar{x} \cdot \sigma_x) \cdot \sigma_x$$

i.e.

$$\bar{z} = (\bar{x}^3 + 3\bar{x} \cdot \sigma_x^2) \qquad \text{(xv)}$$

A slightly different σ_z value is obtained by using the partial derivative method on equation (xv). We had

$$\bar{z} = \bar{x}^3 + 3\bar{x}\sigma_x^2$$

$$\therefore \frac{\partial \bar{z}}{\partial x} = 3\bar{x}^2 + 3\sigma_x^2$$

Thus

$$\sigma_z^2 = (3\bar{x}^2 + 3\sigma_x)^2 \cdot \sigma_x^2$$

or
$$\sigma_z = 3\sigma_x(\bar{x}^2 + \sigma_x^2) \quad \text{(xvi)}$$

We note that equation (xvi) would yield a slightly greater value for σ_z than equation (xiv), but since it has been assumed that $\sigma_x \ll \bar{x}$, this difference will be negligible.

Case 5

$Z = f(x)$, say $z = ax^2 + bx + c$;
$x = $ normally distributed variable; and $a, b, c = $ constants.

Using the appropriate expressions from Table 5.1 (sums and products of correlated variables), we obtain

$$\bar{z} = a(\bar{x}^2 + \sigma_x^2) + b\bar{x} + c \quad \text{(xvii)}$$

and

$$\sigma_z^2 = \sigma_x^2 . (2a.\bar{x} + b)^2 + 2a^2 . \sigma_x^4 \quad \text{(xviii)}$$

Using the partial derivative method on equation (xvii), we obtain for σ_z,

$$\sigma_z^2 \simeq \sigma_x^2(2a.\bar{x} + b)^2 \quad \text{(xiv)}$$

which is the same as (xviii) with the omission of the last small term.

5.5 CORRELATION COEFFICIENTS

Before embarking on lengthy statistical calculations, field and/or experimental data should first be examined as to whether there really exists any relationship, or correlation between such data. For example:

(i) Is there any correlation between lung cancer and smoking? Some opinions still deny such correlation.
(ii) Is there any correlation between the number of cars and the number of accidents on the roads?
(iii) Is there any correlation between the birth rate and the number of television sets during the last 25 years?
(iv) Is there any relationship between the gross national product and the number of suicides per annum?
(v) Do the number of certified insane bear any relationship to the number of doctors practising psychiatry?
(vi) What relationship, if any, is there between steel and oil production in Britain or in the world?

A few of the techniques available to assess correlation coefficent r are given below:

5.5.1 Discrete events or table of data

From statistical theory (Moroney, 1964), the correlation coefficient between two sets of data may be estimated from the equation:

$$\left. \begin{array}{l} r = \sum_1^n [(x_i - \bar{x})(y_i - \bar{y})] \Big/ \left[\sum_1^n (x_i - \bar{x})^2 \cdot \sum_1^n (y_i - \bar{y})^2 \right]^{1/2} \\ \text{or} \\ r = \frac{1}{n} \sum_1^n (x_i - \bar{x})(y_i - \bar{y}) \Big/ \sigma_x \cdot \sigma_y \end{array} \right\} \quad (5.10)$$

where

r = correlation coefficient

n = number of data pairs (x and y)

$\bar{x}$ and $\bar{y}$ = mean values of x_i and y_i respectively

σ_x and σ_y = standard deviations of x_i and y_i respectively

The value of r ranges from $(-1) \to 0 \to (+1)$: when

$r = +1$ the variables are directly correlated

$r = -1$ the variables are inversely correlated

$r = 0$ the variables are not at all correlated

An example of tabulated data which has been examined for correlation is given in Table 5.2. The result indicates a high degree of correlation based on formal statistical theory.

Moroney (1964) cautions against blind use of such statistical estimates. He emphasizes the need for common sense by posing the question:
"The number of radio licences over a period of 20 years correlated extremely well with the number of people certified insane over the same period... who would be so bold as to say that one is the cause of the other?"

5.5.2 Method of paired comparisons

A useful chart, Fig. 5.1, was described by Hill (1960). By comparing successive pairs of data, "agreements" and "disagreements" can be noted. The 95% probability of +ve and −ve correlations can easily be established from the log–log graphs in Fig. 5.1. The simplicity of the method is best illustrated with a typical example (see Table 5.3). This procedure has also been used to check the result calculated in Table 5.2. Both points (P1) and (P2) are shown on the graph, Fig. 5.1.

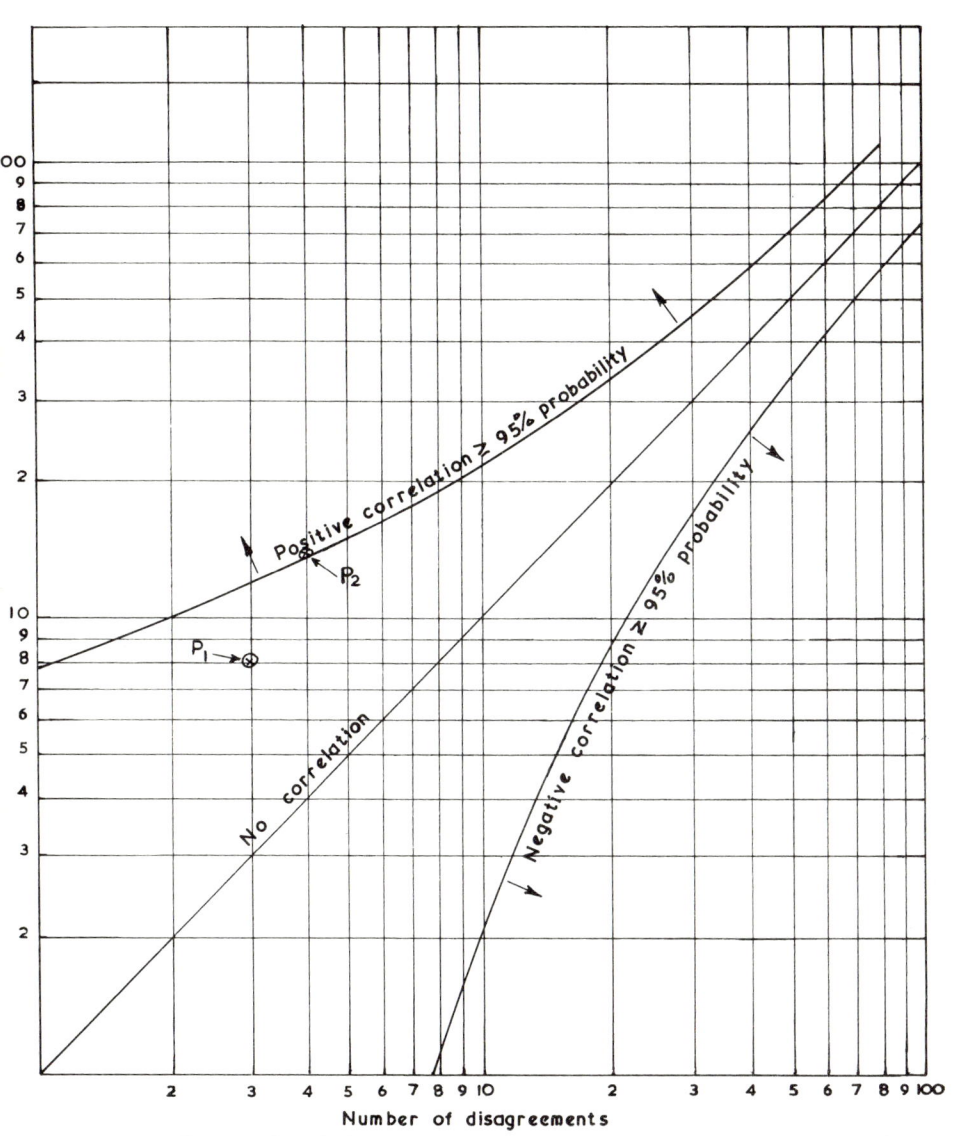

Fig. 5.1. Correlation chart for ten or more pairs of data.

Table 5.2. *Example*: Establish correlation coefficient from data for x and y.

(1) x	(2) y	(3) $(x-\bar{x})$	(4) $(y-\bar{y})$	(5) $(x-\bar{x})(y-\bar{y})$	(6) $(x-\bar{x})^2$	(7) $(y-\bar{y})^2$
0	0·00	−5·5	−1·35	+7·42	30·25	1·823
1	0·56+	−4·5	−0·79	+3·55	20·25	0·624
2	1·11+	−3·5	−0·24	+0·84	12·25	0·058
3	0·72−	−2·5	−0·63	+1·53	6·25	0·397
4	1·28+	−1·5	−0·07	+0·10	2·25	0·005
5	1·44+	−0·5	+0·09	−0·04	0·25	0·008
6	1·41−	+0·5	+0·06	+0·03	0·25	0·004
7	1·81+	+1·5	+0·46	+0·69	2·25	0·212
8	1·60−	+2·5	+0·25	+0·62	6·25	0·063
9	1·78+	+3·5	+0·43	+1·51	12·25	0·185
10	1·92+	+4·5	+0·57	+2·57	20·25	0·325
11	2·55+	+5·5	+1·20	+6·60	30·25	1·44

$\bar{x} = \dfrac{\Sigma x}{N} \quad \bar{y} = \dfrac{\Sigma y}{N}$
$= 5·5 \quad = 1·35$

$\Sigma_{(5)} = 25·42 \quad \Sigma_{(6)} = 143·0 \quad \Sigma_{(7)} = 5·144$

$\Sigma_{(6)} \times \Sigma_{(7)} = 735·6$

$$\therefore r = \frac{25·42}{\sqrt{735·6}} = 0·937.$$

With x in ascending order y values are compared consequently and marked +ve = agreement and −ve = disagreement. This yields 8 agreements and 3 disagreements. Point **P1** on correlation chart confirms calculated value of r.

5.5.3 Continuous function variables

It is of some interest to demonstrate that correlation coefficients can also be established for continuous function variables (Haugen, 1968. These derivations are reprinted by permission of John Wiley and Sons, Inc.).
(i) Consider the function

$$x = z + y \qquad (i)$$

where z and y are correlated through the equation

$$z = x + y \qquad (ii)$$

x and y being independent variables. From Table 5.1, we have for (i),

$$\sigma_x = (\sigma_z^2 + \sigma_y^2 + 2r\sigma_z\sigma_y)^{1/2} \qquad (iii)$$

We wish to determine the value of r in this case. Using an analogy of a triangle we see that

$$A^2 = (B.\sin\theta)^2 + (C - B\cos\theta)^2$$

5. THE ALGEBRA OF RANDOM NUMBERS

Table 5.3. Bush in hole (after Hill, 1960).

(a)	(b)	Diagreement[a]	Agreement
130	45		
120	40		+
115	35		+
110	30		+
100	45	−	
95	40		+
90	35		+
80	30		+
70	35	−	
60	25		+
55	20		+
50	25	−	
40	20		+
35	15		+
30	10		+
25	20	−	
20	15		+
10	10		+
0	5		+
Total = 19 pairs		$\Sigma = 4$	$\Sigma = 14$

[a] If any successive values are equal, allocate $(+\tfrac{1}{2})$ and $(-\tfrac{1}{2})$ in each column.

Check whether there is any correlation between two sets of measurements:
  (a) dimensional interference between bush outside diameter and hole inside diameter (inches $\times 10^{-6}$).
  (b) force to extract bush from hole (lbf).

Method
  (1) Arrange (a) readings in descending order.
  (2) Compare each (b) reading with number above it.
  (3) Decreases = (+ve) and increases = (−ve).
  (4) Add all (+ve) = agreements and all (−ve) = disagreements.
  (5) Plot $\Sigma(+ve)$ v. $\Sigma(-ve)$ on Correlation Chart. See point **P2**.

i.e.
$$A^2 = B^2 + C^2 - 2BC . \cos \theta \qquad \text{(iv)}$$

Comparing (iii) with (iv), we note that
$$2r . \sigma_z . \sigma_y \equiv -2\sigma_z . \sigma_y . \cos \theta$$
or
$$r \equiv -\cos \theta$$

i.e.

$$r \equiv \left(-\frac{\sigma_y}{\sigma_z}\right) \quad \text{(v)}$$

```
        A (=σx)
         /|
      C / |
   (=σz)/ | B (=σy)
       /Θ |
      /___|
```

(ii) Similarly, for

$$x = z - y$$

one can show that

$$r \equiv \left(+\frac{\sigma_y}{\sigma_z}\right) \quad \text{(vi)}$$

Example 5.9 A survey of population densities and corresponding infant survival in a number of localities might give the results shown in Table 5.4.

Table 5.4

Locality	Infant survival (%)	Population density
Langside	23·8	14·6
Longtown	42·2	4·2
Walling	38·2	7·0
Sandhamn	38·8	2·0
Berclay	17·8	23·6
Tilwell	37·4	2·5
Lidingo	28·3	12·2
Denham	37·0	4·5
Beverley	10·8	27·5
Wellboro	33·6	5·2
Bilstow	46·3	5·0
Wilmslow	17·1	14·8
Washton	40·0	1·4
Ormholm	33·6	7·2
Neasden	30·8	6·3
Kirkside	33·3	3·1
Harlem	42·9	1·3
Dockland	43·4	2·2

Carry out a correlation analysis and determine the correlation coefficient.
(a) Use equation (5.10).
(b) Use Hill's (1960) method.

5. THE ALGEBRA OF RANDOM NUMBERS

(iii) Consider function $x = z \cdot y$, \hfill (vii)

where

$\quad z$ is related to y through equation

$$z = \frac{x}{y} \text{ and } x \text{ and } y \text{ are independent variables} \quad \text{(viii)}$$

From Table 5.1 we have: for (vii)

$$\bar{x} = \bar{z} \cdot \bar{y} + r \cdot \sigma_z \cdot \sigma_y \quad \text{(ix)}$$

and

$$\sigma_x = [(\bar{z})^2 \cdot \sigma_y^2 + (\bar{y})^2 \cdot \sigma_z^2 + 2r \cdot \sigma_z \cdot \sigma_y \cdot \bar{z} \cdot \bar{y}]^{1/2} \quad \text{(x)}$$

for (viii)

$$\bar{z} = \frac{\bar{x}}{\bar{y}} \quad \text{(xi)}$$

and

$$\sigma_z = \frac{1}{\bar{y}} \left[\frac{\bar{x}^2 \cdot \sigma_y^2 + \bar{y}^2 \cdot \sigma_x^2}{\bar{y}^2 + \sigma_y^2} \right]^{1/2} \quad \text{(xii)}$$

From (xi) we obtain,

$$\bar{x} = \bar{z} \cdot \bar{y} \quad \text{or} \quad \bar{x}^2 = \bar{z}^2 \cdot \bar{y}^2 \quad \text{(xiii)}$$

and from (xii)

$$\sigma_z^2 \cdot \bar{y}^2 \cdot (\bar{y}^2 + \sigma_y^2) = (\bar{x}^2 \cdot \sigma_y^2 + \bar{y}^2 \cdot \sigma_x^2) \quad \text{(xiv)}$$

substituting (xiii) into (xiv), we get

$$\sigma_z^2 \cdot \bar{y}^2 (\bar{y}^2 + \sigma_y^2) = (\bar{z}^2 \cdot \bar{y}^2 \cdot \sigma_y^2 + \bar{y}^2 \cdot \sigma_x^2)$$

or

$$\sigma_x^2 = \sigma_z^2 \cdot \bar{y}^2 + \sigma_z^2 \cdot \sigma_y^2 - \bar{z}^2 \cdot \sigma_y^2 \quad \text{(xv)}$$

Comparing (xv) with (x), we obtain

$$\sigma_x^2 = (\bar{z}^2 \cdot \sigma_y^2 + \bar{y}^2 \cdot \sigma_z^2 + 2r \cdot \bar{z} \cdot \bar{y} \cdot \sigma_z \cdot \sigma_y) = (\sigma_z^2 \cdot \bar{y}^2 + \sigma_z^2 \cdot \sigma_y^2 - \bar{z}^2 \cdot \sigma_y^2)$$

or

$$2\bar{z}^2 \cdot \sigma_y^2 + 2r \cdot \bar{z} \cdot \bar{y} \cdot \sigma_y \cdot \sigma_z = \sigma_z^2 \cdot \sigma_y^2$$

i.e.

$$r = \frac{\sigma_z^2 \cdot \sigma_y^2 - 2\bar{z}^2 \sigma_y^2}{2\bar{z} \cdot \bar{y} \cdot \sigma_z \cdot \sigma_y}$$

or

$$r \simeq \left[-\frac{\bar{z}}{\bar{y}} \frac{\sigma_y}{\sigma_z} \right] \quad \text{(xvi)}$$

(iv) Consider function $x = z/y$, where z and y are correlated through equation $z = x \cdot y$, x and y being independent variables.

We proceed as for case (iii) above, using Table 5.1. The correlation coefficient for z relative to y, in this case, is

$$r = \left[+\frac{\bar{z}}{\bar{y}} \cdot \frac{\sigma_y}{\sigma_z} \right] \tag{xvii}$$

5.6 CONFIDENCE LIMITS

A second criterion to examine is how meaningful the results of a statistical analysis may be, i.e. with what degree of confidence can we make statements about the data available. The problem frequently arises in engineering, because we invariably deal with practical limits of sample sizes. What we need, therefore, is a method of determining the measure of confidence with which we can make a statement about the whole population or batch, from an inspection of a limited sample.

If one uses the "normal" (Gaussian) distribution as a basis of analysis, one recalls that parameter variations are assumed to range from $-\infty$ to $+\infty$. In engineering practice variations in tolerances, of any measurements, range between finite limits. The proportion of the whole population (or batch) which may be expected to lie within such limits was shown earlier to be:

Tolerance limit	Percentage of batch within the limits
$K = \pm 1\sigma$	68·26
$K = \pm 2\sigma$	95·44
$K = \pm 3\sigma$	99·73

This may be understood to mean that if we examined the whole population, i.e. infinite batch size, we could say with 100% confidence that 68·26%, 95·44% and 99·73% of the whole batch would be expected to lie within the limits of $\pm 1\sigma$, $\pm 2\sigma$ and $\pm 3\sigma$ respectively. Inspecting smaller samples, we clearly cannot be equally sure in predicting the probability of the whole batch.

Figure 5.2 shows a set of curves relating confidence levels against percentage population prepared by Chamberlain (1964), for a range of completely random samples smaller than infinite (= whole population or batch),

$$\text{the constant } K = \frac{\text{mean size} - \text{lower limit size}}{\text{standard deviation}}$$

The following example illustrates the use and interpretation of these curves:

5. THE ALGEBRA OF RANDOM NUMBERS

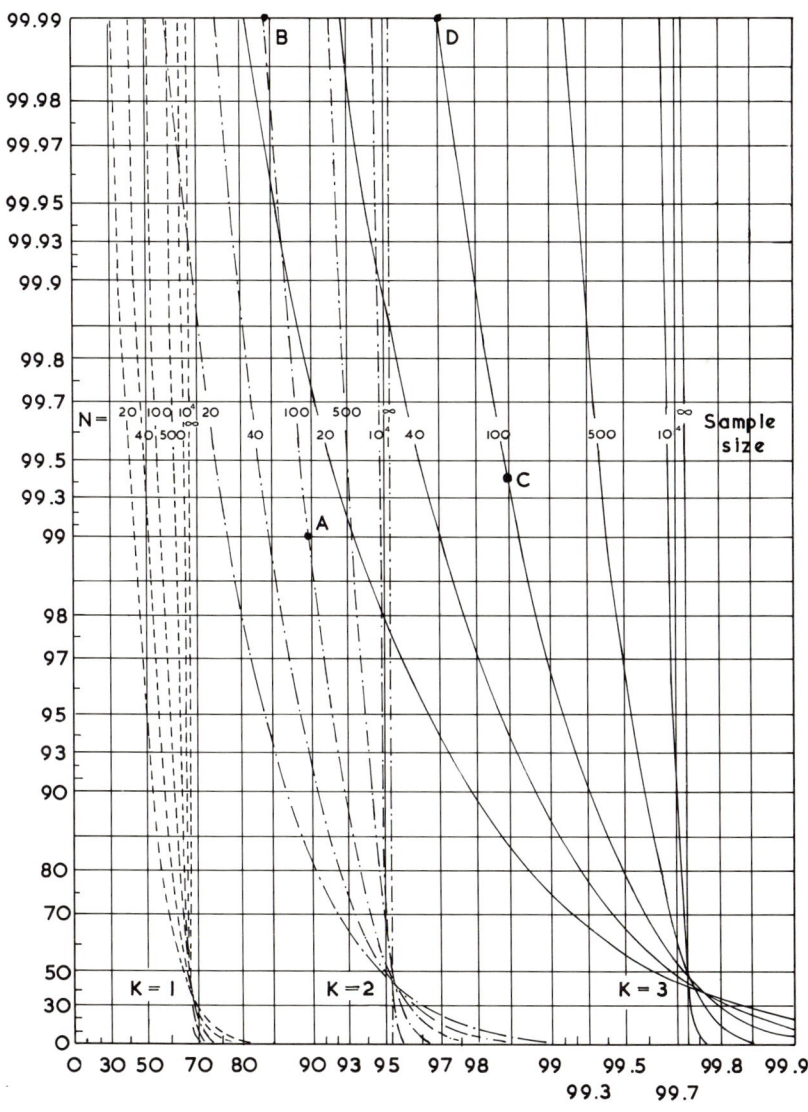

Fig. 5.2. Confidence level v. sample size and population proportion.

Example Quarter-inch B.S.W. screws are to be produced within length tolerance of $2\frac{1}{4}$ in. $\pm \frac{1}{8}$ in. What percentage of the whole production of 100 000 screws may be expected to lie within the specified tolerances if inspection is carried out on a random sample of 100 screws only? A standard deviation of

$+\frac{1}{16}$ in. is assumed in the first instance.

(a) $\quad K = \dfrac{2\frac{1}{4} - 2\frac{1}{8}}{\frac{1}{16}} = 2 \,(= 2 \times \text{standard deviation})$

If all 100 000 screws were inspected we would have 100% confidence that some 95·4% of all screws would lie within the tolerance limits specified. However, if we only inspect 100 screws, we note from Fig. 5.2 that at point A we could be 99% confident that 90% of all screws would be within the specified tolerance, or at point B we could be 99·99% confident that 85% of all screws would satisfy the specified limits.

(b) If the standard deviation was not given, we may use the convention that

$$\sigma = \pm\tfrac{1}{3} \text{ (tolerance limits)}$$

i.e.

$$K = 3(\sigma = \pm\tfrac{1}{24}\text{ in.})$$

Then from the inspection of 100 screws, we obtain from Fig. 5.2, that at point C we could be 99·4% confident that 98·5% of all screws would fall within the tolerance limits, and at point D we could be 99·99% confident that 97·0% of all screws would lie within the upper and lower tolerance lengths.

5.7 EXAMPLES (for solutions see pp. 369–381)

Example 5.1 Find the standard deviation for the following functions, assuming all the variables are independent

(i) $\phi = A \cdot y \cdot \log x$
(ii) $\phi = A \cdot \log \cdot x$
(iii) $\phi = c \cdot y \cdot \sin x$
(iv) $\phi = c \cdot \sin x$

Example 5.2 Given independently varying strength and cross-section of a tension rod, what will be the corresponding permissible variation in tensile load?

$$(\bar{A};\, \sigma_A) = (10\,\text{in}^2;\, 0{\cdot}40\,\text{in}^2)$$
$$(\bar{f};\, \sigma_f) = (5 \times 10^4\,\text{psi};\, 3 \times 10^3\,\text{psi})$$

N.B. The engineering designer should be aware that the mean strength of a component varies with the cross-sectional area; e.g. castings, fatigue strength, hardened and rolled sections, etc. The above assumption is, therefore, an idealized approximation.

5. THE ALGEBRA OF RANDOM NUMBERS

Example 5.3 When a metal rod constrained at its ends was heated from (50; 2·5)°F to (100; 3)°F the stress induced was measured to be (10 500; 650) p.s.i.

If the Young's modulus of the material varied as $(30 \times 10^6; 1·05 \times 10^6)$ p.s.i., determine the random variation of the coefficient of thermal expansion $(\alpha; \sigma_\alpha)$.

Example 5.4 Using statistical analysis determine the probable variation in stiffness of the cantilever flat spring considered in Example 2.8. Compare the results with those obtained using the theory of errors.

Example 5.5

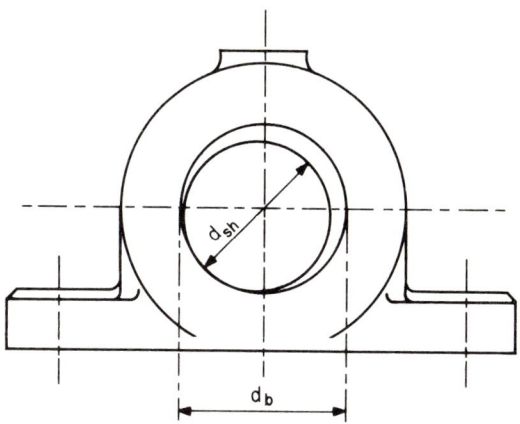

Shafts of diameter (1·048 in.; 0·0013 in.) are fitted randomly into bearing blocks with bush-diameters of (1·059 in.; 0·0017 in.). Assuming "normal" tolerance distributions, what is the probability that the diametral clearance (δ) between a shaft and a bush would be

(a) $\delta \leqslant$ zero   and (b) $\delta \leqslant 0·0010$ in., minimum running clearance

Example 5.6 Two heat exchangers in series, each with fluctuating capacity,

$$\bar{H}_1 = \bar{H}_2 = 5500 \text{ BTU/hour}$$
$$\sigma_{H_1} = 500 \text{ BTU/hour}$$
$$\sigma_{H_2} = 700 \text{ BTU/hour}$$

What is the probability that the heat absorbed by the airstream $\geqslant 12\,750$ BTU/hour?

Example 5.7 In Example 2.10 (p. 42) the variation in transmissible torque of a shrink fit was examined. Assuming the "central limit theorem" applies in this case, what is the probability that the torque will be 50% and 30% less than the mean value?

Example 5.8 In an electric circuit two resistors, $(R_1; \sigma_{R_1})$ and $(R_2; \sigma_{R_2})$ are in parallel. Determine the variation in the resultant resistance of the circuit. For numerical values take

$$(R_1; \sigma_{R_1}) = (1000\,\Omega;\ 300\,\Omega)$$
$$(R_2; \sigma_{R_2}) = (2000\,\Omega;\ 600\,\Omega)$$

REFERENCES

Chamberlain, R. G. (1964). *Journal of Product Engineering* Sep. 28.
Haugen, E. B. (1968). "Probablistic Approaches to Design." John Wiley and Sons, Chichester.
Hill, W. J. (1960). *Journal of Product Engineering* Feb. 15.
Moroney, M. J. (1964). "Facts from Figures." Penguin, Harmondsworth.

Chapter 6

Variabilities of Material Properties

There is as yet only a limited amount of information on statistical variations of material properties. Designers are familiar with national specifications which often quote upper and lower limit values, and they usually use the conservative figures to ensure their design estimates are on the safe side. With increasing restrictions on cost and the demands for reliability and quality assurance, this conventional approach is becoming increasingly unsatisfactory. Some work on the establishment of statistical data has already been started in the U.S.A. and these are published by the Material Properties Data Center, Traverse City, Michigan.

Professor E. B. Haugen at the University of Arizona, U.S.A., has also pursued work in this area. Tables 6.1–6.7 and Fig. 6.1 below represent part of his collated data (Haugen, 1975).

Table 6.1 gives typical examples of static strength of metals. The mean and standard deviation values for the ultimate and yield strengths are quoted for a number of widely used metals.

Table 6.2. Variabilities in strength properties are presented as "coefficients of variation". It is noted that the values range from 5% to 8%, the fatigue endurance limit having the widest spread. This is further clarified in Fig. 6.1.

Table 6.3. Little published information exists that can describe the variability of the modulus of elasticity (E) of metals. The few examples given illustrate that the variability may be small, but not insignificant.

Table 6.4. Dimensions of manufactured components will vary according to the production processes used. The accuracy, or tolerance variability of various processes, are well known. Some of these are shown in this table. Assuming random variations and a normal distribution, the tolerances may be taken to represent $\pm 3 \times$ (standard deviation).

Table 6.1. Typical variations in the strength of metals (Haugen, 1975).

Material	Condition	Ultimate tensile strength S_u (ksi)		Tensile yield strength S_y (ksi)	
		μ	σ	μ	σ
2024-T3 Aluminium	(Bare sheet and plate) < 0·250 in.	73	3·7	54	2·7
6061-T6 Aluminium	½ in. sheet	46	1·9	42	2·9
7075-T6 Aluminium	¼ in. plate; bare	85	3·4	76	3·8
A231B-0 Magnesium	¼ in. plate; longitudinal	41	3·8	24	3·8
Ti-6A1-4V Titanium alloy	Sheet and bar; annealed	135	6·8	131	7·5
304 Stainless steel	Round bars; annealed (0·50–4·6 in.)	85	4·1	38	3·8
ASTM-A7 Steel	½ in. Plate	66	3·3	40	4·0
AISI 1018	Cold drawn round bar (0·75–1·25 in.)	88	5·7	78	5·9

Table 6.2. Variability in materials properties (Haugen, 1975)

	Coefficient of variation
Tensile ultimate strength of metallic materials	0·05
Tensile yield strength of metallic materials	0·07
Endurance limit of steel	0·08
Brinnel hardness of steel	0·05
Fracture toughness of metallic materials	0·07

μ = arithmetic mean
σ = standard derivation
Coefficient of variation = σ/μ

6. VARIABILITIES OF MATERIAL PROPERTIES

Table 6.3. Statistical values for the modulus of elasticity of metals (Haugen, 1975)

	Mean × 10^6 (psi)	Coefficient of variation, C
Steel	30·0	≈0·03
Aluminium	10·0	≈0·03
Titanium	14·7	≈0·09
Nodular iron cast	23·1	≈0·04

Table 6.4. Variability of tolerances (Haugen, 1975)

Process	Tolerance (±) (inches)
Flame cutting	0·060
Sawing	0·020
Shaping	0·010
Broaching	0·005
Milling	0·005
Turning	0·005
Drilling	0·010
Reaming	0·002
Hobbing	0·005
Grinding	0·001
Lapping	0·0002
Stamping	0·010
Drawing	0·010

Table 6.5 lists standard tolerances on the thickness of cold drawn carbon plates. The standard deviation may again be assumed equal to one sixth of the tolerance range.

Table 6.6. At least four parameters—all of which are random variables—determine the characteristics of helical springs. A summary of the variability in design factors relating to helical spring performance is given.

Table 6.7 contains a few statistical data on welded and bolted connections. For design purposes Haugen (1975) suggests the use of a coefficient of variability of 0·015 for the strength of such joints.

Statistical data on bolted and rivetted joints are given by Fisher and Struick (1974). They distinguish "non-slip" and "slip" type joints. The latter are very

Table 6.5. Cold-rolled carbon steel sheet, thickness[a] tolerances ($\pm T$) (Haugen, 1975)

Width (inches)	Thickness (inches)									
	0·2299 0·1875	0·1874 0·1800	0·1799 0·1420	0·1419 0·0972	0·0971 0·0822	0·0821 0·710	0·0709 0·0568	0·0567 0·0509	0·0508 0·389	0·0388 0·0344	0·0343 0·0314	0·0313 0·0255	0·0254 0·0195	0·0194 0·0142	0·0141 and less
Over 6 to 12 incl.	—	—	—	—	—	—	—	0·005	0·005	0·004	0·004	0·003	0·003	0·002	0·002
Over 12 to 15 incl.	0·008	0·007	—	—	0·006	0·006	0·006	0·005	0·005	0·004	0·004	0·003	0·003	0·002	—
Over 15 to 20 incl.	0·008	0·008	0·007	0·006	0·007	0·007	0·006	0·006	0·005	0·005	0·004	0·003	0·003	0·002	—
Over 20 to 32 incl.	0·009	0·009	0·008	0·007	0·007	0·007	0·006	0·006	0·005	0·004	0·004	0·003	0·003	0·002	—
Over 32 to 40 incl.	0·009	0·009	0·009	0·008	0·008	0·007	0·006	0·006	0·005	0·004	0·004	0·003	0·003	0·002	0·002
Over 40 to 48 incl.	0·010	0·010	0·010	0·008	0·008	0·007	0·007	0·006	0·005	0·004	0·004	0·003	0·003	0·002	0·002
Over 48 to 60 incl.	—	—	0·010	0·010	0·008	0·007	0·007	0·006	0·005	0·004	0·004	—	—	—	—
Over 60 to 70 incl.	—	—	0·011	0·011	0·009	0·009	0·008	0·007	0·006	0·005	0·005	—	—	—	—
Over 70 to 80 incl.	—	—	—	0·012	0·012	0·009	0·008	—	—	—	—	—	—	—	—
Over 80 to 90 incl.	—	—	0·012	0·012	0·010	—	—	—	—	—	—	—	—	—	—
Over 90	—	—	0·012	0·012	—	—	—	—	—	—	—	—	—	—	—

[a] Thickness is measured at any point on the sheet not less than $\frac{3}{8}$ in. from an edge.

6. VARIABILITIES OF MATERIAL PROPERTIES

Table 6.6. Variability in helical springs
(Haugen, 1975)

Parameter	Statistical values
Wire diameter, d	$\sigma_d = 1.58 \times 10^{-3} \cdot \sqrt{d}$ (hard drawn carbon steel) $\sigma_d = 1.31 \times 10^{-3} \cdot \sqrt{d}$ (music wire)
Spring diameter, D	$\sigma_D = (48 \times 10^{-4}) \cdot D$ for $d < 100$ mils[a] $\sigma_D = 1.90 \times 10^{-3} \cdot D/\sqrt{d}$ for $d \geqslant 100$ mils
Number of coils, N	$\sigma_N = \frac{1}{12}$
Shearing modulus of elasticity, G	$C_G = 0.02 =$ coefficient of variation

[a] One mil = 0.001 in.

Table 6.7. Statistical data on welded and bolted connections
(Haugen, 1975)

	μ_{mean}	Standard deviation, σ
Shear strength/tensile strength, longitudinal fillet welds (ksi)	0.769	0.097
Shear strength of longitudinal fillet welds (ksi)		
E60	56.4	6.2
E70	63.5	5.7
Slip coefficient (bolted joint) (no special treatment)	0.336	0.707
Shear strength/tensile strength, A325 and A490 bolts (ksi)	0.625	0.033

dependent upon the nature of the interface and the elasticity of the jointed members. Typical results from tests before slipping commenced were:

(i) Steel plates with mill scale surfaces cleaned off:
mean slip coefficient $\bar{\mu} = 0.336$
standard deviation $\sigma_\mu = 0.07$
∴ coefficient of variation $C = 0.208$

(ii) Steel plates with mill scale surfaces grit blasted:
$\bar{\mu} = 0.493$
$\sigma_\mu = 0.074$
∴ $C = 0.15$

(iii) For other surface treatment (see Table 5.1 in Fisher and Struick, 1974) the coefficients of variation ranged from 0·084 to 0·223.

Further useful data on bolted and welded joints are also given by Kulak (1972).

Table 6.8. The coefficients of variation are generally less than 10%, the deviations of yield strength being rather greater than those for the ultimate strength (Haugen, in press).

Figure 6.1 shows the change in dispersion of fatigue strength with the number of stress cycles applied. At the endurance limit (for steels) the standard deviation σ_{SE} is given as equal to $0 \cdot 08 \mu_{SE}$. It is of interest to compare this deviation of the mean endurance limit value with that given in Section 3.3 (Stulen et al., 1956).

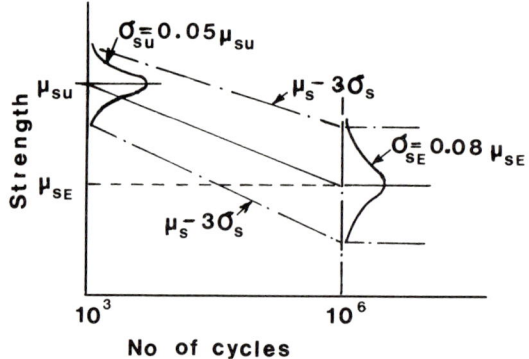

Fig. 6.1. Statistical variation of fatigue strength v. number of cycles.

Using equations (3.6) and (3.7) and Table 3.2, we can estimate the f'_e values for various reliability levels. The following results are obtained:

Reliability level	Z value	f'_e	$\sigma = \dfrac{\bar{f_e} - f'_e}{Z}$
90·00	1·282	0·896	$0 \cdot 0811 . \bar{f_e}$
99·00	2·3025	0·816	$0 \cdot 0799 . \bar{f_e}$
99·90	3·085	0·750	$0 \cdot 0804 . \bar{f_e}$

The close similarity between Haugen's and Stulen's deviation values may not be wholly accidental.

Where upper and lower limits of strength are available, the *Military Standardization Handbook* recommends the use of the lower values for design and notes (in its Section 9.2.6) that "Past experience with many alloy

6. VARIABILITIES OF MATERIAL PROPERTIES

systems indicates that the distributional form of room temperature strength observations is usually normal (Gaussian)". Thus for statistical estimates of reliability, one may use

$$\bar{S}_{ult} = \frac{\text{upper limit} + \text{lower limit}}{2}$$

and

$$\sigma = \frac{\text{upper limit} - \text{lower limit}}{6}$$

Note: By using $\pm 3.\sigma$ as representing the "sure-fit" tolerance limits, we make the assumption that
  (i) 99·73% of all values would fall within the given range;
  (ii) 0·135% may be expected to be less than the lower limit;
  (iii) 0·135% may be expected to be greater than the upper limit.
If quality control rigorously adheres to the "sure-fit" limits, the usable population is then confined within the tolerance range. This implies that we are using a truncated distribution of Gaussian form and we can correct the probability estimates by dividing the results obtained from standard tables by 0·9973.

An important source of information on mechanical properties of engineering materials is being developed by the Mechanical Properties Data Center (MPDC) established in 1979 at Battelle, Columbus Laboratories, Ohio, U.S.A. This centre maintains a constantly growing data base containing already over 1 500 000 well-defined test results on more than 5400 metals and alloy compositions. In addition to giving results of the tests cited, source, test conditions, material form, heat treatment and related information are fully documented.

A number of computer programs have been developed for the retrieval and analysis of these data:
1. Automatic Interaction Detector (AID), which allows the examination of interrelationship of up to 140 predictors.
2. SSP, an integrated system of computer programs for statistical analysis.
3. TENSIL, COMPR, BEARG programs compute, tabulate and print a number of quantities related to various mechanical properties, such as number of data, average and standard deviation.
4. SEVRAL compares material lots statistically to determine whether they can be combined or are statistically different.
5. SC4060 is a plotting routine for such comparisons as strength v. a selected variable.
6. REGRES allows a mathematical model to be selected and serves as an input–output medium to the standard regression program in the Battelle computer.

Table 6.8. Static strength of materials (Haugen, in press)

Material		Remarks	Condition	Ult. tensile, S_u (ksi)			Tensile yield, S_y (ksi)		
				Mean	σ	Sample	Mean	σ	Sample
2014-T651	AL-alloy	Plate L-T	Ambient temperature	69.0	1.82	20	63.0	1.35	20
7039-T6	AL-alloy	Plate −320°F	0.50–1.50 L.T.	78.9	0.56	8	65.9	0.94	8
C 1035	Carbon steel	Round bars	Hot rolled, 1–9 in. dia.	88.4	9.5	913	49.5	5.36	899
C 1035	Carbon steel	Round bars	Cold drawn	92.1	6.1				
Low carbon steel		Sheet, draw quality	Hot rolled, 0.075 in.	47.4	1.26	205	34	2.25	215
5Cr–Mo–V alloy	Low alloy steel	5Cr–1.5Mo–0.4V–0.35C	¾ in. dia. bar, aircraft steel	296.4	1.84	25	240.8	3.10	25
Ni-Cr-Mo alloy	Low alloy steel	0.28C–2Ni–0.8Cr–0.35Mo	Normalized, quenched and tempered	161.1	4.97	44	148.1	4.98	44
Type 301	Stainless steel	Sheet and strip; austenitic	Cold worked, 20–28%	161.4	9.05	182	118.4	5.55	182

Table 6.8.

	Material	Remarks	Condition	Ult. tensile, S_u (ksi)			Tensile yield, S_y (ksi)		
				Mean	σ	Sample	Mean	σ	Sample
Type 302	Stainless steel	Sheet 0·003 in.	Annealed	90·4	3·01	45	42·4	3·53	45
Type 347	Stainless steel	Forging	− 100 °F	118·0	3·34	23	34·5	1·15	187
			+ 70 °F	81·0	1·45	18	29·8	1·15	187
			+ 600 °F	56·3	1·33	24	21·3	1·15	187
			+1200 °F	43·6	0·61	22	18·7	1·15	187
InConel 718	Super alloy	Welded sheet	Solution treated at 1950 0·085–0·095 in.	173·0	2·39	15	141·2	2·34	15
SAE 70310 (11K alloy)	Heat resistant alloy	26% Cr–20% Ni	As cast	72·4	5·12	183	43·5	2·90	366
CA-15	Corrosion resistant steel		Casting	107·0	5·64	23	80·7	4·31	23
Nodular iron	Cast iron	As cast	Auto crankshafts	99·3	6·14	125	74·9	4·62	125

7. RATREG and LINRRG are regression analysis programs for least-square linear regression or various ratios of properties v. thickness of the materials.

The dissemination and utilization of information are provided by the *Aerospace Structural Metals Handbook* and the *Structural Alloys Handbook*. Both handbooks are regularly updated to reflect new data or established materials and information on newly developed alloys.

The storage and retrieval of test data on thousands of metals and alloys has necessitated a material classification identification system which recognizes that many "equivalent materials" are identified by a variety of association or society designations and trade names. The identification system developed provides a common code number for "chemically equivalent" metals and alloys. An Alloy Cross Index is utilized by designers, engineers, purchasing agents and others involved in the acquisition, application and selection of metals and alloys. The index lists some 22 000 designations in alphanumeric sequence (Part I) and also combines like material under a common code number (Part II).

REFERENCES

Fisher, J. W. and Struick, J. H. A. (1974). "Guide to Design Criteria for Bolted and Rivetted Joints." John Wiley and Sons, Chichester.
Haugen, E. B. (1975). "Probablistic Design." *Machine Design* May 15, 1975.
Haugen, E. B. (in press). "Probablistic Mechanical Design." John Wiley and Sons, Chichester.
Kulak, C. I. (1972). "Safety Reliability of Metal Structures." A.S.C.E.
Material Properties Data Centre, Traverse City, Michigan, U.S.A.
Mechanical Properties Data Center (MPDC), Battelle, Columbus Laboratories, Columbus, Ohio, U.S.A.
Military Standardization Handbook (1973). "Metallic Materials and Elements for Aerospace Vehicle Structures." Naval Publications Form Center, Philadelphia.
Stulen, F. B., Cummings, H. N. and Schulte, W. C. (1956). "Conference on Fatigue of Metals." I.Mech.E.

Chapter 7

Reliability

The experienced engineering designer, while aiming for increased accuracy and perfection in his work, has always been conscious of the limitations in human knowledge, the unreliability of data about materials, and the approximations inherent in modelling and mathematical methods. Aware too that he will bear the blame for poor performance or "failures", he tends to be cautious and prefers to err on the side of safety. Unfortunately, demands for lower costs, higher performance, enhanced efficiency, lower weight-to-power ratio, more sophisticated computer-aided analytical techniques as well as the demand for higher reliability assurance have increased the pressure on engineering designers to reduce the margin of safety of his design.

The old-fashioned "factor of safety", which in reality represented a factor of ignorance of actual operating conditions and true performance capabilities, was used to provide a sensible reserve against such unknowns and was usually based on years of accumulated in-service experience. Today engineers are sentient that reliability is associated with risk taking, cost levels, life patterns and life expectancy.

The first statistical theory of reliability was formulated by Lusser in 1952. Lusser noted that mechanical equipment showed histories of "failures", defined by the fall-off of performance outside specified limits, which could be represented by typical mortality curves. Figure 7.1 shows the characteristics of a Lusser mortality curve.

7.1 COMPONENT RELIABILITY

Typical factors that can influence true strength and actual load (stress) are illustrated in Fig. 7.2. The Warner block diagram clearly illustrates the difference between apparent safety margins based on mean values and the much reduced actual safety margin that may well exist in practice. Whereas the block diagram leaves the impression that such an actual safety margin appears to exist, the conversion of this histogram into a continuous function, e.g. a normal distribution (see Section 7.4), indicates that this too may be deceptive.

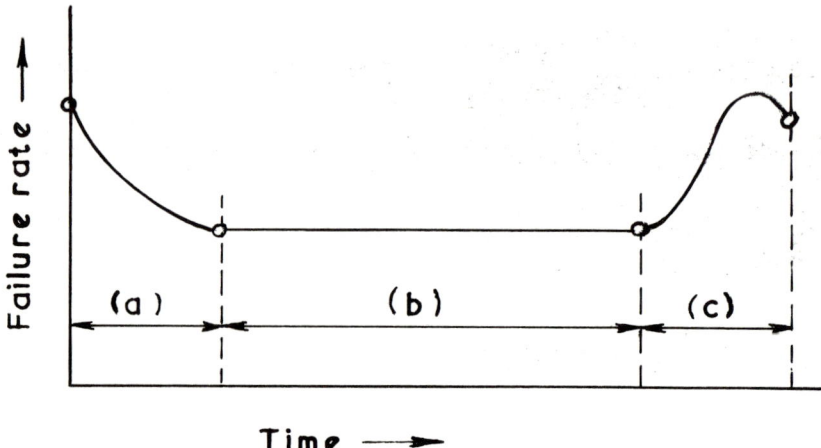

Fig. 7.1. (a) *Infant mortality.* Associated with the running-in period. In biology influenced by growing strength of infant. (b) *Constant hazard.* Accidental (random) failures caused by dirt, unexpected loads, and temperatures, human errors, etc.

The failure rate, λ is defined as

$$\lambda_t = \frac{\text{no. of failures per unit time at time } t}{\text{no. of items surviving at time } t}$$

(c) *Wear-Out.* After a useful life, wear and degradation will finally cause failure. Preventative maintenance can prolong useful life of a machine.

It is clear from Figs 7.2 and 7.3 that:
(a) If a sure-fit model (histogram) is used an actual, albeit much reduced, margin of safety, may be available.
(b) For continuous (e.g. normally distributed) strength and load characteristics a failure region will always exist, and this is represented by the overlap of the two distribution curves. The overlap area represents the probability of "failure" of the unit or machine.

7.1.1

It is of interest, at this juncture, to obtain some feel for the relative significance of conventional factors of safety and associated reliability levels. Consider, for example, a typical mechanical engineering structure such as the jib-crane shown in Fig. 7.4. The tie-rod is a forged steel component with a diameter of $(30 \pm 1\cdot2)$ mm and the material strength may range from 30 000 to 54 000 p.s.i. If the applied load equals $(30\,000 \pm 9600)$ lbf,
  (i) what factor of safety (SF) do we have, and
  (ii) what is the probability of overstressing the rod?

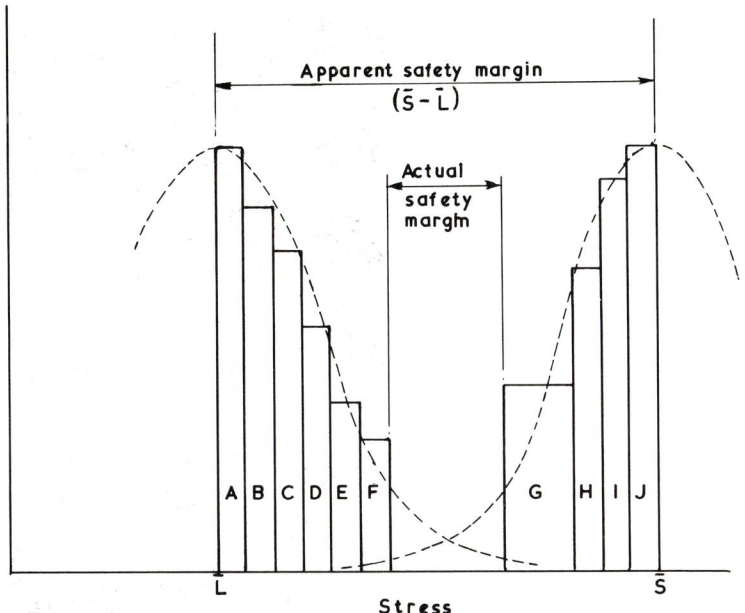

Fig. 7.2. "Warner" diagram. Comparison of actual and apparent safety margins.
A. Stress raisers; B. Unknown loads; C. Internal or residual stress; D. Wear; E. Production variation; F. Corrosion; G. Fatigue; H. Undetected defects; I. Normal production variations; J. Advertising.

(a) For conventional calculation of an SF, we use the mean values of load, area and strength:

$$\bar{P} = 30\,000 \text{ lbf}$$
$$\bar{A} = \frac{\pi}{4}\left(\frac{30}{25 \cdot 4}\right)^2 = 1 \cdot 095 \text{ in}^2$$
$$\bar{S} = 42\,000 \text{ p.s.i.}$$

The applied stress is equal to

$$\frac{\bar{P}}{\bar{A}} = \frac{30\,000}{1 \cdot 095} = 27\,397 \text{ p.s.i.}$$

and the SF is equal to

$$\frac{42\,000}{27\,390} = 1 \cdot 533$$

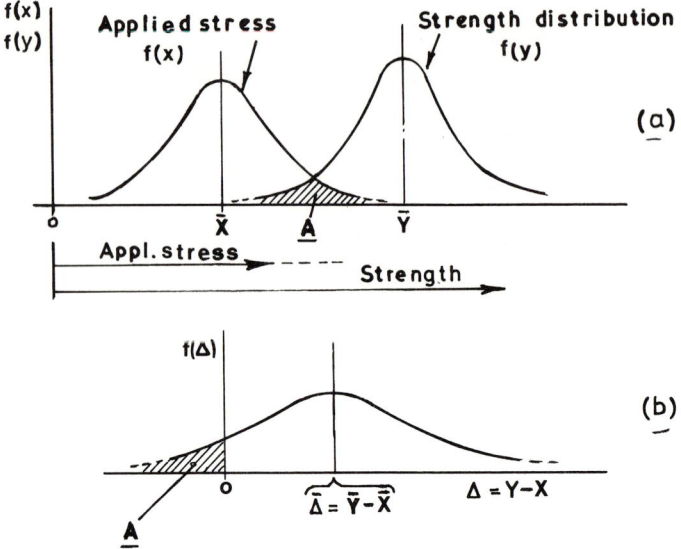

Fig. 7.3. Reliability: strength v. stress. Area $A(a) = A(b)$ = probability of failure, i.e. when stress $\geqslant$ strength

where

$$\bar{\Delta} = \bar{y} - \bar{x}$$

$$\sigma_\Delta = \sqrt{\sigma_y^2 + \sigma_x^2}$$

$$R = \frac{1}{\sqrt{2\pi}} \int_{-z_0}^{+\infty} e^{-\frac{1}{2}(z^2)}\, dz = \text{reliability of non-failure}$$

$$z_0 = \frac{0 - \bar{\Delta}}{\sigma_\Delta} = -\left(\frac{\bar{\Delta}}{\sigma_\Delta}\right)$$

Therefore probability of failure = Area $A = 1 - R$. (Values of R and F obtained from standard tables.)

A check on the minimum SF is also advisable since all three parameters can vary:

$$A_{min} = \frac{\pi}{4}\left(\frac{25 \cdot 8}{25 \cdot 4}\right)^2 = 1 \cdot 01 \text{ in}^2$$

and

$$P_{max} = 39\,600 \text{ lbf,}$$

$$S_{min} = 30\,000 \text{ p.s.i.}$$

These yield,

$$SF_{min} = \frac{30\,000}{39\,208} = 0 \cdot 765$$

i.e. for worst conditions we could expect failures due to overstressing.

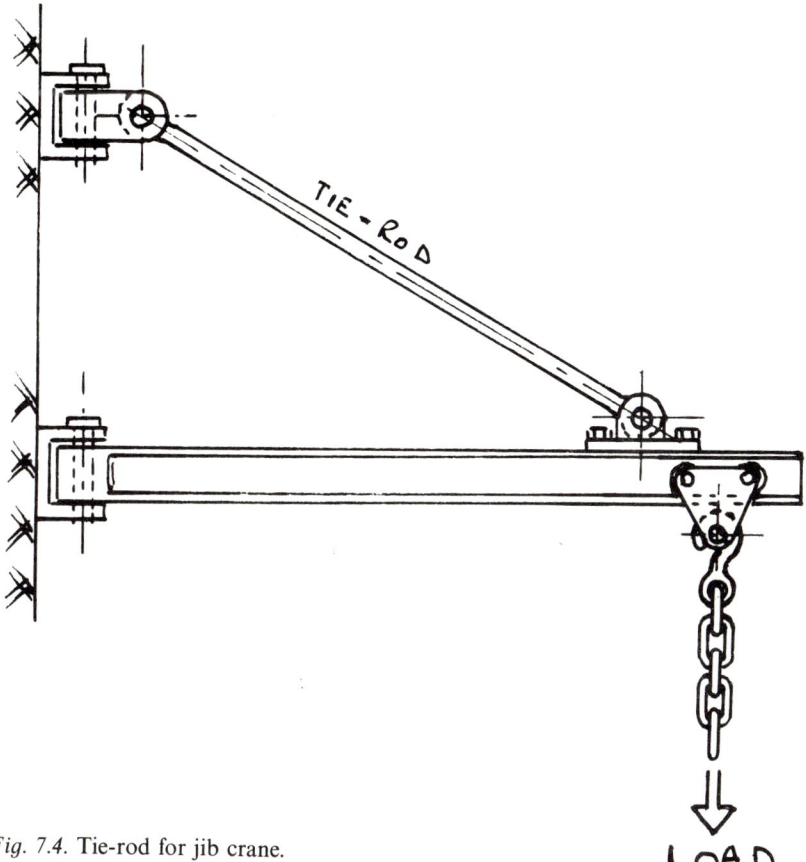

Fig. 7.4. Tie-rod for jib crane.

(b) Assuming the three parameters have "normal" distributions, we may estimate the likelihood of such overstressing to occur. Now,

$$(\bar{P};\ \sigma_p) = (30\,000;\ 3200)\,\text{lbf}$$
$$(\bar{S};\ \sigma_s) = (42\,000;\ 4000)\,\text{p.s.i.}$$
$$(\bar{d};\ \sigma_d) = (30;\ 0\cdot4)\,\text{mm}$$

For the diameter values we obtain for the area

$$\bar{A} = 1\cdot095\,\text{in}^2 \quad \text{and} \quad \sigma_A = 0\cdot0292\,\text{in}^2$$

Thus the applied stress is

$$\bar{f} = \frac{\bar{P}}{\bar{A}} = 27\,397\,\text{p.s.i.}$$

and
$$\sigma_f = \frac{1}{(\bar{A})^2}[(\bar{P})^2 \cdot \sigma_A^2 + (\bar{A})^2 \cdot \sigma_P^2]^{1/2}$$

i.e.
$$\sigma_f = \left(\frac{1}{1\cdot 095}\right)^2 [(30\,000)^2 \cdot (0\cdot 0292)^2 + (1\cdot 095)^2 \cdot (3200)^2]^{1/2}$$

or
$$\sigma_f = 3012\cdot 5 \text{ p.s.i.}$$

The mean difference between strength and stress is now
$$\bar{\Delta} = \bar{S} - \bar{f} = 42\,000 - 27\,397$$
$$\bar{\Delta} = 14\,603$$

Also
$$\sigma_\Delta = [\sigma_s^2 + \sigma_f^2]^{1/2}$$
$$= [(4000)^2 + (3012\cdot 5)^2]^{1/2};$$

i.e.
$$\sigma_\Delta = 5008 \text{ p.s.i.}$$
$$\therefore Z = \frac{0 - 14\,603}{5008} = -2\cdot 916$$

From probability tables (see Table 4.1), we obtain, for $Z = 2\cdot 916$
$$R(Z) = 99\cdot 826\%$$

or
$$F(Z) = 0\cdot 174\% \text{ probability of overstressing}$$

Thus if the resultant distribution of strength/stress difference has a normal distribution, the chances of overstressing are very remote. This illustrates that the danger of failure as indicated by the adverse minimum factor of safety for extreme conditions gives an exaggerated and excessively alarming impression.

7.1.2

Another illuminating example is that of a strut member in a roof-truss. The geometry and loading of the truss is shown in Fig. 7.5.

The following information is available:
(i) (AB) = 240 inches = 2(DC),
(ii) length tolerance for (DC) = $\pm \frac{3}{8}$ inches,
(iii) tube diameter tolerance = $\pm 3\cdot 42\%$,
(iv) tube thickness = $\frac{1}{50}$. (tube diameter),
(v) Young's modulus, $E = 30 \times 10^6$ p.s.i.

7. RELIABILITY 139

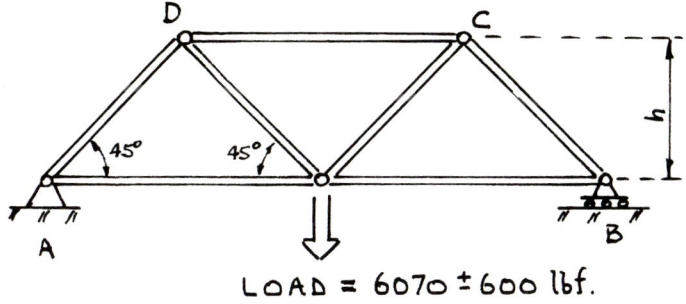

Fig. 7.5. Roof truss.

Required, to estimate the tube diameter for (DC),
(a) by conventional deterministic method using a safety factor of 1·50 on the crippling load, and
(b) by probabilistic method to give a reliability level of 0·99997 against buckling failure.

(a) *Deterministic calculations*

A force analysis of the truss structure yields a mean compressive load of 6070 lbf on member (DC). Allowing for loading tolerances the maximum load may be expected to be $= 6070 + 600 = 6670$ lbf.
The critical load to be catered for is then

$$P_{\text{crit}} = 1 \cdot 5 \times 6670 = 10\,000 \text{ lbf} \qquad \text{(i)}$$

The critical stress is

$$f_{\text{crit}} = \frac{P_{\text{crit}}}{A} \quad \text{(compressive)} \qquad \text{(ii)}$$

Using Euler's criteria for a column with both ends hinged, we have

$$P_{\text{crit}} = \frac{\pi^2 \cdot E \cdot I}{l^2} \qquad \text{(iii)}$$

$$\therefore f_{\text{crit}} = \frac{\pi^2 \cdot E}{l^2} \cdot \frac{I}{A}$$

or

$$f_{\text{crit}} = \frac{\pi^2 \cdot E}{l^2}(\rho^2)$$

where ρ = radius of gyration. For a tube (diameter $\gg$ thickness)

$$\rho^2 = \left(\frac{r^2}{2}\right) \qquad \text{(iv)}$$

140   APPROXIMATE METHODS IN ENGINEERING DESIGN

From equation (iii) we can determine I; viz.
$$I = \pi \cdot r^3 \cdot t = \pi \cdot r^4 \cdot (\tfrac{1}{25})$$
Thus,
$$10\,000 = \frac{\pi^2 \cdot 30 \cdot 10^6 \cdot \pi \cdot r^4}{(DC_{max})^2 \cdot 25} \qquad (v)$$

Now, from truss geometry $DC = \tfrac{1}{2}(AB)$, i.e.
$$DC = (120 \pm \tfrac{3}{8}) \text{ in.,}$$
$$\therefore l_{max} = 120 \cdot 375 \text{ in.}$$

Inserting this value into (v), we obtain
$$r^4 = 3 \cdot 895$$
and
$$r = 1 \cdot 405 \text{ inches (mean radius)} \qquad (vi)$$

(outside diameter of tube $\simeq 3''$) and tube wall $\simeq 0 \cdot 056''$ (or $0 \cdot 06''$). The corresponding critical stress is
$$f_{crit} = \frac{10\,000}{\pi \cdot d \cdot t} = \frac{10\,000 \times 50}{\pi \cdot (2 \cdot 810)^2}$$
i.e.
$$f_{crit} = 20\,156 \text{ p.s.i.} \ (\simeq 9 \text{ ton } f/\text{in}^2)$$

(b) *Probabilistic calculations*

(i) Given that $P = (6070 \pm 600)$ lbf and taking sure-fit tolerance range to be $\equiv \pm 3\sigma$, then
$$3\sigma_p = 600 \quad \text{or} \quad \sigma_p = 200 \text{ lbf} \qquad (vii)$$

(ii) The column length is given as $l = (120 \pm 0 \cdot 375)$ inches. Thus, again
$$\sigma_l = 0 \cdot 125 \text{ inches} \qquad (viii)$$

(iii) The tube diameter is unknown, but the diametral tolerance is specified as $\pm 3 \cdot 42\%$, i.e.
$$\sigma_r = 0 \cdot 0114 \bar{r} \qquad (ix)$$

(iv) A reliability level of $R(z) = 0 \cdot 99997$ is required. From standard probability tables this corresponds to $z = 4$ standard deviations.

(v)
$$I \simeq \pi \cdot r^3 \cdot t = \frac{\pi \cdot r^4}{50} = 0 \cdot 1257 \cdot r^4 \qquad (x)$$
and
$$A = \pi \cdot d \cdot t = \frac{4\pi r^2}{50} = 0 \cdot 251 \cdot r^2 \qquad (xi)$$

7. RELIABILITY

(vi) To estimate the maximum allowable stress ($=$ crippling stress) we use equation (iv)

$$S = \frac{\pi^2 . E}{(l/\rho)^2} = \frac{\pi^2 . E . r^2}{2 . l^2}$$

Entering numerical values, we obtain

$$\bar{S} = \frac{\pi^2 . 30 \times 10^6 . \bar{r}^2}{2 \times (120)^2} = 10\,280 . (\bar{r})^2 \tag{xii}$$

or

$$\bar{S} = 148 \times 10^6 . \left(\frac{\bar{r}}{\bar{l}}\right)^2 \tag{xiia}$$

(vii) Next we have to determine the standard deviation of the maximum allowable stress. We apply the partial derivate method to equation (xiia)

$$\left(\frac{\partial \bar{S}}{\partial r}\right) = 148 \times 10^6 . \frac{2\bar{r}}{(\bar{l})^2}$$

$$\left(\frac{\partial \bar{S}}{\partial l}\right) = 148 \times 10^6 . \frac{(-2) . (\bar{r})^2}{(\bar{l})^3}$$

Thus

$$\sigma_s = \left[(148 \times 10^6 \times 2)^2 . \left(\frac{\bar{r}}{(\bar{l})^2}\right)^2 . \sigma_r^2 + (148 \times 10^6 \times 2)^2 . \left(\frac{(\bar{r})^2}{(\bar{l})^3}\right)^2 . \sigma_l^2\right]^{1/2}$$

or

$$\sigma_s = 296 \times 10^6 . \left[\left(\frac{\bar{r}}{(\bar{l})^2}\right)^2 . \sigma_r^2 + \left(\frac{(\bar{r})^2}{(\bar{l})^3}\right)^2 . \sigma_l^2\right]^{1/2} \tag{xiii}$$

From equations (xii) and (ix), we have

$$\sigma_l = 0.125; \qquad \sigma_l^2 = 1.565 \times 10^{-2}$$
$$\sigma_r = 0.0114 . \bar{r}; \qquad \sigma_r^2 = 1.30 . (\bar{r})^2 . 10^{-4}$$

Therefore,

$$\sigma_s = \frac{296 \times 10^6}{(120)^2} . (\bar{r})^2 . \left[\frac{1.30 \times 10^{-4} \times (120)^2 + 1.565 \times 10^{-2}}{(120)^2}\right]^{1/2}$$

or

$$\left.\begin{array}{r}\sigma_s = 308.4 . (\bar{r})^2 \\ \\ \bar{S} = 10\,281 . (\bar{r})^2\end{array}\right\} \tag{xiv}$$

and from before

This is the maximum allowable critical stress level against buckling.

142   APPROXIMATE METHODS IN ENGINEERING DESIGN

(viii) It is now necessary to evaluate the *actual* compressive stress generated by the load (6070 ± 600) lbf.

From equation (vii) we have

$$(\bar{P}; \sigma_p) = (6070; 200) \text{ lbf},$$

and from equations (xi) and (ix), we obtain

$$\bar{A} = 0.251 . (\bar{r})^2$$
$$\therefore \sigma_A = 0.502 . \bar{r} . \sigma_r$$

but

$$\sigma_r = 0.0114 \bar{r}$$

therefore

$$\sigma_A = 0.005\ 72 . (\bar{r})^2$$

Thus,

$$(\bar{A}; \sigma_A) = (0.251; 0.005\ 72).(\bar{r})^2 \qquad \text{(xv)}$$

From equation (ii), and generally,

$$\bar{f} = \frac{\bar{P}}{\bar{A}} = \frac{6070}{0.251.(\bar{r})^2} = \frac{24\ 154}{(\bar{r})^2} \text{ lbf/in}^2$$

Referring to Table 5.1, we obtain

$$\sigma_f = \frac{1}{(\bar{A})^2} [(\bar{P})^2 \cdot \sigma_A^2 + (\bar{A})^2 . \sigma_p^2]^{1/2},$$

and thus

$$\sigma_f = \frac{1}{(0.251)^2} \cdot \frac{1}{(\bar{r})^2} \cdot [(6070)^2 . (0.005\ 72)^2 + (0.251)^2 . (200)^2]^{1/2}$$

which yields

$$\sigma_f = 841.5 . \left(\frac{1}{\bar{r}}\right)^2$$

Thus

$$(\bar{f}; \sigma_f) = (24\ 154; 841.5) . \left(\frac{1}{\bar{r}}\right)^2 \qquad \text{(xvi)}$$

(ix) Comparing allowable stress (maximum) with the induced stress, equations (xiv) and (xvi), we get

$$(\bar{\Delta}; \sigma_\Delta) = \left[\left(10\ 281 . \bar{r}^2 - \frac{24\ 154}{(\bar{r})^2}\right); (\sigma_s^2 + \sigma_f^2)^{1/2}\right]$$

so

$$\bar{\Delta} = \left(\frac{1}{\bar{r}}\right)^2 . [10\ 281 . (\bar{r})^4 - 24\ 154] \qquad \text{(xvii)}$$

and

$$\sigma_\Delta = \left[(308\cdot 4 . (\bar{r})^2)^2 + \left(841\cdot 5 . \left(\frac{1}{\bar{r}}\right)^2\right)^2\right]^{1/2}$$

$$= \left(\frac{1}{\bar{r}}\right)^2 [95\,048\cdot 9 . (\bar{r})^8 + 708\,122\cdot 3]^{1/2}$$

which yields

$$\sigma_\Delta = \left(\frac{10}{\bar{r}}\right)^2 . [9\cdot 505 . (\bar{r})^8 + 70\cdot 81]^{1/2} \qquad \text{(xviii)}$$

For reliability requirements, equation (x), we know that $z = 4$, i.e.

$$4 = \frac{0 - (1/\bar{r})^2 [10\,281 . (\bar{r})^4 - 24\,154]}{(10/\bar{r})^2 . [9\cdot 505 . (\bar{r})^8 + 70\cdot 81]^{1/2}}$$

Squaring and simplifying this yields,

$$0 = 1\cdot 0418 . (\bar{r})^8 - 4\cdot 967 . (\bar{r})^4 + 5\cdot 7207$$

or

$$(\bar{r})^2 - 4\cdot 768 . (\bar{r})^4 + 5\cdot 491 = 0$$

for which

$$\bar{r} = 1\cdot 267$$

Thus,

and $\left. \begin{array}{l} \bar{d} = 2\cdot 534 \text{ inches} \\ \bar{t} = 0\cdot 051 \text{ inches} \end{array} \right\}$ for $R(z) = 99\cdot 997\%$

These dimensions are smaller than the values obtained for a safety factor of 1·5, which were

and $\left. \begin{array}{l} d = 2\cdot 81 \text{ inches} \\ t = 0\cdot 056 \text{ inches} \end{array} \right\}$ SF $= 1\cdot 5$

7.1.3

We know that for given loading conditons the factor of safety of a component may generally be improved by increasing its size and/or using a material of greater strength. Similar actions are usually also necessary to enhance reliability. Invariably the cost of a component increases. The following example (Haugen, 1968) clearly demonstrates the relationships between these three criteria.

Consider a typical design problem. What would be the "optimum" diameter of a tension bar for the following conditions?*

---

* This example is based on one published by E. B. Haugen and P. H. Wirsching in *Machine Design*, 29 May 1975, and partly used with their permission.

(a) The material is to be SAE 4130;

$$= (156\,000;\ 430)\,\text{p.s.i.}$$

(b) The applied load will be

$$P = (6000;\ 90)\,\text{lbf}$$

(c) The bar is to be of circular cross-section produced with a radial tolerance of

$$\Delta r = \pm 0{\cdot}015\,.\,\bar{r}$$

(d) A cost factor relationship has been ascertained to be

$$(\text{CF}) = 0{\cdot}25 + 1{\cdot}884(\bar{r})^2.$$

It is required

(i) to estimate a range of mean diameters to give safety factors varying from, $\eta = 0{\cdot}9$ to $\eta = 2{\cdot}0$;
(ii) to evaluate the corresponding "reliability" levels;
(iii) to calculate the cost factors of the various sizes of bar.

Since the intention is to cover a range of bar sizes, it will be useful to establish general correlations and write a computer program.

(i) *Factors of safety*. The general equation, based on mean parameter values, is

$$\text{SF} = \frac{156\,000}{(600/A)} = 26\,.\,\pi\,.\,r^2$$

or for the computer program

$$N = 26\,.\,\pi\,.\,R^2 \tag{i}$$

(ii) To calculate "reliability" it is necessary to compute value of Z

$$Z = \frac{\bar{s}-\bar{f}}{[\sigma_s^2 + \sigma_f^2]^{1/2}} \tag{7.1}$$

where

$\bar{s}$ = mean strength
$\bar{f}$ = mean stress
σ_s = standard deviation of strength
σ_f = standard deviation of stress

Using the numerical values given this results in

$$Z = \left(156 - \frac{6}{(\bar{r})^2}\right) \bigg/ \left[\frac{0{\cdot}0117}{\pi^2\,.\,(\bar{r})^4} + (4{\cdot}3)^2\right]^{1/2}$$

7. RELIABILITY

For the computer program this is re-written as follows:

$$A = 156 - \frac{6}{\pi \cdot (R)^2}$$

$$B = \frac{0 \cdot 0117}{\pi^2 \cdot (R)^4} + 18 \cdot 49 \tag{ii}$$

and

$$Z = \frac{A}{[B]^{1/2}}$$

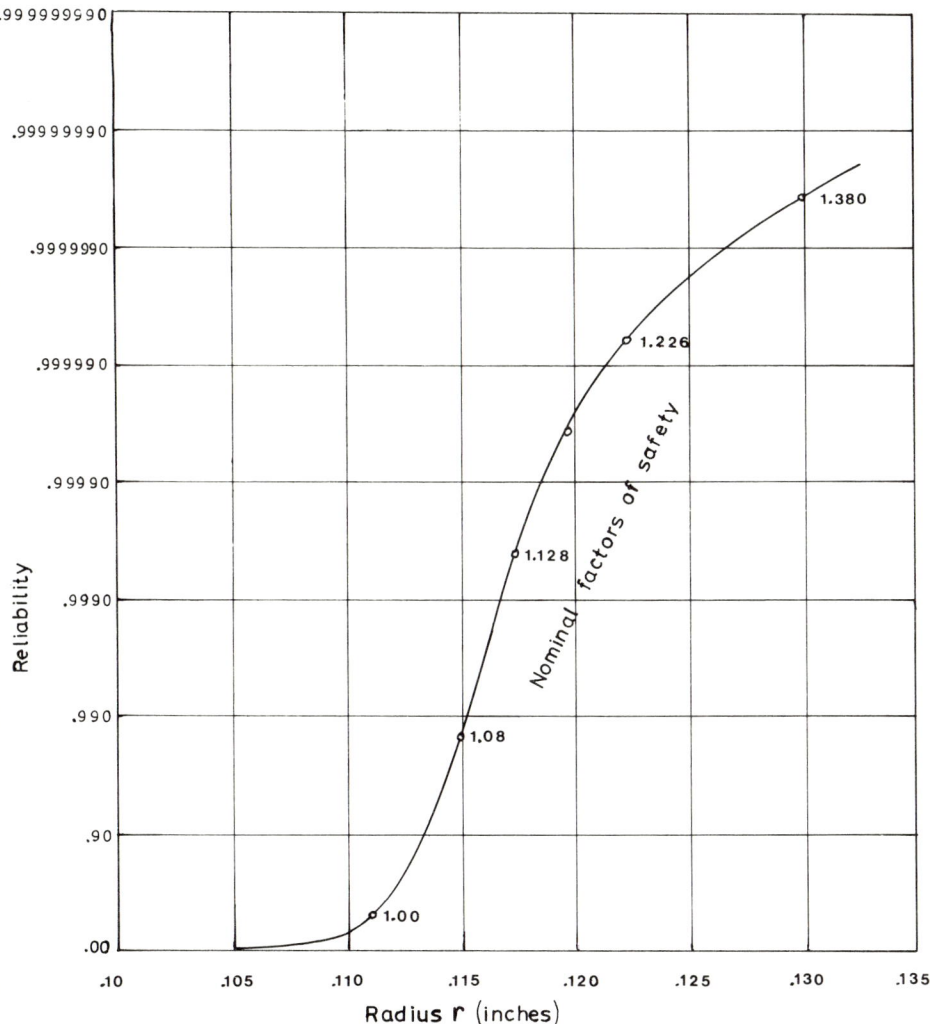

Fig. 7.6. Tension bar. Effect of size (diameter) on reliability and factor of safety.

(iii) Having determined Z, the corresponding reliabilities, R(Z), may be found from standard probability tables (see Table 4.1). For the computer program it is more convenient to use the approximate correlation by Johnson (1978):

and
$$F(Z) \simeq \left[\frac{1 \cdot 29}{|Z|}\right]^{7 \cdot 8125}$$

$$R(Z) \simeq 1 - \left[\frac{1 \cdot 29}{|Z|}\right]^{7 \cdot 8125}$$
(7.2)

which are valid for $0 \cdot 001 \leqslant F(Z) \leqslant 0 \cdot 015$, with less than 2% error. In the computer program the symbols are modified and the equations used read

$$F = \left[\frac{1 \cdot 29}{|Z|}\right]^{7 \cdot 8125}$$
and
$$R1 = (1 - F) \tag{iii}$$

(iv) The cost factor for each tube diameter (or radius) is calculated from the relationship given, i.e.

$$\text{cost factor} = 0 \cdot 25 + 1 \cdot 884 \cdot (\bar{r})^2$$

which for the computer program is written as

$$C = 0 \cdot 25 + 1 \cdot 884 \cdot (R)^2 \tag{iv}$$

(v) The computer read-outs were:

R	N	Z	R1	C
0·110	0·988 345	+0·356 774	0·3590	0·272 796
0·115	1·080 24	−2·305 12	0·989 274	0·274 916
0·120	1·176 21	−4·750 13	0·999 962	0·277 130
0·125	1·276 27	−6·989 01	0·999 998	0·279 438
0·130	1·380 42	−9·035 02	1·0	0·281 840
0·135	1·488 64	−10·9027	1·0	0·284 336
0·140	1·600 96	−12·6068	1·0	0·286 926
0·145	1·717 35	−14·1618	1·0	0·289 611
0·150	1·837 83	−15·5817	1·0	0·292 390
0·155	1·9624	−16·8792	1·0	0·295 263

Figure 7.6 is a plot of Rl versus the tube radius with the nominal safety factors also shown. It can be seen that although the safety factor increases beyond $R = 0 \cdot 125$ inches, the reliability level hardly alters, so it must be decided whether it is economical to increase the bar size beyond this value. The

relationships between factors of safety and reliability levels will be discussed further in Section 7.2.

7.2 RELIABILITY AND SAFETY FACTORS

In the examples of the preceding section comparisons were made between the conventional factors of safety and the probability of non-failure. The latter were based on the assumption that the "normal" distribution of the parameters involved was approximately valid. It was noted that reliability values were extremely high even for quite moderate levels of factors of safety. Clearly some redefinition of safety factors is needed to allow for the difference between distributed and deterministic parameters.

7.2.1

Shigley (1972) introduced a modified "statistical factor of safety" (n) which expresses the degree of safety available beyond any specified reliability (R). For example,

$$\text{if} \quad \bar{s} = 60\,000 \text{ p.s.i.} \quad \text{and} \quad \sigma_s = 4000 \text{ p.s.i.}$$
$$\text{and} \quad \bar{f} = 5000 \text{ p.s.i.} \quad \text{and} \quad \sigma_f = 1000 \text{ p.s.i.}$$

then equation (7.1) yields

$$z = \frac{\bar{s}-\bar{f}}{[\sigma_s^2 + \sigma_f^2]^{1/2}} = \frac{60\,000 - 50\,000}{[(4000)^2 + (1000)^2]^{1/2}}$$

or

$$z = 2\cdot 425$$

From standard probability tables the corresponding value of reliability is

$$R(2\cdot 425) = 0\cdot 9935$$

Now, Shigley notes that there is *no* reliability beyond $R(z) = 99\cdot 35\%$, i.e. that this reliability has a safety factor value of unity only. If a statistical safety factor is required then equation (7.1) becomes

$$z = \frac{\bar{s} - n.\bar{f}}{[\sigma_s^2 + \sigma_f^2]^{1/2}} \tag{7.3}$$

or

$$n = \frac{1}{\bar{f}} \cdot [\bar{s} - z.(\sigma_s^2 + \sigma_f^2)^{1/2}] \tag{7.4}$$

Equation (7.3) cannot easily be used for determining component size. It is more appropriate to go through an iterative procedure by guesstimating dimensions, calculating $\bar{f}$ and σ_f values, selecting the (Z) value corresponding to a desired reliability level R(Z), and thence calculate the statistical safety factor (n) from equation (7.4). Typical values of R(Z) and (Z) are:

R(%)	50	90	95	99	99·9	99·99
Z	0	1·285	1·645	2·325	3·08	3·899

A number of examples are given by Shigley, but it should be noted that the author has made a mistake by calculating $R = 2$ (value of "A" from his Table A.10). The correct answers are obtained from $R = (\text{``}A\text{''} + 0·50)$.

7.2.2

Carter focuses attention on the parameters which influence failure rates (λ). He designates equation (7.1) as the "safety margin", viz.

$$\text{SM} = \frac{\bar{S} - \bar{L}}{[\sigma_S^2 + \sigma_L^2]^{1/2}} \tag{7.1a}$$

and, if the number of load applications equal n, each application representing a given time interval, say $(k \cdot t)$ then

$$\lambda(n) = \frac{f(n)}{R(n)} \tag{7.5}$$

where
$\bar{S}$ = mean strength
$\bar{L}$ = mean load (or stress)
σ_S and σ_L = standard deviations of strength and load respectively

Plotting curves of failure rate v. safety margin (SM), Carter (1973) showed that for SM $\geqslant 6$ the failure rates were negligible (Fig. 7.7). However, as the safety margin is reduced the failure rate increases at an exponential rate; it can increase a hundred-fold for only a unit increase in the safety margin. The implication drawn by Carter from these relationships is that "to achieve failure rates within a factor of 2, i.e. 100% error, it would be necessary to estimate the safety margin to an accuracy close on 0·10". This is clearly impossible; it represents the heart of the mechanical engineering problem and gives, in Carter's words, "the direct lie to the claim that the failure rate, or reliability, can be estimated at the design stage". He therefore concludes that since the failure rate cannot be estimated analytically with any reasonable degree of accuracy, it must be determined experimentally, thus underlining the importance of a planned and properly conducted development programme. It is, of course, necessary to note that the situation is critical only where the safety margin values tend to be low, i.e. where stringent weight or cost criteria are

7. RELIABILITY

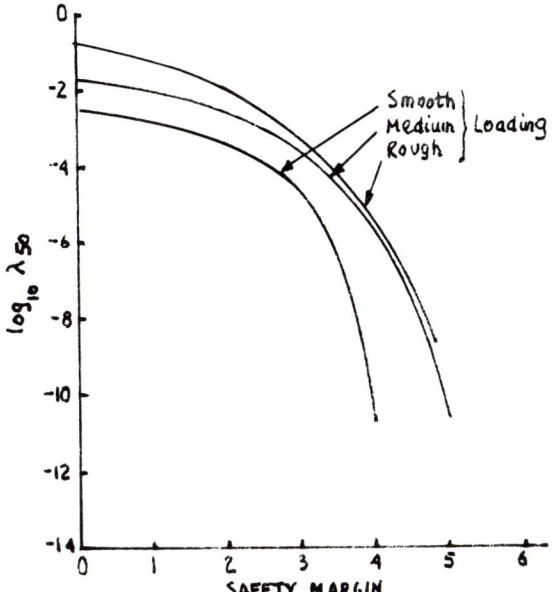

Fig. 7.7. Failure rate v. safety margin at 50 applications of load. (After Carter, 1973, by permission of Macmillan and Co.)

being imposed, or where the loading, or operating conditions cannot be tightly defined or controlled.

We note from equation (7.1a) that the safety margin is dependent upon (a) the difference between the mean strength and mean stress (or load) and (b) the standard deviations of both strength and stress, i.e. the larger the resultant deviation the smaller the SM value.

While it is generally possible to control the mean strength and its variance by careful design, rigorous material specifications, and good quality control of manufacturing processes (at a price, of course), it is much more difficult to control the nature of loading, i.e. operator usage, service and maintenance conditions. Carter (1973) therefore distinguishes between the effects of "smooth" (small variance) and "rough" (large variance) loadings (see Fig. 7.8). Plotting failure rates (λ) against loading roughness (σ_L/σ_S), he notes (Fig. 7.9) that (i) an early life failure pattern is indicative of smooth loading and low safety margins and (ii) there is only a rudimentary early life failure rate with rough loading both with low and high safety margins. The latter conditions are more common with mechanical engineering equipment. The shape of the conventional "bath-tub" curve is clearly dependent upon the "smoothness" or "roughness" of the load/strength variance ratio.

Generally there are two major unknown areas which control failure rates for mechanical equipment: one is the deterioration in strength with usage and

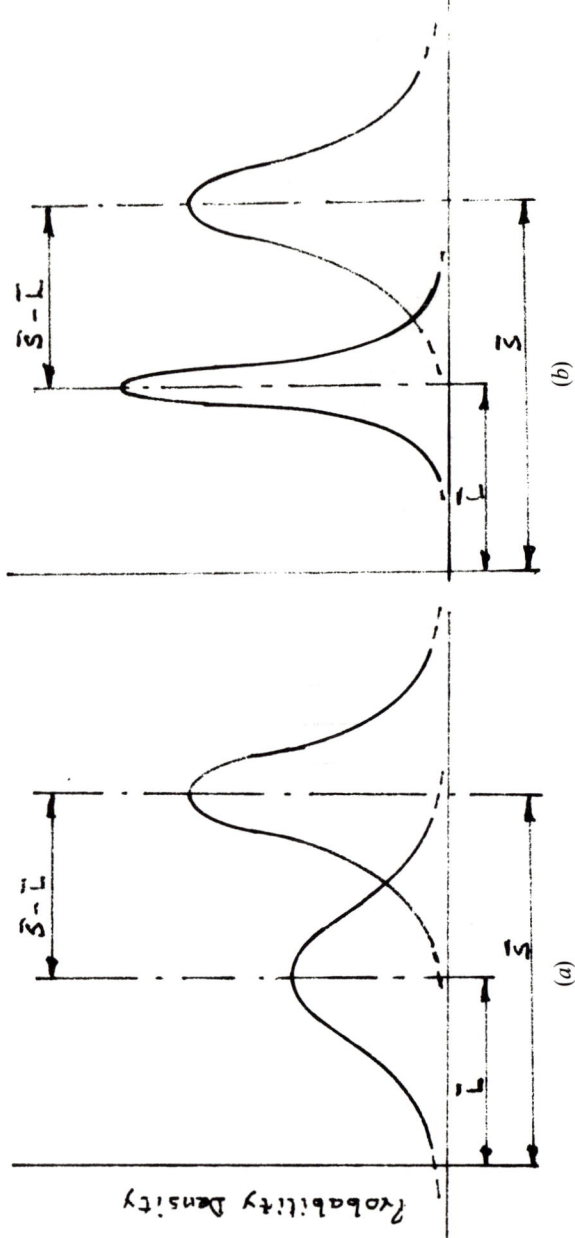

Fig. 7.8. Distributions of load and strength. (a) Rough load. (b) Smooth load. (After Carter, 1973, by permission of Macmillan and Co.)

7. RELIABILITY

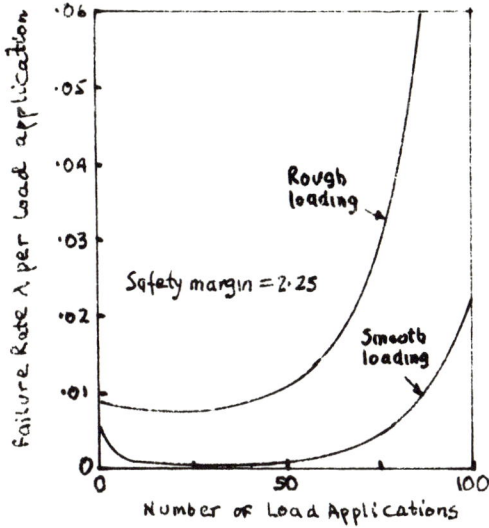

Fig. 7.9. Failure rate v. number of applications of load at constant strength and medium roughness. (After Carter, 1973, by permission of Macmillan and Co.)

time, and the other the uncertainty of the levels and variations in loading. Both are involved in "quality assurance" which is therefore much more complex than the relatively simple "quality control", used merely as a manufacturing tool.

7.3 RELIABILITY OF ASSEMBLIES AND SYSTEMS

Most machines or structures are built up from a number of individual components or parts, each of which contributes to the overall performance, reliability and durability of the whole assembly. If one can establish, by calculations, testing or from field data, the reliability characteristics of the individual parts—simulating as accurately as possible their function *in situ*—it would be useful to be able to predict the performance of the whole system.

Some mathematical (statistical) relationships have been established, but they must be treated with caution and understanding. The performance of a system may be no better than its weakest part, or it may be much more complex and versatile than simply the sum of the performances of the individual parts. These truths are simply illustrated by the following examples.

(i) Consider three electric bulbs of three different colours: white, red and blue. Each bulb can give two basic signals, i.e. on = 1 and off = 0. Since all signals are the same, we have from the individual bulbs a total of 4 signals. If

the three bulbs are placed in a simple formation, one can obtain the following combinations of signals:
(1) All off
(2) White on
(3) Blue on
(4) Red on
(5) (White + blue) on
(6) (White + red) on
(7) (Red + blue) on
(8) (White + red + blue) on

So, whereas the simple sum of the individual lights were 4, the system is capable of giving 8 different signals.

(ii) An anchor chain can certainly allow a distance between the anchor and the ship, but its load capacity is no greater than that of its single (weakest) link.

7.3.1 In-series system configuration

Every machine system has a number of items, or parts linked together like a chain, e.g.

piston/gudgeon-pin/connecting-rod/crankpin/ crank web/crankshaft/coupling, and so on.

In a bicycle the series consists of the

pedals/shaft/large sprocket/chain/small sprocket/ wheel hub/spokes/wheel rim/tyre.

Such systems can be described diagrammatically as in Fig. 7.10.

Fig. 7.10. In-series system configuration.

Conditions assumed are:
(a) Failure of any one part is independent of the failure of any other part in the system.
(b) Failure of any one part causes the failure of the whole system.

If p_i = probability of successful operation (= reliability) of part (i), the overall reliability of the whole system then is

$$R_{system} = p_1 \cdot p_2 \cdot p_3 \cdots p_n$$

or

$$R_{system} = \prod_{i=1}^{i=n} p_i \qquad (7.6)$$

If the reliability of each component in series is nearly the same (e.g. several

7. RELIABILITY

links in a chain), equation (7.6) becomes

$$R_{\text{system}} = (\bar{p})^n \qquad (7.7)$$

where $(\bar{p})$ = mean reliability of all components.

This implies that the reliability of a system consisting of components all in series equals the product of the reliabilities of the individual parts, and that this overall system reliability is *less* than the individual reliabilities. As the number of components increase so the probability of failure of the whole system increases. This is clearly indicated in Fig. 7.11.

For 80% overall system reliability the average individual probability of failure would have to be:

Number of components	Average probability of failure
10	1/45
50	1/225
100	1/450
200	1/900
300	1/1350
400	1/1800

The above approach is highly simplified; the in-series system is discussed further in Section 7.3.6, where Carter's (1973) concepts of "rough" and "smooth" loadings are considered.

7.3.2 Parallel system configuration (see Fig. 7.12) Consider first a simple stand-by system in which only one unit (machine, or component) is in operation at any one time, and when it fails it is immediately replaced by a stand-by unit in parallel. Denoting $q_i = (1 - p_i)$ as the probability of failure of the ith unit, the reliability of the system until all the stand-by units have failed is

$$R_{\text{system}} = (1 - F_{\text{system}}) = 1 - (1 - p_1)(1 - p_2)(1 - p_3) \ldots (1 - p_n)$$

or

$$R_{\text{system}} = 1 - \prod_{i=1}^{i=n} (1 - p_i)$$

or,
$$\qquad (7.8)$$

$$R_{\text{system}} = 1 - \prod_{i=1}^{i=n} q_i$$

If the various stand-by units have nearly similar reliability levels, equation (7.8) can take the form

$$R_{\text{system}} = 1 - (\bar{q})^n,$$

154   APPROXIMATE METHODS IN ENGINEERING DESIGN

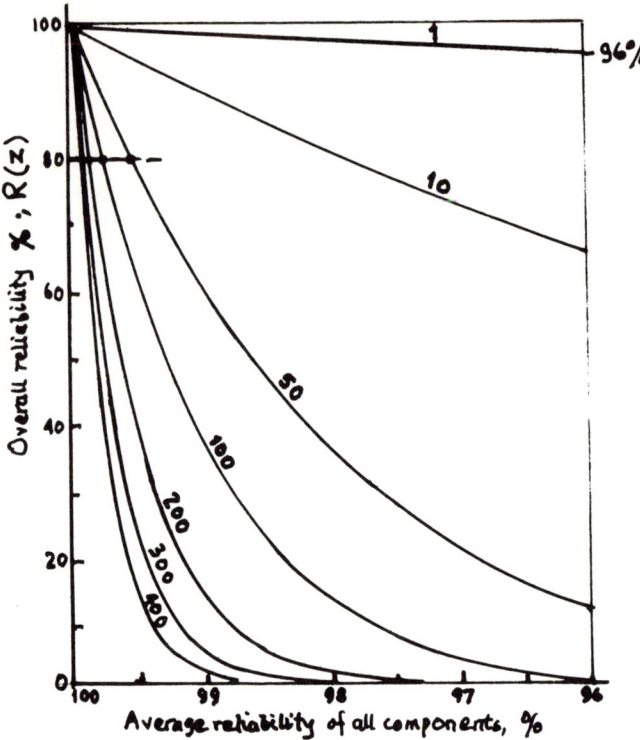

Fig. 7.11. Overall reliability of in-series system. (After Carter, 1972, by permission of Macmillan and Co.)

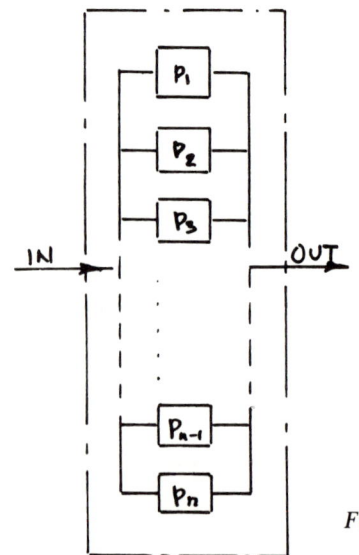

Fig. 7.12. Simple stand-by parallel system.

or
$$R_{system} = 1-(1-\bar{p})^n \qquad (7.9)$$
and
$$F_{system} = (\bar{q})^n$$

It is clear that the probability of failure is reduced as the number of parallel stand-by units is increased. Conversely the reliability of such a parallel system is greater than that of the individual units.

7.3.3 Several in-series sub-systems in parallel (see Fig. 7.13)

In this arrangement there are several (k) complete in-series sub-systems acting as stand-bys. Each of the in-series systems contain (n) components or units.

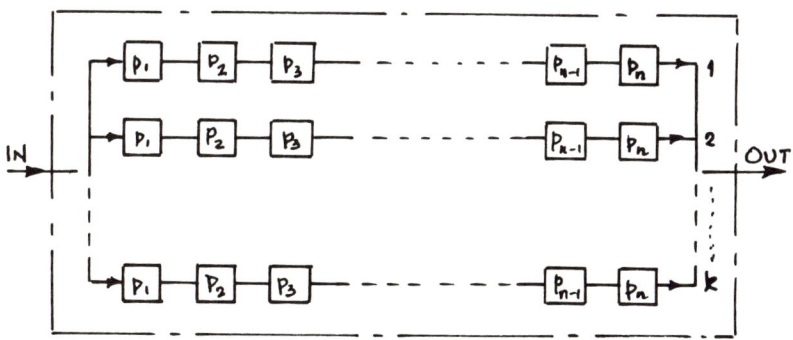

Fig. 7.13. Several in-series subsystems in parallel.

Considering each sub-system as operating one at any one time, the reliability of one such a system is (from equation (7.6))

$$R_n = \prod_{i=1}^{i=n} p_i$$

The probability of failure of each in-series system, therefore, is

$$F_n = 1 - \prod_{i=1}^{i=n} p_i \qquad (7.10)$$

For (k) such in-series systems in parallel, the probability of failure of the whole system, assuming each stand-by system to have identical reliabilities,

$$F_k = \left(1 - \prod_{i=1}^{i=n} p_i\right)^k$$

or

$$R_k = 1 - \left(1 - \prod_{i=1}^{i=n} \cdot p_i\right)^k \qquad (7.11)$$

7.3.4 Several parallel sub-systems in series (see Fig. 7.14)

An alternative arrangement to increase overall system reliability is to have a number (k) stand-bys for each component or unit, these (n) units being in series.

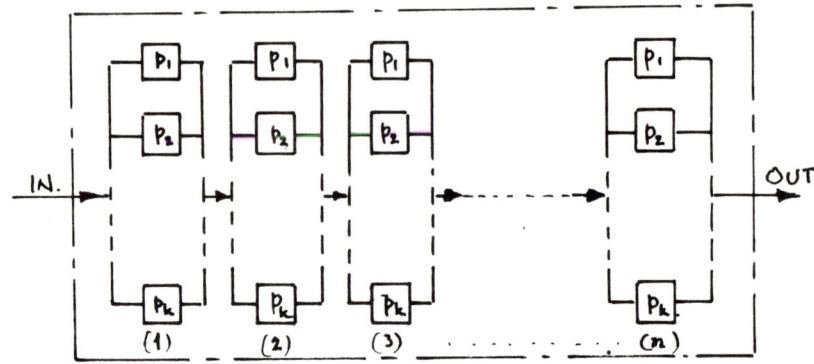

Fig. 7.14. Several parallel subsystems in series.

In this case we have:
Reliability of each parallel stage is

$$R_k = \left(1 - \prod_{i=1}^{i=k} q_i\right)$$

For (n) such stages in series

$$R_n = \prod_{i=1}^{i=n}\left(1 - \prod_{i=1}^{i=k} q_i\right) \tag{7.12}$$

If the reliabilities in each stage are identical, then

$$R_n = \prod_{i=1}^{i=n} (1 - q_i^k) \tag{7.13}$$

and further that if the reliability of each stage is identical then

$$R_n = (1 - q_i^k)^n \tag{7.14}$$

The overall system reliability is thus considerably enhanced.

7.3.5 Active redundancy

Active redundancy implies a system in which all, or several, units in parallel operate simultaneously and the failure of one of the units does not significantly affect the continued operation of the whole system. Examples of such active redundant systems are: multi-engine aircraft; bridge and frame structures; gas, electricity, water, rail and road networks; telecommunication systems, etc. The probability of failure of all parts or units in parallel at any given time can be

7. RELIABILITY

expected to be lower than the failure probability of any one unit. Poisson's distribution may well represent the probability of failure of one, two, three or more units in such a paralle system (before repair and re-coupling of any failed unit). If the average expectation of failure of a unit in such a system is (z), the Poisson's distribution predicts the following probabilities of failure:

$$0 \text{ units failed} = e^{-z}$$

$$1 \text{ unit failed} = z \cdot e^{-z}$$

$$2 \text{ units failed} = \frac{z^2}{2!} \cdot e^{-z}$$

$$3 \text{ units failed} = \frac{z^3}{3!} \cdot e^{-z}$$

$$n \text{ units failed} = \frac{z^n}{n!} \cdot e^{-z}$$

If, for example, the average failure rate $z = 0 \cdot 61$, then:

Number of failed units =	0	1	2	3	4	etc.
Probability of failure =	0·543	0·331	0·101	0·021	0·003	etc.

Thus the probability of failure of the whole system is reduced from 0·543 for one engine to 0·003 only for a system with four engines. However, it is obvious that the increase in system reliability by the use of multiple active or stand-by units is expensive.

Jardine (1973) discusses several situations of optimizing redundancy levels against various criteria, such as cost of design effort, cost of manufacture and installation, cost of repairs and down-time, etc. It is useful here to describe one of the cases Jardine examines.

Problem: Determine the optimal number of machines to have an active redundancy arrangement such as to minimize the total cost per unit time of operation and down-time losses.
Model assumed: (i) Failure distribution is of the Poisson form:

$$\text{mean time to failure} = \frac{1}{\lambda},$$

i.e. λ = number of failures per unit time.

158   APPROXIMATE METHODS IN ENGINEERING DESIGN

(ii) Time to repair a failed machine or unit also has a Poisson distribution:

$$\text{mean service time} = \frac{1}{\mu}$$

i.e. μ = number of machines that can be serviced in unit time.

(iii) There are (k) machines in parallel in each stage. A stage fails when the kth machines has failed, all other machines having failed before and not yet repaired and re-connected.

Applying queueing theory, Jardine (1973) relates the down-time of the whole stage to the parameters given as

$$d(k) = \frac{\rho^k}{(1+\rho)^k} \tag{7.15}$$

where

$$\rho = \left(\frac{\lambda}{\mu}\right)$$

(iv) The cost per unit down-time = C_d.

(v) The total cost of operation per unit time for one machine, or unit in a stage = C_0.

(vi) The total cost of running (k) machines in a stage is equal to

$$\left[\begin{pmatrix}\text{cost per machine}\\ \text{per unit time}\end{pmatrix} \times \begin{pmatrix}\text{number of machines}\\ \text{in parallel in a stage}\end{pmatrix}\right]$$

$$+ \left[\begin{pmatrix}\text{cost of down-time}\\ \text{per unit time}\end{pmatrix} \times \begin{pmatrix}\text{proportion of unit-time}\\ \text{stage has failed}\end{pmatrix}\right]$$

or

$$\text{total cost} = n \cdot C_0 + d(k) \cdot C_d \tag{7.16}$$

Example: Let

$\lambda = 20$ failures per unit time

$\mu = 20$ machines serviced in unit time

$C_0 = £100$ = cost of operating one machine per unit time

$C_d = £500$ = cost of unit down-time of whole stage

From equation (7.15),

$$d(k) = \left(\frac{\rho}{1+\rho}\right)^k = \left(\frac{1}{1+1}\right)^k = (\tfrac{1}{2})^k \tag{a}$$

From equation (7.16), we have

$$\text{total cost} = k \cdot 100 + (\tfrac{1}{2})^k \cdot 500 \tag{b}$$

7. RELIABILITY

Entering various values of (k), one can tabulate the results as below:

k	d(k)	Total cost (£)
1	0·50	350
2	0·25	325
3	0·125	375·5
4	0·0625	436·25
5	0·031 25	518·125

It can be seen that the optimum number of machines in parallel in each stage, to minimize total cost per unit time of operation and down-time losses, is 2 machines.

For further examples and analysis of more complex systems the reader is referred to Jardine (1973).

7.3.6 System models

The simple models of in-series and in-parallel configurations have seldom been confirmed in practice. Chaddock (Carter, 1972) studied a number of weapons for which ample field data were available. He concluded that "No such correlations existed... The level of reliability of a system depends primarily on the weakest link in the most severe operating conditions."

Other investigators have confirmed this conclusion and have noted the variability and scatter in the strength and duty of equipment. Taking these factors into account Carter (1972) extended his analysis of systems and reasoned as follows:

(i) If stress due to load < strength, component (or unit) survives.
(ii) If stress due to load > strength, component (or unit) fails.

He distinguishes three typical situations (Fig. 7.15)

(1) $Smooth \simeq$ constant load ($\bar{L}$) (i.e. minimum scatter) and variable strength $= (\bar{s}, \sigma_s)$

$$\sigma_L \ll \sigma_s$$

(2) $Rough$ = variable load ($\bar{L}; \sigma_L$) and steady $\simeq$ constant strength ($\bar{s}$)

$$\sigma_L \gg \sigma_s$$

(3) Variable load and strength;

$$\sigma_L \simeq \sigma_s$$

These considerations lead to modification of the relationships for systems (cf. 7.3.1 and 7.3.2). Figure 7.16 (from Carter, 1972) clearly shows the variation in system reliability with the number of units in series depending upon the

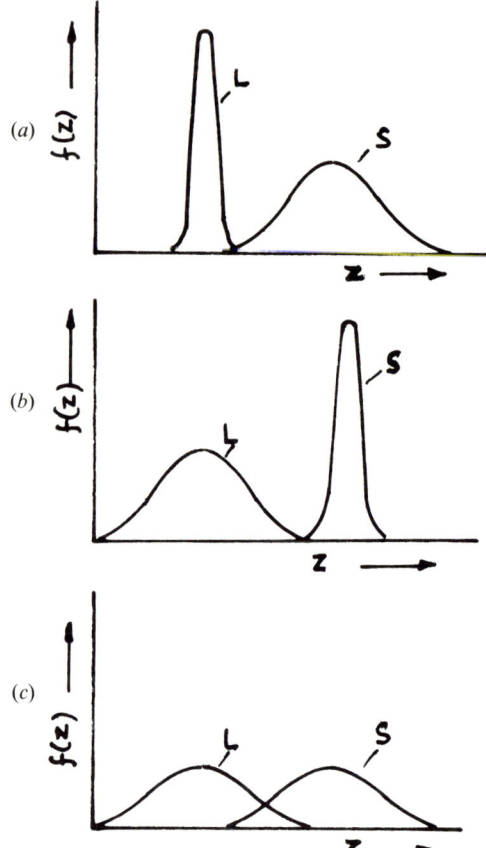

Fig. 7.15. "Smooth" ($\sigma_L \ll \sigma_S$), "rough" ($\sigma_L \gg \sigma_S$) and "intermediate" loadings ($\sigma_L \approx \sigma_S$). (After Carter, 1973, by permission of Macmillan and Co.)

smoothness or roughness of the load/strength relationship. It is noted that

for case (a) (smooth loading) $R_{\text{system}} \simeq (\bar{R})^n$
for case (b) (rough loading) $R_{\text{system}} \simeq \bar{R}$ ($=$ weakest link)
for case (c) (load and strength variable) $(\bar{R}) > R_{\text{system}} > (\bar{R})^n$

Meaningful numerical reliability data for mechanical equipment is still very scant. Individual companies do gather and analyse their field reports, but these are usually not available for general study. The in-house information is used for design and manufacturing modifications on a more or less *ad hoc* and intuitive basis. As yet there are scant correlative data between accelerated test procedures and service experience. At best, as far as the author is aware, computer-controlled simulated test programs developed at the National

7. RELIABILITY

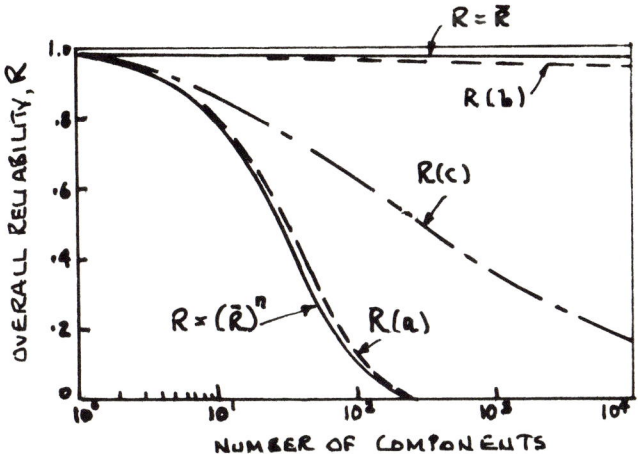

Fig. 7.16. Overall reliability of systems for (*a*) smooth, (*b*) rough and (*c*) intermediate loadings. (After Carter, 1973, by permission of Macmillan and Co.)

Engineering Laboratories for full-scale equipment come closest to real-life situations for any particular equipment being tested.

Example 7.1 Consider the braking system of a front-wheel driven car. The system may be represented by a block-diagram (Fig. 7.17).

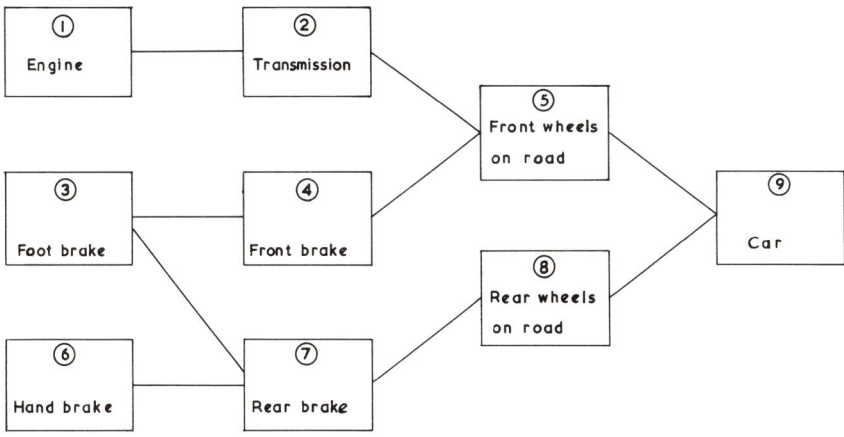

Fig. 7.17. Block diagram of braking system of a front-wheel driven car.

Let the individual "component" reliabilities be:

(1) Engine = p_1
(2) Transmission = p_2

Braking effectiveness assumed to be only, say, = 0·60. (foot brake)

(3) Hydraulics = p_3
(4) Front brakes = p_4
(5) Front wheels on road = p_5
(6) Handbrake system = p_6;   effectiveness on moving car is, say, only 0·50 (foot brake)
(7) Rear brakes = p_7
(8) Rear wheels on road = p_8

(N.B. rear tyres on road less effective than front wheels, say, 0·90. (front wheels) (i.e. $p_8 = 0·90 . p_5$).)

The reader may insert his own individual reliability values in order to estimate the overall braking reliability of the car. (For a method of solution see p. 382.)

7.4 THE ROLE OF DESIGN IN ACHIEVING RELIABILITY

"There is nothing in *design* for reliability that is not contained in good design itself" (Carter, 1972).

To achieve a high level of reliability the designer must consider the following:
(1) The need to use safety margins between 3 and 6 standard deviations for all important parameters, e.g. load/strength; lift/drag; cost/price, etc.
(2) The need to diagnose the critical item(s). If necessary to apply fail-safe techniques or introduce redundancy.
(3) The need to establish the secondary function(s) any item or part may have to perform.
(4) The use of well-tried and tested parts and materials, rather than new and uncertain ones.
(5) New designs or parts should be carefully tested and matched into system.
(6) Always aim for simplicity of design. "What isn't there, cannot fail!"
(7) Service and maintenance procedures should be planned for. They are included in the concept of "quality assurance".
(8) Reliability costs money; value analysis and cost-saving procedures should not be introduced at the expense of reliability.
(9) Manufacturing specifications and routes are important; they may include special inspection and proof-test procedures and even the careful selection of preferred manufacturer(s).
(10) The design must include facilities for correct inspection and any test required.
(11) Effects of handling, transport and storage should be taken into account.

(12) Always make allowance for human fallibility.
(13) Assign meaningful reliability values whenever possible and feasible. These highlight important features and stimulate critical thought and contingency ideas.
(14) Analyse feedback information from field and service experience and data and use these for modifications and improvements. Plot Pareto-type curves (Fig. 7.18) (Carter, 1972).

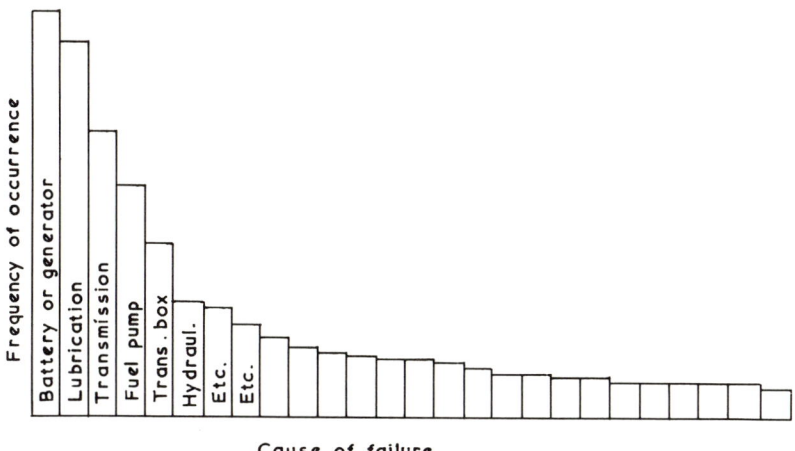

Fig. 7.18. The Pareto curve. (After Carter, 1972, by permission of Macmillan and Co.)

In this context it is of interest to present a number of points from Chapter II, "Responsibility of the Design Office for Quality, Reliability and Cost", in the O.E.C.D. survey published in Paris in 1967:

(1) Quality cannot be inspected into a product, the responsibility for quality depends on:
   (i) management policy and direction;
   (ii) design specifications;
   (iii) detail drawings and their interpretation;
   (iv) type and condition of plant and machine tools;
   (v) type and condition of tooling and gauges;
   (vi) accuracy of machine setting;
   (vii) training and attention of individuals performing their various tasks.

"Quality is everybody's responsibility for his own activity." (See also the Zero Defect Production System, Chapter 8 in Pronikov (1973).)

(2) Defining a problem (specifications) is, perhaps, the most important task in the design process.

(3) The designer holds the key in all questions concerning quality of a product, i.e. superiority in
   (i) function and performance;
   (ii) quality and reliability;
   (iii) price competitiveness;
   (iv) delivery and after service.
(4) Cost is usually directly related to complexity; simplicity is most difficult to achieve in design.

The concluding summary reads:
(1) (a) The designer himself must determine the rules and the measurable values to enable the workshops to manufacture the product according to the best quality criteria.
   (b) The designer must hold the key position in all questions concerning quality at all stages of the production.
   (c) To maintain the quality, the designer must receive feedback information from all departments concerned with the product.
   (d) Prototypes and products must be tested for reliability in different operational and environmental conditions, and the results communicated to the designer.
   (e) The designer must be informed about all requirements of the product so that he can establish the master specification for the design project
   (f) It may be advisable to hold regular meetings between the design staff and the inspection and service department, to maintain and improve the quality and the reliability of the product.
(2) (a) It is essential to provide designers and draughtsmen continually with comparative cost information on:
      (i) new production techniques and processes;
      (ii) costs of materials and costs of bought-out items;
      (iii) cost comparisons of alternative designs;
      (iv) actual manufacturing costs in the organization.
   (b) To design and detail for consistent quality at controllable costs, the channels in which to direct research, education and training are:
      (i) simplification in the functioning and construction of the product;
      (ii) reduction in the number of component parts;
      (iii) reduction to a minimum of the number of processes and operations;
      (iv) maximum utilization of materials and the minimum of material removal operations;
      (v) application of new materials and processes;

7. RELIABILITY

(vi) standardization and variety reduction of materials and components;
(vii) avoidance of unnecessary requirements for accuracy of manufacture;
(viii) selection of appropriate quality of surface finish;
(ix) consideration of the use of commonly purchased articles to avoid manufacture;
(x) consideration of all available sources of supply for all bought-out items

(c) It may be advisable to hold regular meetings between the design staff and the Estimating and Cost Department to check on estimated and actual costs.

(d) Cost consciousness must be developed and stimulated in the design staff by the management. Value analysis, properly applied by suitable committees, can make the greatest contribution towards cost reduction and will even discover completely new ways of performing a specific product function.

Summarizing, Carter (1972) says: "The achievement of high reliability at the design stage is mainly the application of engineering common sense coupled with a meticulous attention to trivial details. While not a subject superficially of great attraction to academics, it should be recognised that a systematic approach and a sound appreciation of fundamentals of engineering are vital to success. It is, therefore, essentially an intellectual activity."

The essence of what Carter says is true, but nevertheless well-defined reliability information on mechanical equipment is sadly lacking, both in general engineering literature as well as in information from engineering companies about their products. It is anticipated that the 1980s will bring considerable development in this area.

It would appear appropriate to conclude this chapter with a few practical examples of mechanical engineering equipment of poor designs and some in which modified designs have been developed to improve reliability.

7.5 CASE STUDIES

Carter's conclusion that "the achievement of high reliability at the design stage is mainly the application of common sense coupled with a meticulous attention to trivial details" (Carter, 1972) is in fact something of an understatement. Although engineering practice may not always involve complex mathematics, the achievement of high performance and high reliability is often the result of the application of scientific methods to the problems experienced in practice, of a thorough analysis of performance

166   APPROXIMATE METHODS IN ENGINEERING DESIGN

deficiencies, of simulated experiments with refined measurement techniques to determine influences of various factors and causes of non-reliability, and of patient development work.

A number of practical case studies have been selected from a variety of industries. Apart from the varying degree of complexity of the equipment described, the case studies underline the interplay between engineering science and engineering practice, and the intense intellectual demands made on the professional engineers. The task they perform cannot be said to be simply the "application of common sense" nor the "meticulous attention to trivial details".

Case Study 7.5.1

Figure 7.19 shows a fairly conventional two-pulley bottom block complete with a standard crane hook. On careful examination the reliability of the

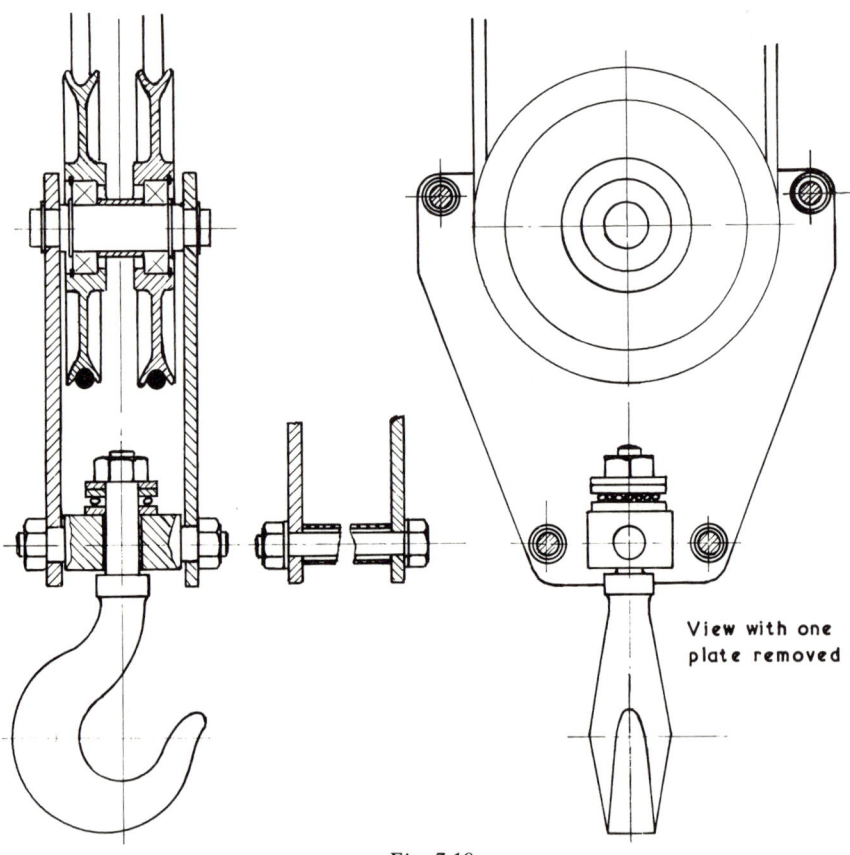

Fig. 7.19.

7. RELIABILITY

design leaves much to be desired. The following shortcomings are noted:
(1) The nuts that hold the hook and cross-bar are not secured.
(2) Both shaft and pulleys can rotate and this necessitates both inner and outer bearing races to be interference fitted.
(3) All rolling element bearings are open to dirt ingress, and no lubrication facilities are provided.
(4) The ropes can come off the pulleys, when slack, and could loop between them on to the shaft.

Figure 7.20 incorporates a number of modifications to overcome the above defects and improve the reliable working of the equipment.
(1) The hook and cross-bar have been firmly secured and both are running in shell-bearings.
(2) The pulley-shaft has been secured by two anchor plates which prevent its rotation.

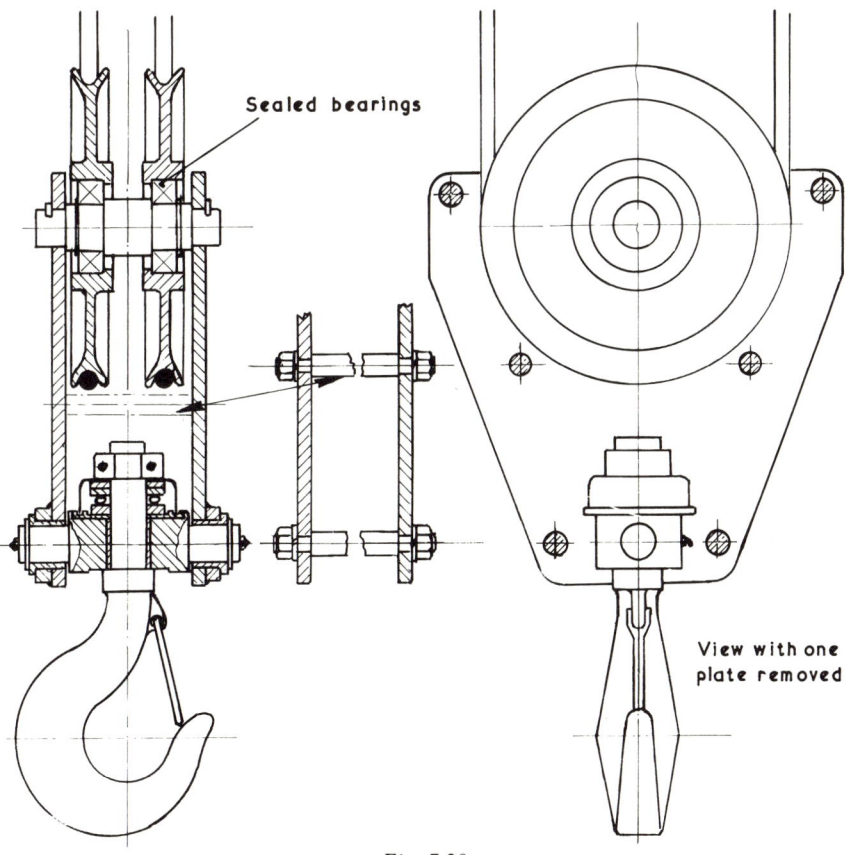

Fig. 7.20.

168   APPROXIMATE METHODS IN ENGINEERING DESIGN

(3) All rolling element bearings have been sealed and grease packed. Lubrication is also provided for the shell bearings.
(4) Two tie-bars have been added and positioned so close to the periphery of the pulleys to prevent the ropes coming out of their grooves.
(5) A latch has been added to the hook to give extra security to a load sling or chain link.

Case Study 7.5.2

This case study of a large industrial fan illustrates some shortcomings in the original design with regard to details and especially with respect to maintenance and serviceability. Figure 7.21 shows various features of the fan casing, the impeller mounting and the shaft-bearing assembly.

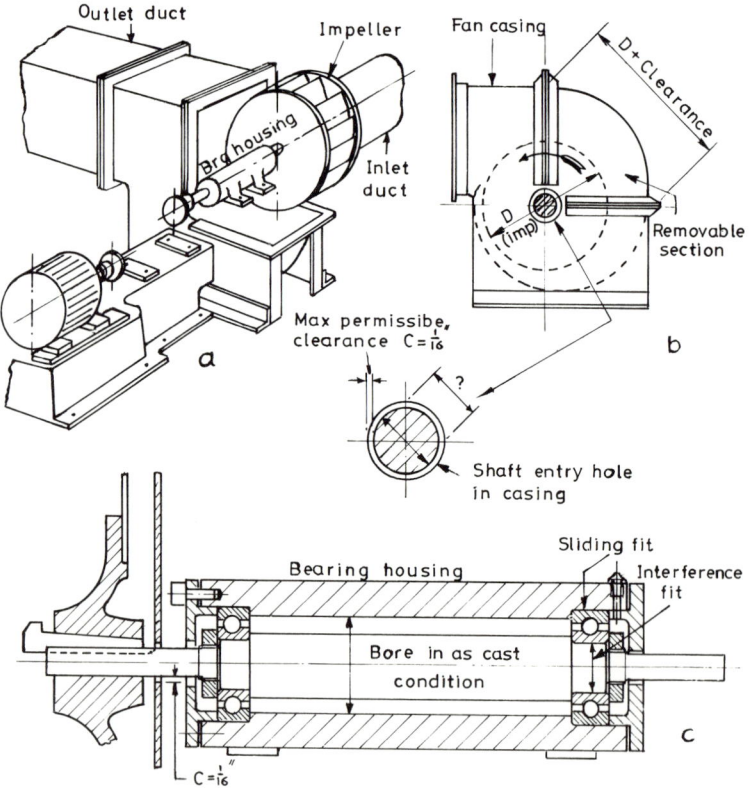

Fig. 7.21. (With acknowledgements to A. E. Seaborn, University of Salford.)

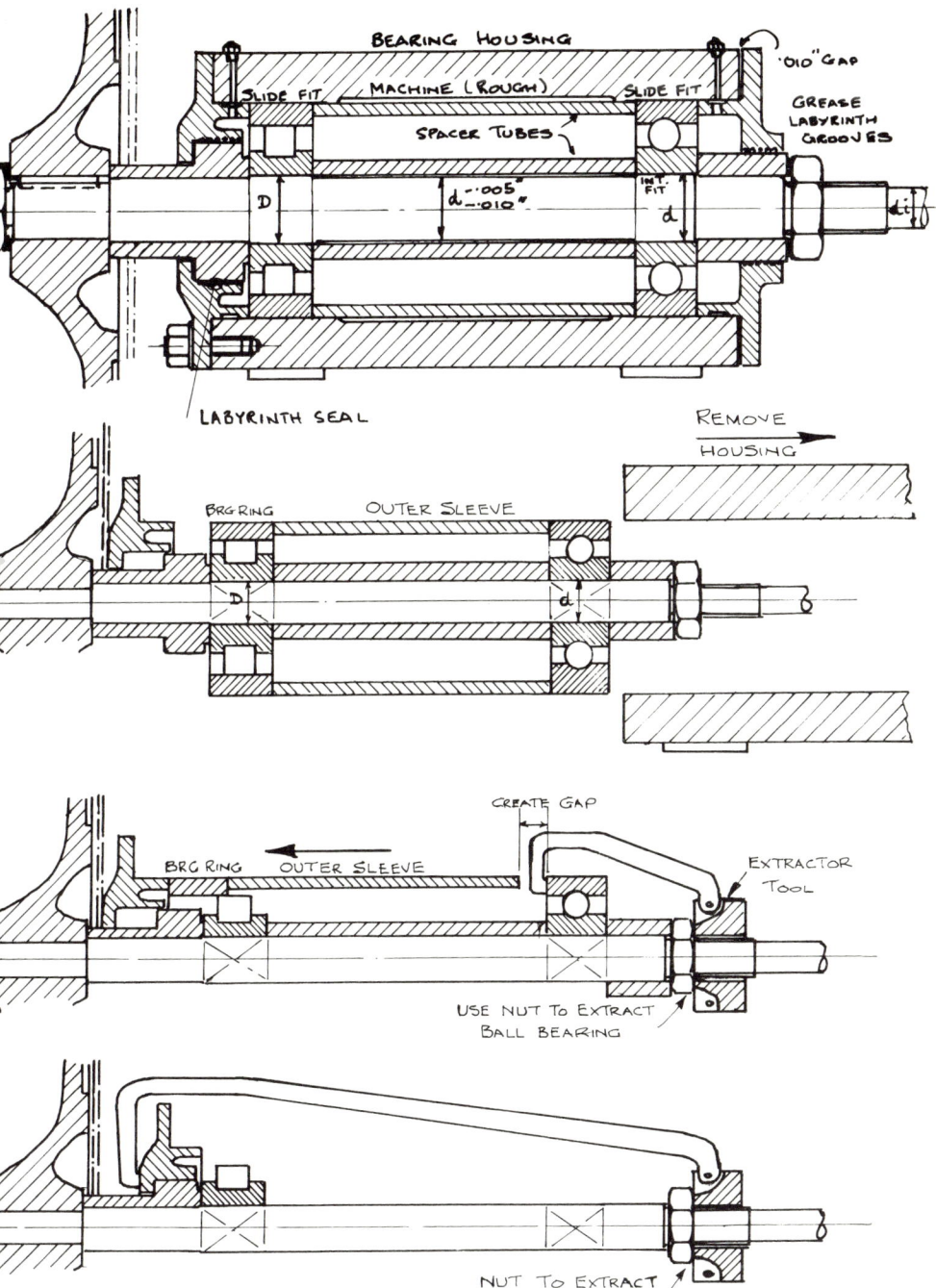

Fig. 7.22. (With acknowledgements to A. E. Seaborn, University of Salford)

170   APPROXIMATE METHODS IN ENGINEERING DESIGN

Deficiencies
(1) The 90° volute section must be sized so that the complete impeller and shaft assembly can be inserted and withdrawn as shown in (A). If the maximum permissible clearance between shaft and hole in the casing-wall is not to exceed $\frac{1}{16}$ in (to avoid excessive air leakage) the shaft will not clear the 90° cut-out.
(2) The gib-head of the taper key overhangs the end of the shaft. When trying to remove the key from the shaft (with drift and hammer) the gib-head would tend to bend.
(3) The bearing housing cannot be removed from the shaft without opening the fan and withdrawing the impeller. This restricts the servicing and replacement of bearings.
(4) A high stress raiser can be expected at the impeller end of the shaft, where the nut clamps the ball-bearings on the shaft.
(5) Location of ball bearings axially both on shaft and housing is likely to cause pre-loading of the races due to the effect of tolerances and any thermal differential expansion.
(6) The space between bearing housing and fan wall is too small, making it impossible to remove the set-screws securing the end flanges.
(7) The outer bearing race diameters at the ends of the housing have to be bored out separately. This makes it difficult to achieve perfect concentricity.
(8) The end seal on the shaft is inadequate for retaining oil or grease.

Modifications (see Fig. 7.22)
(1) Make holes in casing wall larger and cover with split plates.
(2) Use set-in key and end nut and tab-washer to secure impeller to shaft.
(3) The outer bearing diameters in housing bored right through in one operation. Tubular spacers may be used for axially clamping the races.
(4) Using one roller and one ball bearing obviates axial pre-loading.
(5) Stress concentrations at high bending stress section of shaft avoided by using practically plain shaft.
(6) Space between fan-wall and bearing end-flange increased, using studs and nuts. These enable bearing housing to be withdrawn from the shaft without opening up the fan.

Case Study 7.5.3

A stores lift, operating between two floors only, consisted of a cage attached at the rear to a four-wheeled trolley running in two vertical channels (Fig. 7.23). The ascent and descent of the cage was by means of a long vertical lead-screw driven by a 3-phase electric motor through a reduction gear unit.

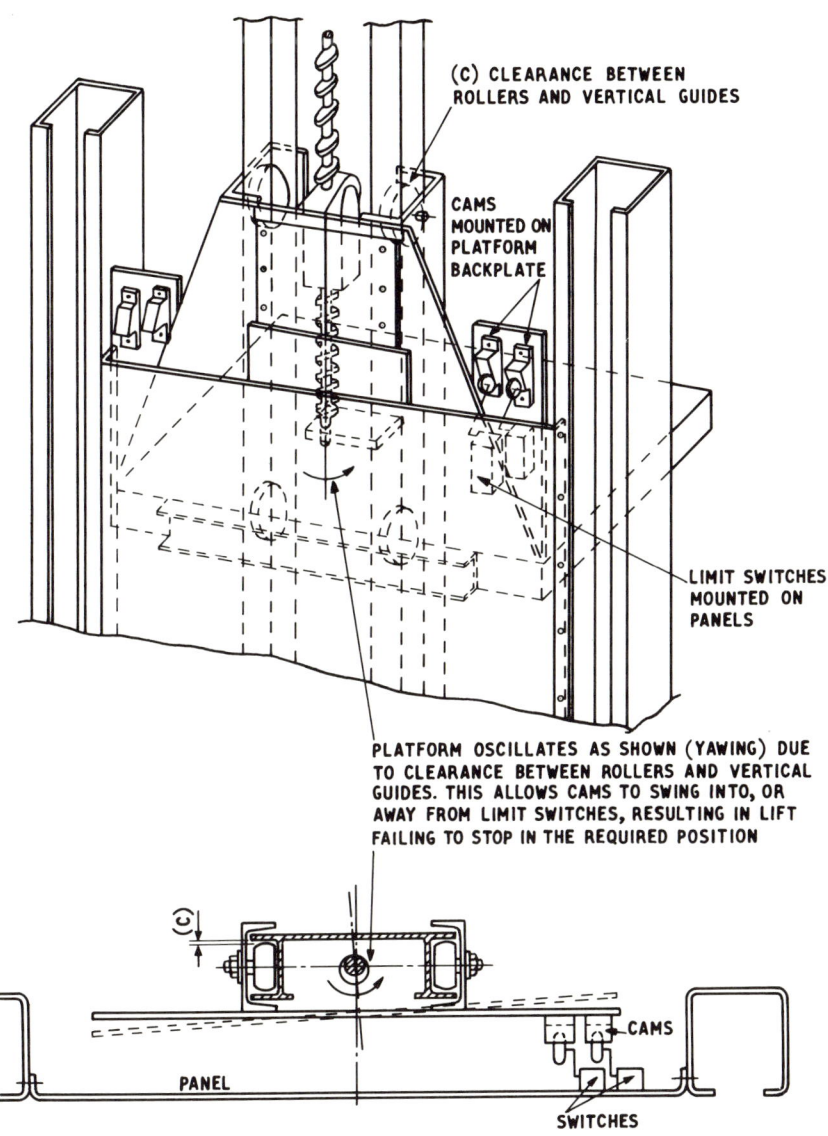

Fig. 7.23.

The lead-screw actuated a composite "nut-assembly" consisting of a housing, a cylindrical brass nut and a free-wheel clutch controlled safety nut (Fig. 7.24). The latter can turn against a light frictional resistance of the coil-clutch on the upward movement of the cage, but is prevented from doing so on the downward movement of the nut assembly.

The cage hangs on the nut-assembly by hooks welded to a plate bolted to the back of the cage (Fig. 7.24). There was thus a positive upward drive but the cage followed the nut-assembly down by its own gravity.

The cage was stopped at the top and bottom positions by means of pairs of limit switches (the second one for safety in case of overrun). The switches were mounted on the stationary lift casing and operated by preset linear cams bolted to the cage (Fig. 7.23).

This original design and construction had a number of unreliable and unsafe features:
(1) If for any reason the cage was obstructed in its descent, the nut-assembly would continue its downward journey leaving the cage hanging on the obstruction. Removal of the obstruction, without calling the nut-assembly back, would allow the cage a free fall, with serious consequences (such an accident had actually been experienced).
(2) The trolley running in rolled channel sections naturally needed a fair amount of clearance for the four wheels. This caused the cage to oscillate approximately about the centre of the lead-screw as it was raised or lowered. It can be seen from Fig. 7.23 that this swinging of the cage caused the linear cams to move in and out relative to the rollers of the limit switches, thus making it extremely difficult to get reasonably accurate stopping at both top and bottom positions.
(3) In an attempt to overcome the first noted deficiency, the suppliers screwed two bolts to the back of the nut-assembly through clearance holes (and end-clearance) in the cage plate. But this only created a new problem almost worse than the one it tried to overcome. If an obstruction prevented the cage descending, the spindle would continue to drive the nut-assembly down until the two bolts were hard against the bottom of their clearance holes; now either these bolts would shear, or if the cage was far enough down, the spindle would buckle due to the high compressive load induced by the full lifting torque capacity of the motor. The possible damage and highly unsafe situation this would engender is fairly obvious.

To improve the safety and reliability of this equipment a nunber of modifications were incorporated by the user:
(1) Referring to Fig. 7.25 the cams and limit switches were mounted at 90° to the back-plate so that the effect of the swinging of the cage was minimized. The cam shapes were made steeper so as to be more quick acting on the switches. The action distance between the first, and the second safety switch was reduced to $\frac{3}{4}$ in. Stopping accuracy within $\frac{1}{16}$ in. was easily achieved.
(2) The clearance holes in the back-plate were elongated, and the two bolts replaced by two studs (Fig. 7.24).

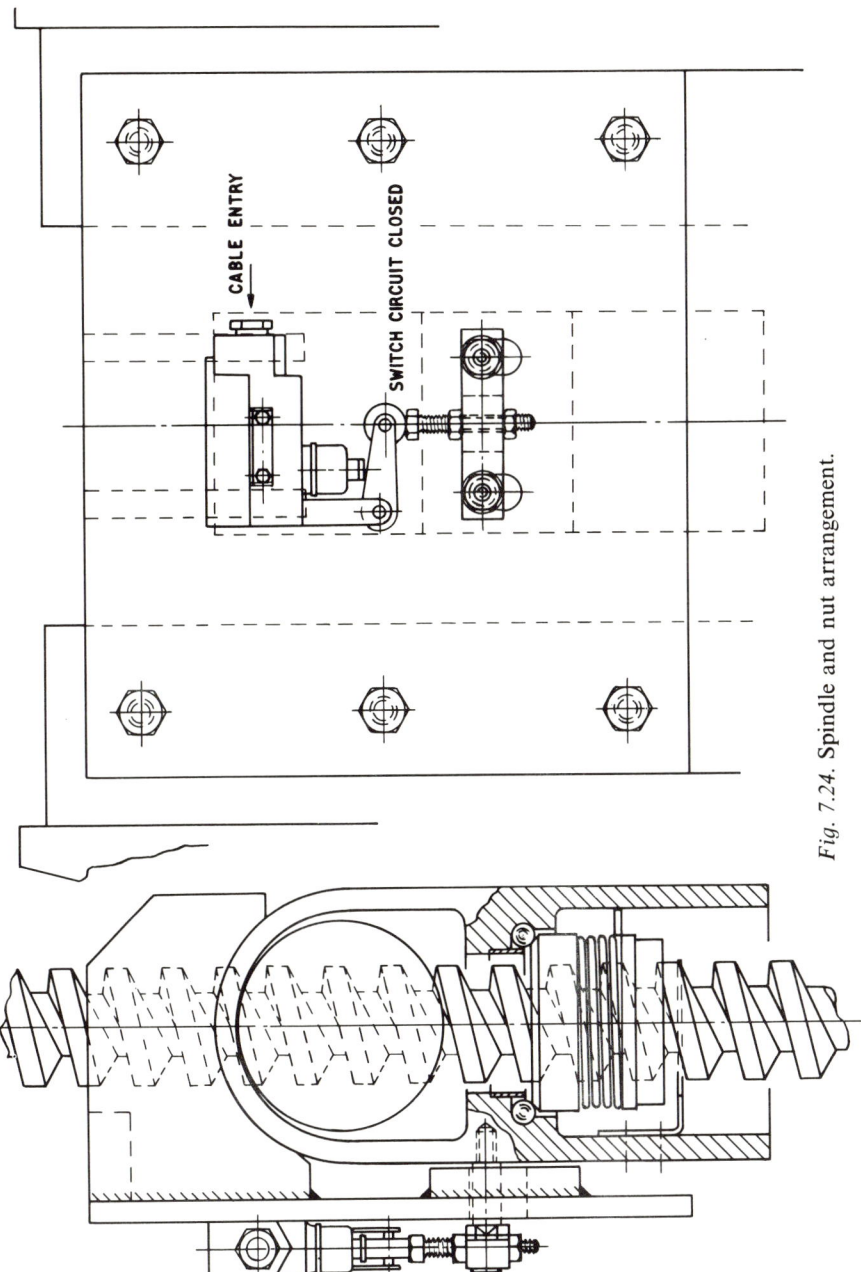

Fig. 7.24. Spindle and nut arrangement.

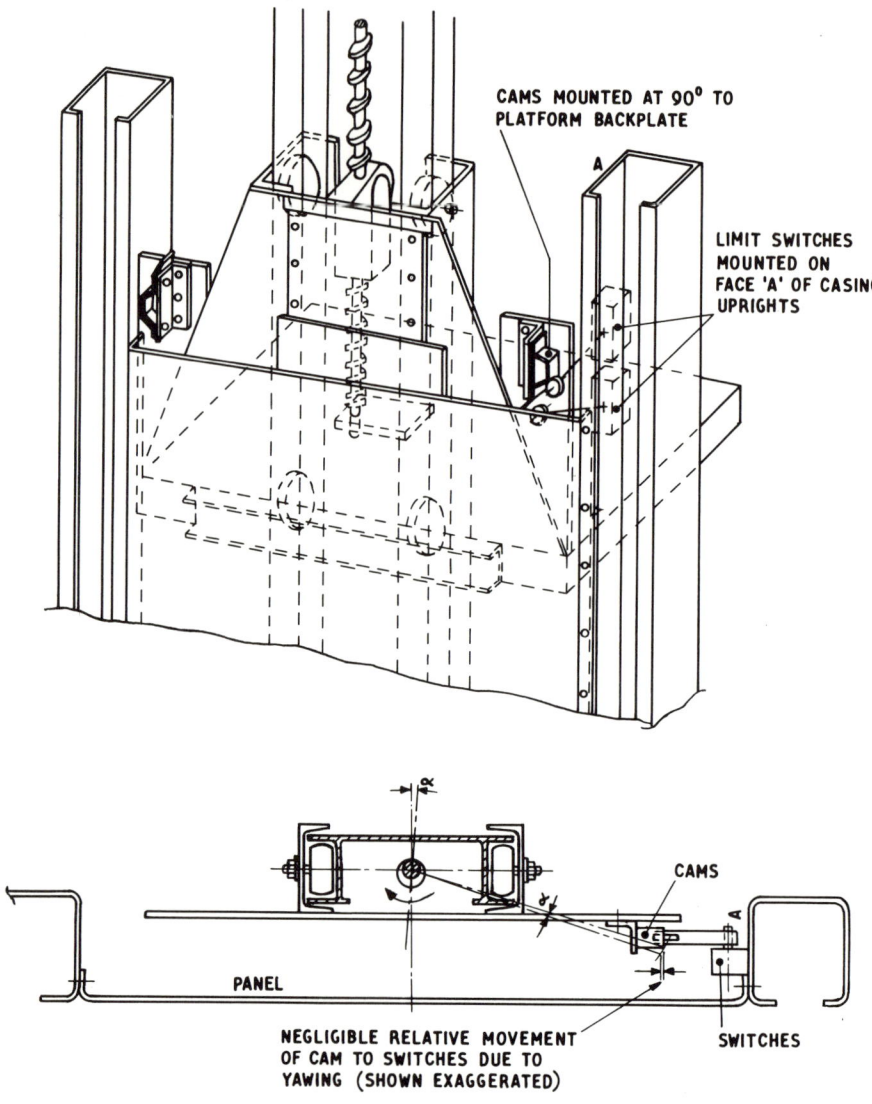

Fig. 7.25.

A bridge piece was secured to the studs outside and clear of the back-plate. The bridge piece carried an adjustable bolt which was arranged to actuate a heavy type micro-switch, bolted to the cage back-plate.

When the cage was hanging firmly on the nut-assembly the micro-switch was set to "closed". In this position the "down" circuit was completed and responded to the call switches inside or outside the cage (Fig. 7.26).

If any obstruction prevented the cage descending, the bridge-piece with the adjustable bolt moved down in the slots and caused the microswitch to "open". The "down" relay circuit would thus be broken and

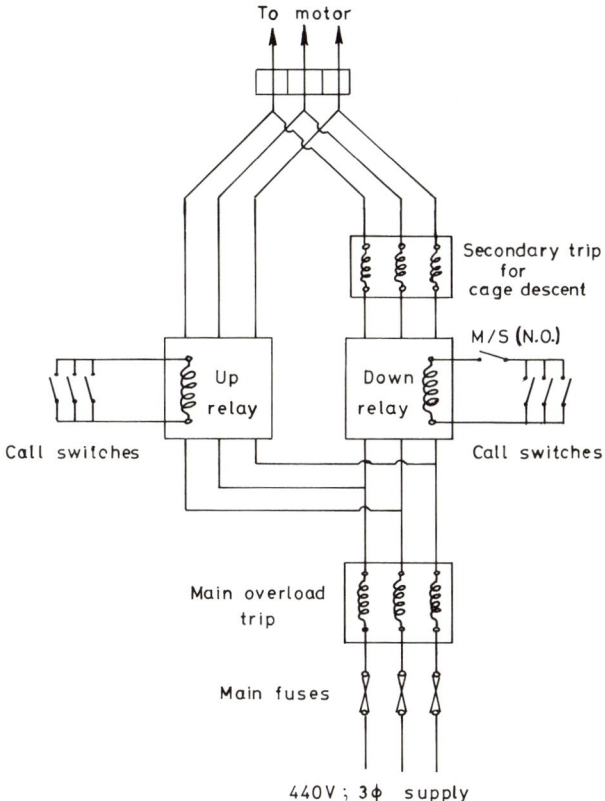

Fig. 7.26. Modified control circuit for cage lift.

the motor stopped. The amount of separation between the cage and nut-assembly was set to about $\frac{1}{16}$ in. Pressing the "up" buttons would relieve the cage off the obstruction and re-close the micro-switch.

(3) As an extra precaution against the possibility of the non-operation of the micro-switch and the two studs being forced down to the bottom of the elongated holes, as assessment was made of
   (a) the maximum downward driving torque required to overcome friction in the system;

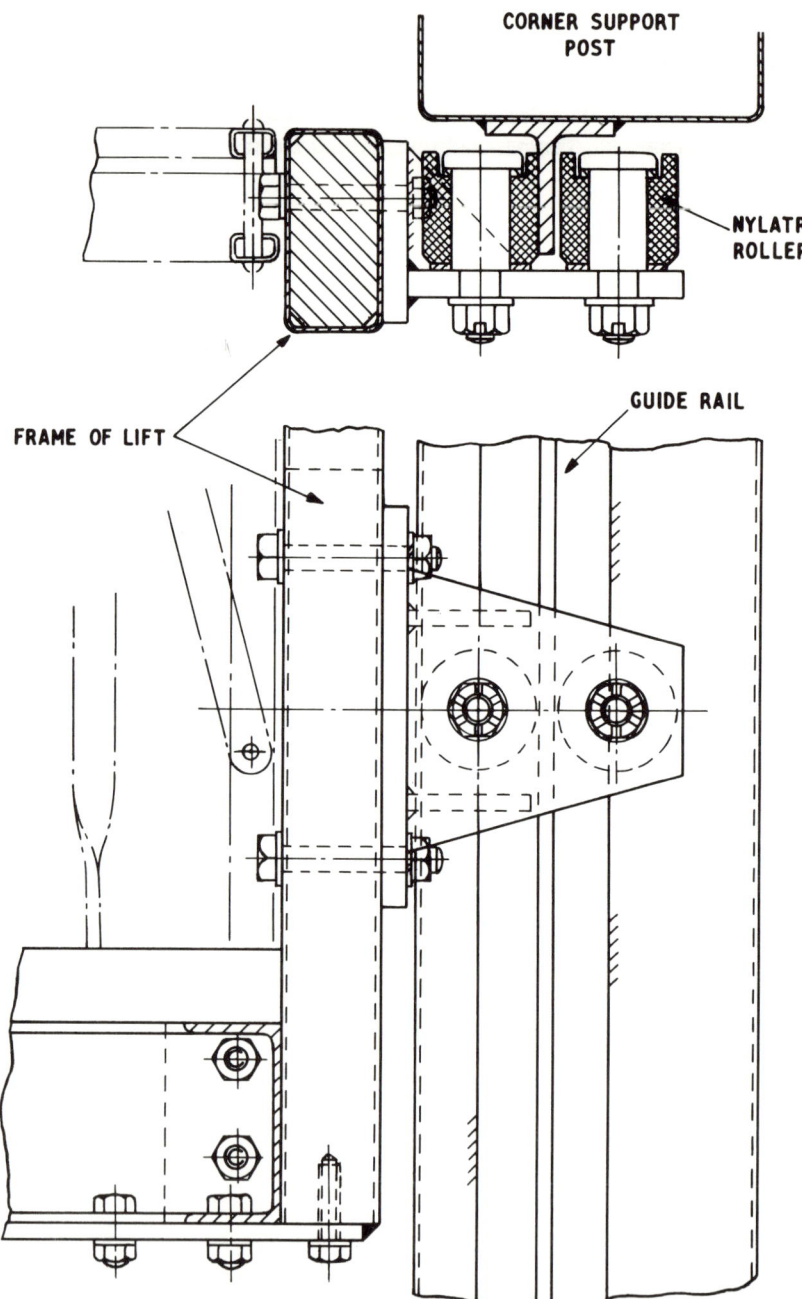

Fig. 7.27. Cage steady.

(b) the maximum torque (compressive load) the lead screw could tolerate with the cage in its lowest position (i.e. longest free length of lead screw) without buckling.

It was found that there was a good margin between (a) and (b), allowing reduced capacity overload cut-outs to be inserted in the "down" circuit (Fig. 7.26).

Thus, should the limit switches fail or should the micro-switch and relays fail to operate on the down-circuit, the secondary overload trips would fall out before the lead-screw would buckle or cause any other serious mechanical damage.

Both the main and secondary overload trips were fitted with dashpots to permit temporary current surges required at motor start-up.

The insurance company's inspector was not happy with the oscillating movement of the cage as it lifted or descended, and insisted this be constrained by additional guiding. Figure 7.27 shows a modification introduced. One corner at the floor of the cage was strengthened by inserting a plug into the hollow vertical tubing and secured by bolting on the underside of the floor. To this stiffened part of the corner a bracket was bolted carrying two Nylatron rollers. These run with a small clearance on both sides of a rolled T section welded to the corner member of the lift enclosure structure.

Case Study 7.5.4

Designing heavy-duty engines for high reliability (by courtesy of Mirrlees Blackstone Ltd)

The development of the K-Major heavy duty diesel engine (Fig. 7.28) by Mirrlees Blackstone Ltd (Henshall and Fletcher, 1976) is an interesting example of the problems posed by the demands for high-level performance, reliability and durability. These engines are used in a variety of installations: in distant water trawlers, in generating stations operating in arctic and tropical environments, on large earth-moving equipment, etc.

A typical power plant installation was at Hamersley in N.W. Australia, some 800 miles from the nearest large city. The major part of the base load was the supply of power to the iron ore pelletizing plant which operates continuously day and night. In addition there was the electricity supply to the township for all its essential services. Heat recovered from the engine's exhausts was to be used to operate the desalination plant which would meet the entire water requirements of the community. Here indeed was a rigorous demand for high reliability of performance. The major specifications were:

(1) minimum maintenance, i.e. low frequency and shortest possible service time;

(2) simple operating procedures (initially only inexperienced staff available);
(3) predictable wear rates; planned maintenance and easy access to worn parts;
(4) fail-safe features; minimum damage from mishaps;
(5) availability of spares and replacements.

Most engineers will agree that one of the surest ways to produce intrinsic reliability in complex machinery is to draw on past successes as well as past failures; and to proceed in a logical evolution from one successful design to another. A great fund of knowledge had been built up from the experience with over 1000 series 1K engines. The company, realizing the importance of in-service experience, had set up a special Field Information Group organized as shown in Fig. 7.29. It was estimated that this essential service added only 0·1% to the company's overhead costs.

For continuous operation of these engines on heavy fuels the following requirements were established as essential for satisfactory service life:
(1) exhaust valve seat temperature should be ⩽ 550 °C to avoid deposits of sodium/vanadium compounds;

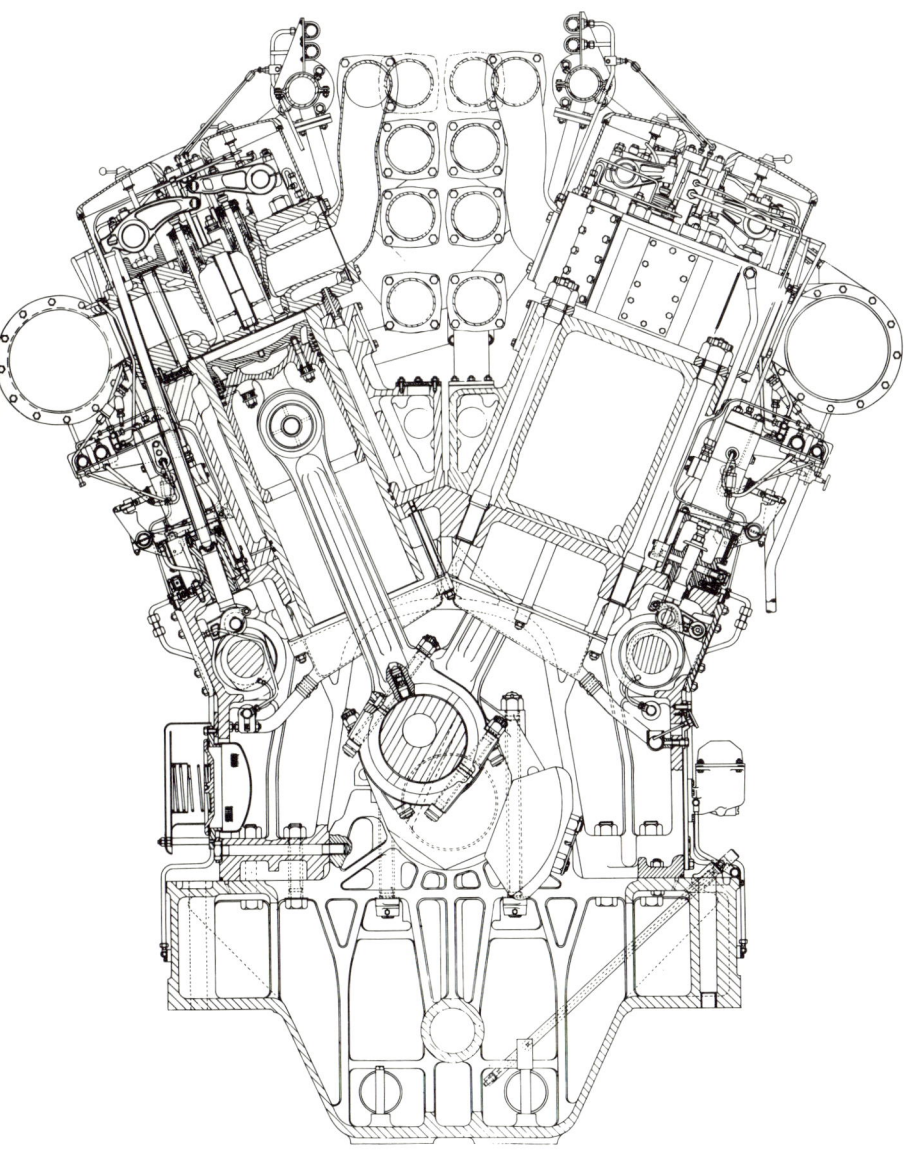

Fig. 7.28. (*opposite*) KV major 18-cylinder generating set installed at the Huntingdon Water Treatment Works of the Western Division, N.W. Water Authority. (*above*) Cross-section.

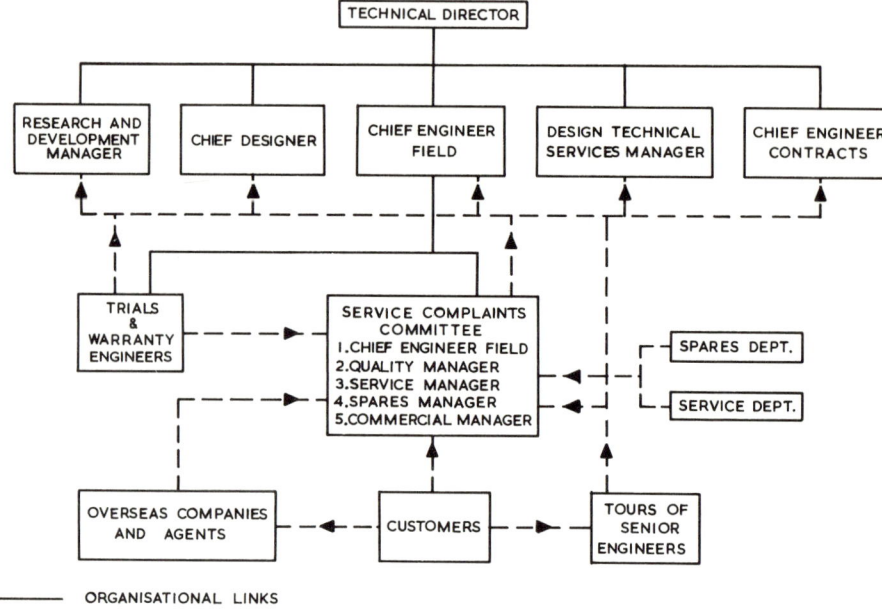

Fig. 7.29. Organization of field information grouping (by courtesy of Mirlees Blackstone Ltd.)

(2) top piston ring temperatures should be $\leqslant 230\,°C$;
(3) injection nozzle tip temperatures to be $\leqslant 180\,°C$;
(4) bearing loads to be well within the capacity of the lining material;
(5) adequate factors of safety on the stressing of all components both thermally and mechanically must be provided.

To achieve high reliability, high output and low cost—generally conflicting demands—sophisticated manufacturing techniques were needed to achieve simple designs. A few of the important design improvements are described below.

Bearings

The thin shell main and large end bearings were originally both tin–aluminium on steel backing. Soft overlays could not be added on these bearings. They were sensitive to dirt ingress and had to run with a minimum film thickness of 2·92 µm. Occasional seizures were reported. Bearings with copper–lead linings could take a soft lead–tin overlay and were, therefore, more tolerant to particle ingress. However, some of these bearings suffered from corrosion damage of the lead phase.

7. RELIABILITY

Development testing in co-operation with bearing manufacturers resulted in the adoption of a 6% tin–aluminium lining able to carry a lead–tin softer overlay plating.

Pistons

The original designs and improved designs of piston are shown in Fig. 7.30. The basic construction of an alloy steel crown with a cantilevered ring belt and intensive oil cooling behind the rings remains unchanged. Top ring groove temperatures were maintained at about 180 °C.

At higher cylinder pressures the stress-concentration at the intersection between the vertical oil drilling and the piston pin bore (original design) became excessive. This problem was overcome by angling the hole so that it intersected the pin bore at about 30° to one side (current design).

Further, steel bushes lined with bearing material were inserted in the piston boss-bore so as to reduce the stress in the cast iron body and avoid high pressure oil on the porous iron surface. Bushes lined with phosphor-bronze gave the highest fatigue strength.

Exhaust valves

Exhaust valve maintenance is probably the most important single item in the reliability of medium-speed engines running on residual fuels.

The original valve cage (a Cr/Mo casting, Fig. 7.31) had a cooling groove machined in its bottom face and a matching groove in a solid stellite seating. The two were joined by electron-beam welding to provide an annular cooling water passage. Development tests proved satisfactory, but subsequent reports from customer service noted cracks appearing in the solid stellite seats. Metallurgical examination diagnosed the cause of these cracks as thermal shock initiated in the water space. This was confirmed in simulated laboratory tests when the cooling water flow was irregular, or erratic.

A modified design was developed which was less sensitive to faulty coolant flow. It was found that a valve seat ring made of EN40B steel (of similar composition to the cast steel cage) faced with stellite on the valve seating tolerated more than 300 cycles of exaggerated heat fluctuations.

The effects of cage cooling and lubrication of the valve stem produced lower temperature levels and a symmetrical distribution in the whole valve.

Expansion bellows (Fig. 7.32)

The exhaust bellows on the earlier engines had experienced cracking through the convolutions.

A series of rig tests was carried out on strain-gauged bellows to assess the effect on stress from internal pressure, vibration, misalignment, expansion and contraction. The tests showed that to resist the normal expansion and

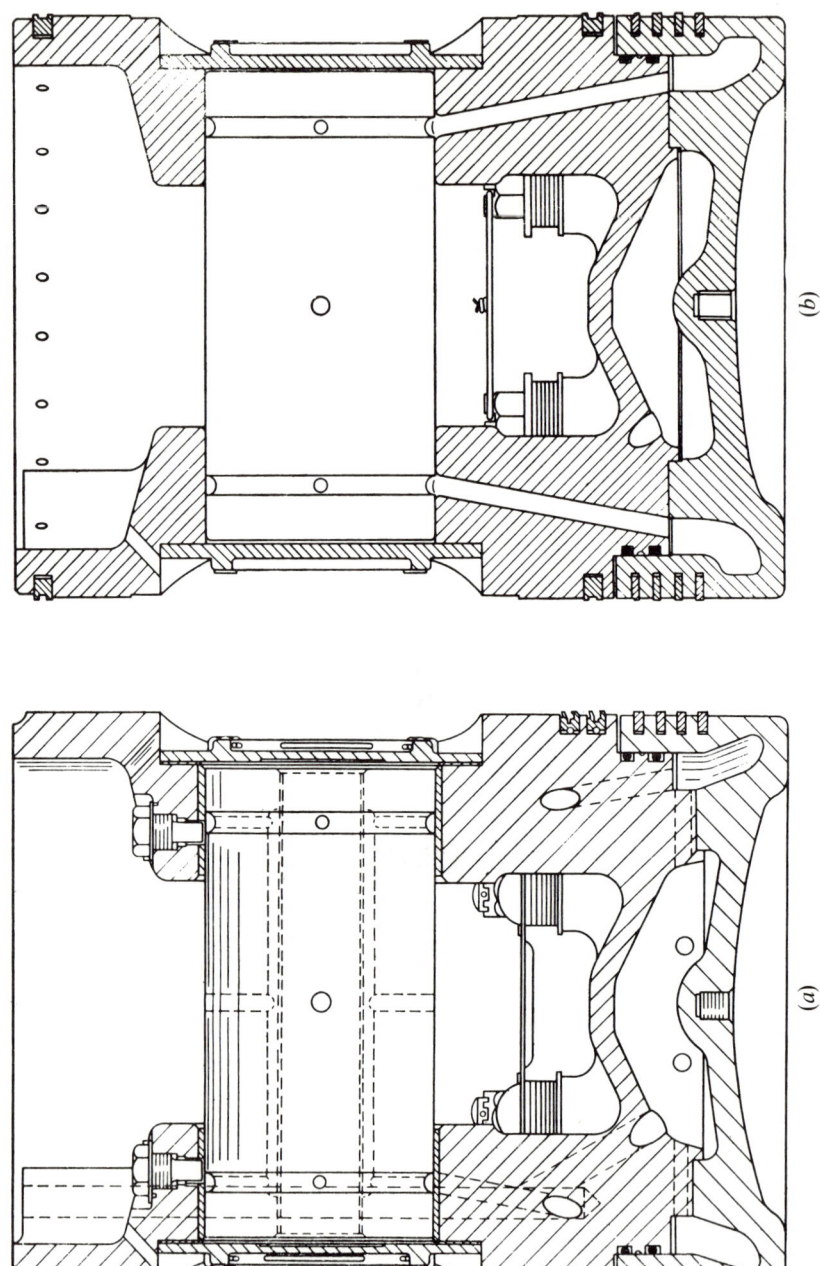

Fig. 7.30. Assembly of two-piece piston: (*a*) original design; (*b*) current design.

7. RELIABILITY

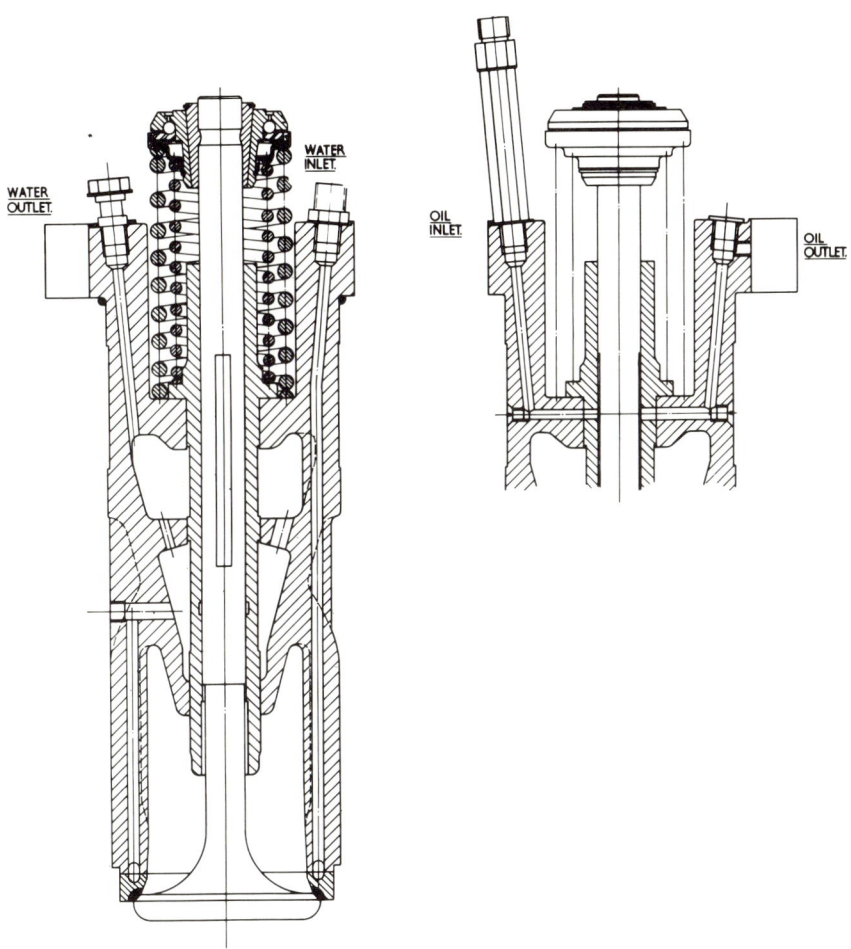

Fig. 7.31. Exhaust valve cage.

contraction the bellows should be thin, but to resist the pulsating internal pressure the material should be as thick as possible. The solution to this problem lay in making the convolutions of two layers of thin stainless steel, and to fit an internal sleeve to attenuate the pulsating internal pressure. This was confirmed by engine tests.

The new bellows is, of course, rather more expensive than the original single-skin type, but this is almost totally compensated for by the elimination of the need for exhaust pipe brackets and rollers due to the self-supporting feature of the close-fitting internal sleeving.

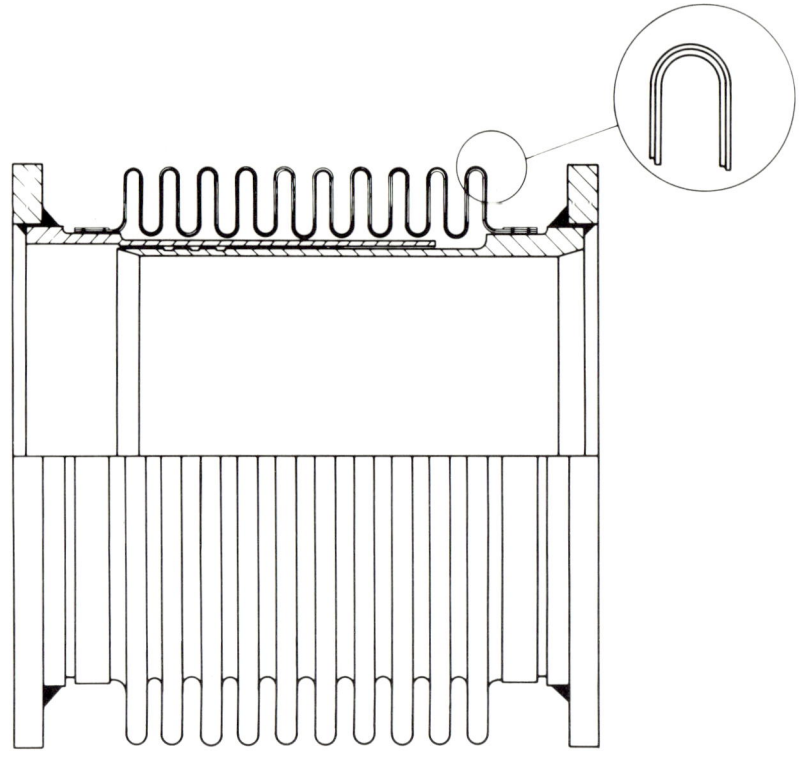

Fig. 7.32.

Case Study 7.5.5

Design and reliability of high-performance gears (by courtesy of GEC, Lynn, U.S.A.)

It is appropriate to start by listing some of the important requirements of gear drives for high-speed applications (Heter and Moore, 1971):
  (1) Highest possible reliability for not less than three years' continuous operation without forced down-time
  (2) High power capacity: 10 000 h.p. and above.
  (3) High speeds: 12 000 r.p.m. and above.
  (4) High pitch line velocities: 30 000 f.p.m. and above.
  (5) Mostly speed increasing.
  (6) Relatively low speed ratios.
  (7) Efficiency.

A number of typical gear type failures, such as pitting, scoring, wear, tooth breakage, etc., are described by Heter and Moore (1971). The authors discuss

7. RELIABILITY

considerations to be given by the designer and the user to minimize the possibility of such failures occurring and thus obtain the operational reliability levels required of high performance gear drives.

Design considerations

Centre distance and face width Selecting the best relationship between centre distance and face width is the very essence of the gear design process:
  (a) Too small a centre distance will make the pinion long and slim, sensitive to twisting and bending and prone to misalignment.
  (b) Too great a centre distance, although reducing the face width, causes high pitch line velocities, increased windage losses, and greater sensitivity to tooth pitch and profile inaccuracies.

Direction of rotation For hydrodynamic reasons, i.e. to avoid choking (foaming and inability to drain), it is preferable for the direction of meshing to be upwards (Fig. 7.33).

Service factor (SF) should be considered as a margin on gear capability to allow for dynamic or cyclical loads which are known, or suspected, to be present in the connected equipment, but not shown in their name plate ratings. The SF is not the extent to which a gear is overdesigned. The American AGMA standards 421.06, entitled "Practice for High Speed Helical and Herringbone Gear Units", recommends: $SF = 1 \cdot 1$ for smooth running machines and $SF = 2 \cdot 5 - 3 \cdot 0$ for reciprocating equipment.

Single v. double helix Single-helical gears are limited to lower power capacities by the axial thrust bearings required. Helix angles between $8°$ and $23°$ are used: below $8°$ the axial cross-overs are too few, and above $23°$ the axial thrust loads become excessive.

Helix angles for double-helical gears usually range between $20°$ and $45°$, with $30-45°$ being most common. Double-helical gears are generally more expensive to produce than single ones but can transmit much higher powers.

Tooth size Large teeth have greater strength than small ones, but certain limiting factors have to be considered.
  (a) For a given pitch line speed, surface sliding velocity and aero-dynamic heating increase with tooth size. This increases the susceptibility to scoring.
  (b) At pitch line velocities above 30 000 f.p.m. the windage power loss increases in proportion to the square of tooth size. These losses can become greater than the mesh and bearing losses put together.

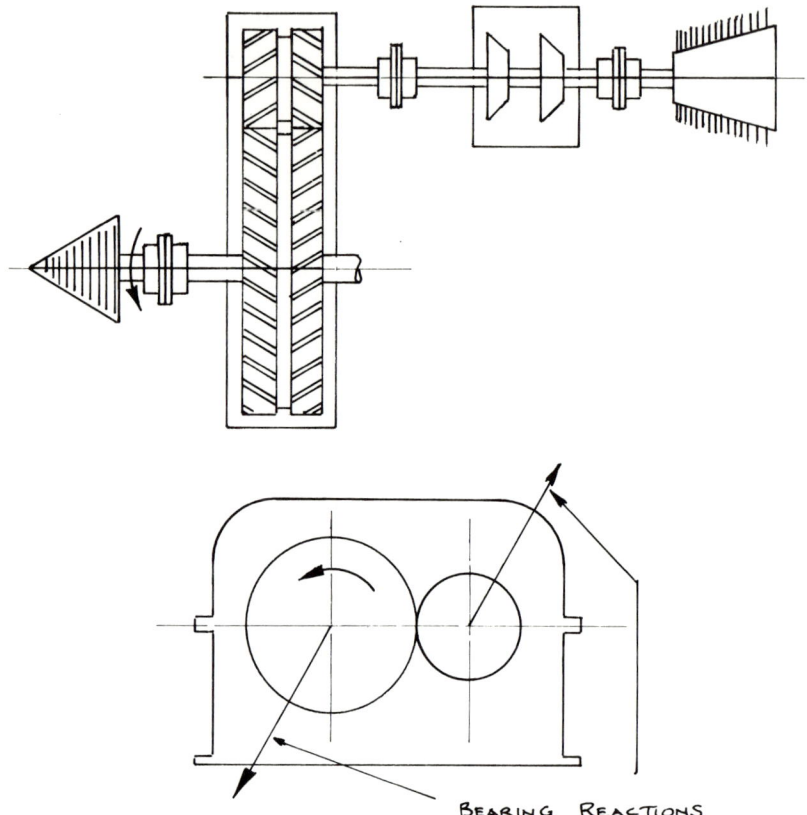

Fig. 7.33. Note. For the rotation shown the pinion must be offset to right to obtain upward mesh.

Bearing size and geometry In general, plain hydrodynamic journal bearings are used. The main load is due to tooth reaction and therefore this should come on only when the hydrodynamic film has been fully developed. Load densities up to 500 p.s.i. are commonly used on high performance gears. For weight-dominated gears, only about half this load density is considered safe, because of starting conditions.

Shaft centralization and stabilization are important to maintain proper meshing. Tilting-pad bearings are considered by many as the ultimate in stable journal bearing design. They have, however, a hidden long-term unreliability in their tendency to wear slightly due to fretting at the trailing contact edge of pads. The wear rates on each pad are unlikely to be exactly equal; parallelism between the gear axes cannot, therefore be maintained.

7. RELIABILITY

Heter and Moore (1971) refer to a "tri-taper" bearing in which "three powerful hydrodynamic wedges" are formed to create stabilized conditions independent of external loading (Fig. 7.34). This bearing design, they claim, has proved superior to the tilting-pad bearing for critical gear applications, where high vibratory loads are present.

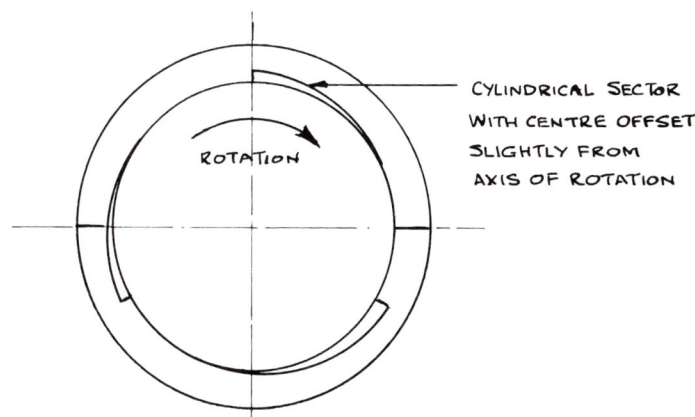

Fig. 7.34. Three-lobe bearing.

Thrust bearings are required to locate the double-helical gears in the casing and to take any axial forces transmitted from external couplings. Only one of the gears, usually the low-speed wheel, is constrained axially (Fig. 7.35). Thrust forces acting on the unconstrained gear (pinion) are taken across the meshing teeth. Such axial loads must be limited to avoid damage to the gear teeth. Using flexible couplings with limited end-floats may completely eliminate the necessity for internal thrust bearings in double-helical gears.

Lubrication of gear teeth In low-speed gears it is normal to inject the oil into the in-going side of the meshing gears, but in high velocity gears the outflow of displaced air from the rapid filling of a space by a tooth prevents the oil reaching the teeth surfaces if injected in this area. In this case a slight negative pressure is actually developed on the disengaging side of the teeth. Thus, for high-speed gears injection of oil is more effective on the out-going mesh. More efficient cooling is also achieved, since the tooth temperatures are highest immediately after meshing.

Efficiency Most measures to improve the efficiency of a high-speed gear drive involve compromises with reliability. For example, reducing the oil supply to

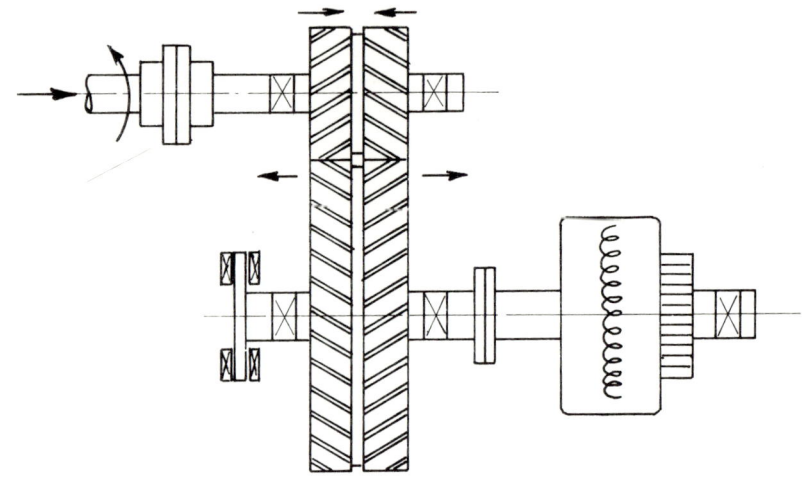

Fig. 7.35. Gear thrust bearing and its effect on tooth load.

gears and bearings can effect a modest saving in power loss, but may result in the following problems:
(1) increasing the risk of scoring;
(2) high bearing temperatures;
(3) misalignment due to temperature effects;
(4) insufficient backlash due to greater thermal expansion of rotating parts relative to casings.

Windage losses can be reduced by using smaller teeth (finer pitch), but at this risk of tooth breakage due to higher root stresses.

An example of a "Design Defect" which caused a power loss of some 100 kW more than had been calculated was the geometry of a cover plate for a vertical offset gear as shown in Fig. 7.36(a). A very high temperature was recorded at X. The rotating pinion and the convergent space between the teeth and the cover-plate acted as a compressor causing the high local temperature and the excessive power loss. The problem was overcome by the installation of a cylindrical baffle plate, as shown in Fig. 7.36(b).

Restrictions in space for oil flow inside the gear casing can often result in power losses due to turbulence. Honeycomb baffles have been found to impede such turbulence and reduce the associated power losses.

User considerations

Information supplied to the Gear Manufacturer on usage, environment, installation, maintenance conditions, etc. should be as comprehensive as possible.

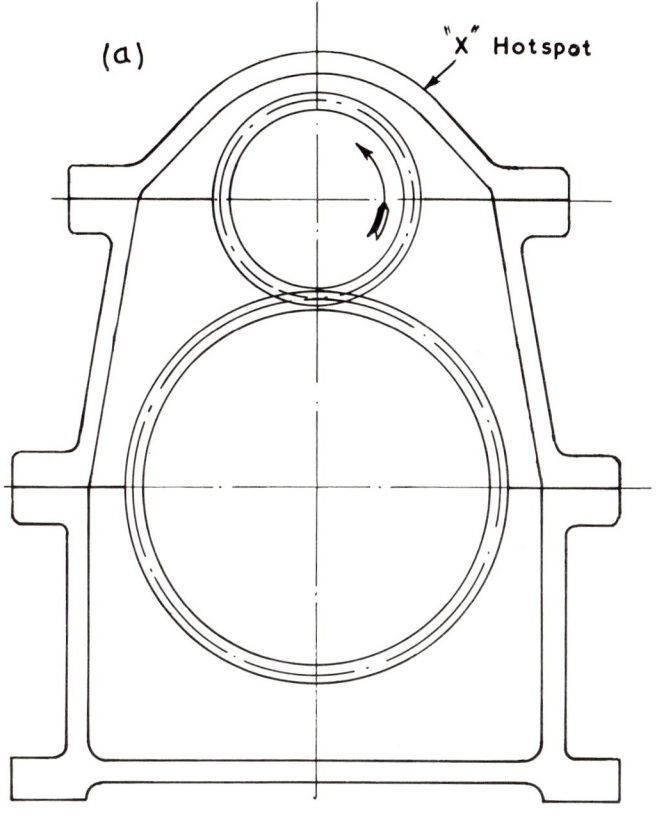
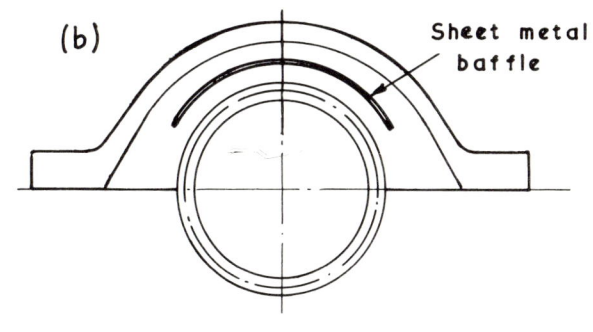

Fig. 7.36.

190   APPROXIMATE METHODS IN ENGINEERING DESIGN

Supporting foundations should be rigid and as symmetrical as possible. Non-resonance and good damping characteristics are important. Reinforced concrete is a good foundation material.

Good running alignment to both input and output shafts are essential.

Torsional and lateral vibration analysis of the whole installed system should be carried out to avoid coincidence with critical frequencies.

Case Study 7.5.6

Improvements in reliability of thermal station pumps (Anderson, 1973; by courtesy of Weir Pumps Ltd and reprinted by permission the Council of the Institution of Mechanical Engineers)

Improvements in multi-stage centrifugal pumps were necessitated by the upgrading of turbo-generators from 30 MW to 600 MW over 30 years. Associated with these power changes were corresponding increases in pressures and temperatures. The re-designs also aimed at improvements from 100% to only 20% stand-by capacities.

Preliminary studies indicated that:
(a) Higher operating speeds would have to be used: this demanded improved mechanical rigidity.
(b) Higher water velocities would be involved: this implied more precise geometries to reduce pressure and power losses.
(c) Higher water velocities would increase corrosive and erosive actions: superior materials would have to be used and keen attention given to running clearances.
(d) recent Power Station requirement specified:
   (i) pumps to run below critical speed with water,
   (ii) pumps to run above critical speed in air or vapour.
(e) Improved reliability was specified: this implied a rotating assembly would have to run true under all normal and emergency conditions, including thermal shocks up to 600 °C per minute. Attention would also have to be given to reducing maintenance down-time.

Major stages of development
(1) Up to 150 °C conventional ring-section pumps were used. The stages were clamped together with long bolts between heavy end covers (Fig. 7.37). These long bolts could become a weakness with thermal shocks causing leakage at joints, or over-stretching of the bolts.

The pumps were driven by electric motors at 2970 r.p.m. For higher outputs and pressures several stages were needed (up to 10 stages in some installations). This resulted in long rotating assemblies with static deflections up to 0·75 mm. Not only did such long shafts involve more frequent maintenance with low critical speeds, but the large deflections

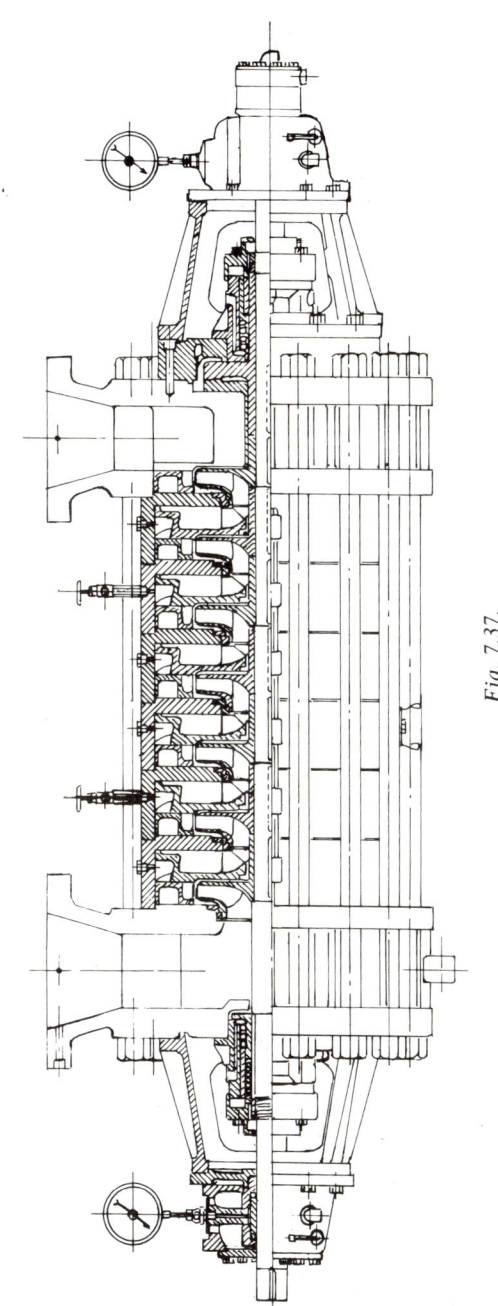

Fig. 7.37.

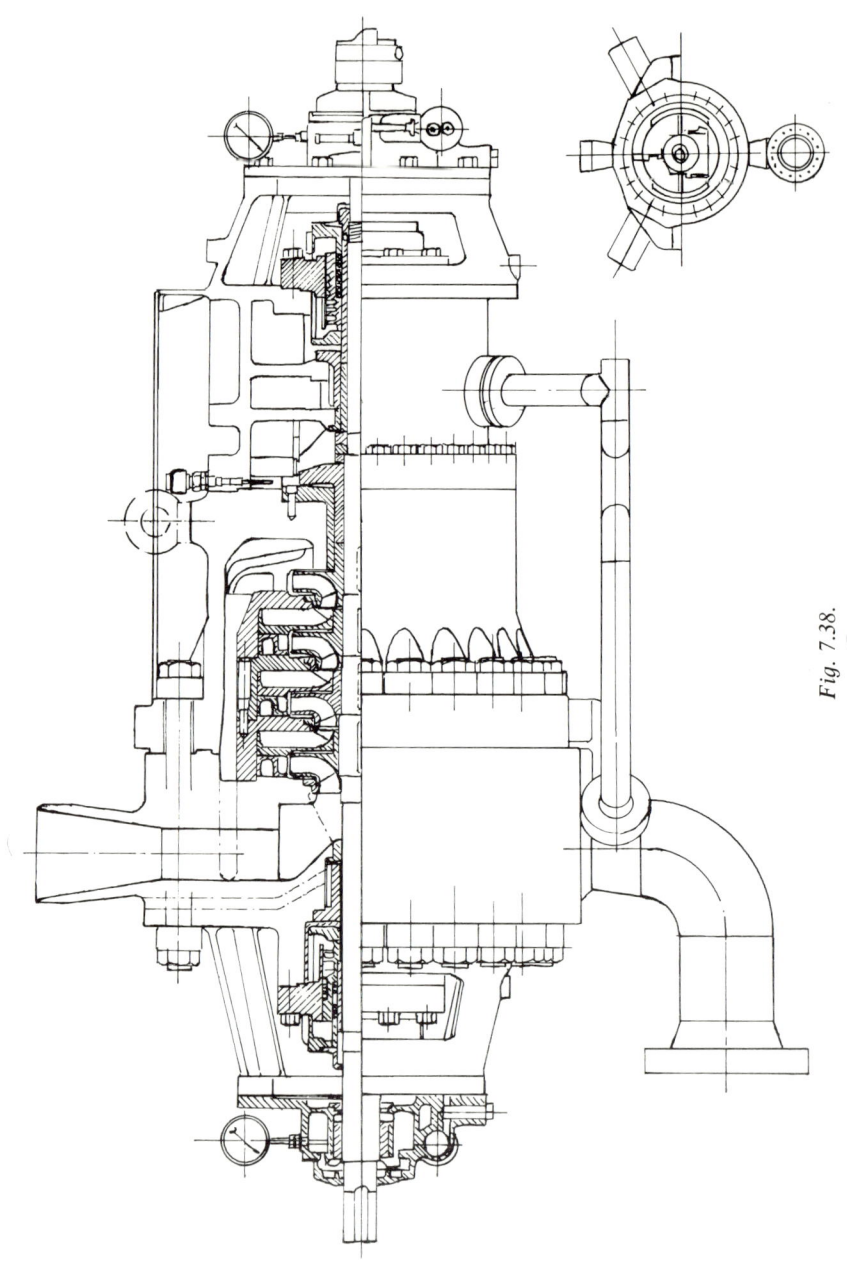

Fig. 7.38.

worked against recommended maximum clearances between the impellers and the radial diffusers.
(2) Thermal problems, long rotating assemblies, limitations in speed and pressure led to the development of "barrel-type" pumps (Fig. 7.38). These machines were designed for outputs of 350–600 MW and ran (turbine driven) at 5000 r.p.m. The features to note were:
   (i) Four stages only were needed, making the rotating assembly shorter and stiffer.
   (ii) Wholly axisymmetrical barrel gave totally symmetrical water flow, heat flow and stress distributions. Thermal shocks could be accepted.
   (iii) Barrel was made from a spun stainless-steel forging, and could therefore be reduced to 50 mm thickness, compared with the 200–250 mm walls of the mild-steel pump casings.
   (iv) Stress variations in the (shorter) bolts were minimized.
(3) The third phase of development was a feed pump for a 660 MW power plant. It was specified that such a pump should:
   (a) not have more than three stages;
   (b) be capable of running dry;
   (c) be so constructed as to enable a major overhaul to take no longer than 8 hours;
   (d) have generally higher reliability under all operating conditions.
   To achieve these objectives the design aimed at:
   (i) extreme rigidity;
   (ii) complete symmetry;
   (iii) wide separation of impeller and diffuser vanes;
   (iv) largest possible clearances between running and stationary surfaces.

The two-stage high output pump developed is shown in Fig. 7.39. Some of the important features are discussed below:

1. **Casing.** This is a single piece wholly symmetrical stainless steel forged unit.

2. **Stage clamping.** Internal bolts for holding the stages together have been eliminated. In the new design shrink-rings are fitted red-hot into trepanned grooves straddling stage one and two. On dismantling they are drilled through and discarded.

3. **Stage alignment.** Conventional spigot and recess for concentric alignment proved inadequate for thermal transient conditions. Instead three radial dowel pins are positioned centrally on the contact faces between the stages, half a dowel being in each stage. Changes in diameters due to temperature differences cannot affect central alignment.

4. **Internal cartridge assembly.** The shaft, impellers, diffusers and bearings

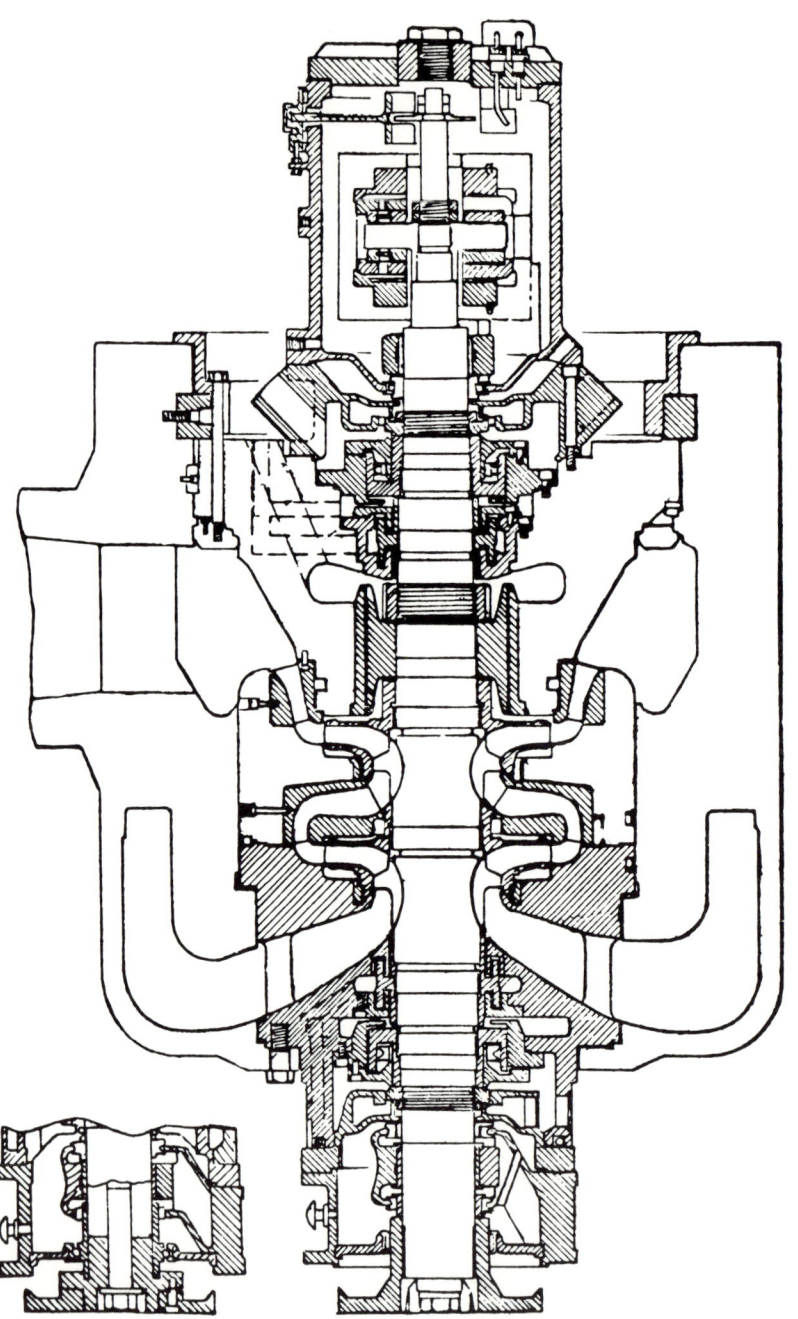

Fig. 7.39. Two-stage high output pump.

are built up as a complete cartridge unit outside the pump casing. One can be extracted from, and another complete one inserted into, the casing within an eight-hour shift. The inlet guide flange is shown bolted to the casing, on the left in Fig. 7.39, whilst the discharge cover is clamped via a specially developed self-seal joint to the segmental shear rings, shown on the right. The forward end of this discharge cover sits over the "balance drum" and has an O-ring seal with the internal diameter of the second stage diffuser.

5. **Impeller geometry and materials.** The through-movement of the water and the rotation of an impeller may be considered analogous to a worm-and-wheel drive, the worm threads being represented by the impeller vanes and the wheel-teeth by the flow of water through the impeller (Fig. 7.40). Thus the

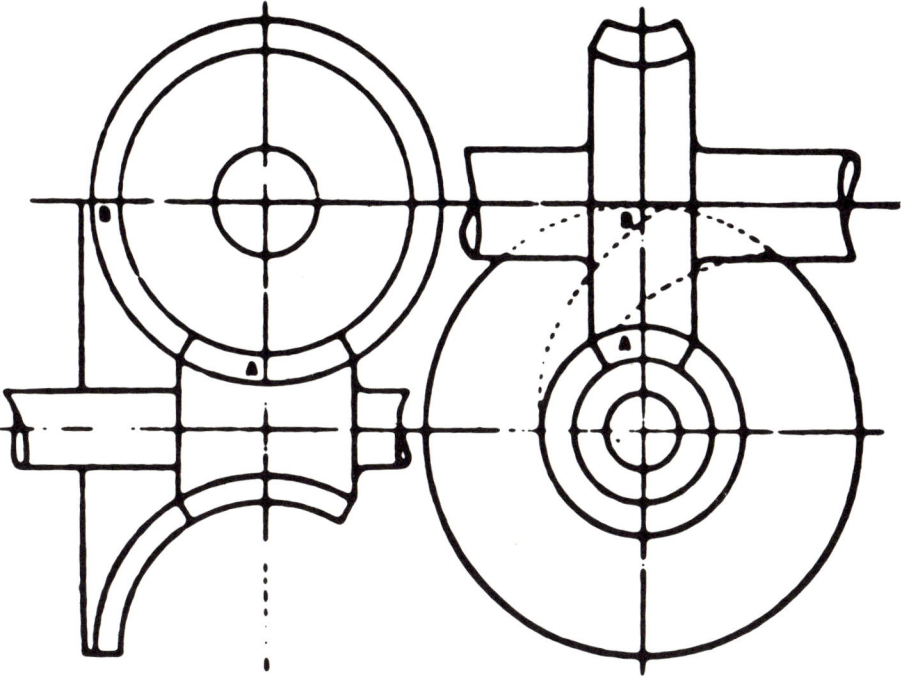

Fig. 7.40. Generating impeller geometry by worm gear analogy.

impeller geometry can be accurately generated by conventional gear-cutting methods. This results in considerable improvement in efficiency and much-reduced cavitation.

Impeller and diffusers are precision cast in 17% Cr + 4% Ni precipitation-hardened stainless steel with excellent corrosion and erosion resistance. Surface finishes as fine as two to three microns are obtained. Each vane surface is carefully inspected to ensure adherence to the tight tolerances specified.

6. **Diffusers.** The diffuser geometries have been radically changed. Whereas the earlier pumps had radial diffusers with rather fine clearances between the impeller tips and the stationary inlet edges of the diffuser blades, the new design uses predominantly axial blades with a 90% toroidal clearance between the rotating and static vanes. This latter feature has been particularly beneficial in obviating pulsing and flow disturbance at the diffuser entry, so that erosion is virtually absent and noise levels considerably attenuated. Cavitation also has been much reduced. The overall effect is a major improvement in the reliability of high-head centrifugal pumps.

7. **Quality assurance**
   (i) *Low cycling fatigue test.* It is estimated that boiler feed pumps would normally experience about 15 000 pressure cycles during their operating life.
      (a) A quarter-scale model of a two-stage pump casing was cycled from zero to hydrostatic test pressure (pressure vessel code) 25 000 times. It was then sectioned and examined. No cracks or deformations were found.
      (b) A 250-mm model of the "self-seal" joint was cycled from zero to 50 MPa 20 times, to 43 MPa 100 times, to 32 MPa 20 000 times, and between ± 1 MPa 2×10^6 times. Thorough examination revealed no defects.
   (ii) *Bursting tests.* Destruction tests on both (a) and (b) above gave failure loads several times higher than working pressures.
   (iii) *Endurance tests.* These were carried out on a 150 mm diameter impeller with associated diffuser in order to determine best geometry and establish the parameters for inlet conditions, etc. The stage head was 1860 m and the input power was 2·3 MW.
   (iv) *Photo-elastic tests.* Photo-elastic models were made in order to minimize any stress-concentrations.

REFERENCES

Anderson, H. H. (1973). "Improvement of Reliability in Thermal Station Pumps." I.Mech.E. Conference Publ. No. 8.

Carter, A. D. S. (1972). "Mechanical Reliability." Macmillan and Co. Ltd, London.

Carter, A. D. S. (1973). "The Bath-tub Curve for Mechanical Components—Fact or Fiction?" I.Mech.E. Conference Publ. 8.

Haugen, E. B. (1968). "Probablistic Approaches to Design." John Wiley and Sons, Chichester.

Henshall, S. H. and Fletcher, N. (1976). "Designing Heavy-duty Engines for High Reliability." Mirrlees Blackstone Ltd.

Heter, E. D. and Moore, M. G. (1971). "Factors Affecting Design and Reliability of High Performance Gears in Process Compressor Trains." Petroleum Mech. Eng. Technology Conf., ASME.

Jardine, A. K. S. (1973). "Maintenance Replacement Reliability." Pitman, London.
Johnson, R. C. (1978). "Mechanical Design Synthesis with Optimisation Applications." Van Nostrand Reinhold Co.
Pronikov, A. S. (1973). "Dependability and Durability of Engineering Products." (Translation by T. T. Furman.) Butterworths, London.
Shigley, J. E. (1972). "Mechanical Engineering Design." McGraw-Hill Book Co., New York.

Chapter 8

Machine Failure

"Just as human life has an expectancy and mortality which can be statistically evaluated, so does each individual machine. Perfect reliability ... is impossible, and the life of a machine, like that of a human, depends upon conception, birth and life." Collacott (1976). This quotation may be interpreted to refer also to health ($\equiv$ efficient operation) and sickness ($\equiv$ malfunction).

8.1 DEFINITION OF MACHINE FAILURE

The Organisation for Economic Cooperation and Development (Paris, 1969) defined failure as "the point at which a given system no longer performs as designed".

Pronikov (1973) is somewhat more explicit, saying that "failure is an event which defines a breakdown in the efficiency of the machine, or in one of its elements". He says further that "the more advanced the design of a machine the fewer the number of these 'legitimate' failures will be and the longer the machine will remain in continuous reliable operation".

8.2 ORIGINS OF MACHINE FAILURES

Failures can have their origins at any stage throughout conception, manufacture and usage.

8.2.1 Design

Deficiencies emanating from the design stage may be due to a number of reasons:
 (i) The true functional needs were not clearly established between the designer and the customer.

8. MACHINE FAILURE

(ii) The designer did not appreciate fully the implications of the specifications; he did not ascertain the up-to-date state of the art, the design data available, and any research and development work needed.
(iii) Requirements to the detailing office, the production department, inspection, erection and testing were not clearly and unambiguously specified.
(iv) Instructions for operation and service, limitations in usage, interlocks, overload relief, and fail-safe and fool-proof features were inadequately defined.
(v) Where low weight and/or low cost are important, safety margins are inevitably reduced. In spite of more rigorous computations and test procedures some risk still has to be taken—but not risk due to ignorance or carelessness.

8.2.2. Manufacture

No two maufactured parts can be exactly alike. This fact is recognized in the provision of tolerances on material properties, geometrical shapes and dimensions, and surface finishes. To ensure that every part or component lies within specified limits would require 100% inspection and the elimination of human error. This is not feasible economically and socially for large-scale production.

Quality assurance consciously accepts reliability levels which are built into the quality control system operated at all stages of the manufacturing process.

8.2.3 Operator errors

Even when clear and well-defined instructions have been provided by a manufacturer, operator usage is still likely to digress from such directions: starting-up procedures are deviated from; continuous operations are often carried on beyond specified limits (to get a bit extra out of the machine); monitoring of performance is lax; maintenance and service programmes are not properly carried out, and so on.

8.2.4 Technical defects

Technical defects can arise from the following factors:
(i) design error, due to lack of reliable data or erroneous interpretation of available information;
(ii) variations in material properties;
(iii) thermal distortions during the manufacturing stages, or later during operation of the machine;

(iv) the effects of vibrations and residual stress relaxation;
(v) corrosion, erosion, and cavitation effects;
(vi) radiation and ageing;
(vii) general errors of judgement at all levels and all stages, from the design brief to customer liaison and service information feedback.

The points above are well summarized by Collacott (1976), who distinguishes five constituent origins of early mechanical failures (Fig. 8.1) and three causes of terminal failures (Fig. 8.2) and notes that "failures do not just happen; there is always a reason, and until one looks for failure points in the grass roots of design and manufacture one cannot hope to see a reduction in these incidents". Reliability and durability are nowadays important statistical concepts. Our pioneer engineers were perfectly aware of the importance of these criteria although they expressed their ideas in different terms, e.g. "There is no safe way of judging anything except by experience" (Sir Henry Royce), and "Most advances in engineering have in fact been made by turning failures into success." (Sir H. Guy) (both quoted by Whyte, 1975).

8.3 CAUSES OF MACHINE FAILURES

Deviations from nominal performance parameters are caused by internal and external disturbances. Pronikov (1973) points out that "whereas classical science considers operating errors resulting from wear, thermal deformation, material faults, etc., as undesirable and even avoidable deviations, modern science, especially cybernetics, views these deviations as natural features of all real systems. Causes of deviations in performance are studied and are generally found to be random in nature leading to scatter in machine performance."

8.3.1 Classification of processes causing performance deviations
(Pronikov, 1973)

(i) Rapid processes. These act and are repeated during each working cycle of a machine, for example, vibrations, variations in friction, fluctuations in working load. They tend to distort the working cycle.

(ii) Medium rate processes. These are generally caused by intermittent operation of a machine, e.g., variations in temperature, variations in moisture content of the environment or raw material, wear of cutting tool. They cause alterations in initial machine parameters and are random in nature.

(iii) Slow processes. A number of well-known processes fall into this category, e.g. stress relaxation and creep, wear and corrosion, erosion, accretion and

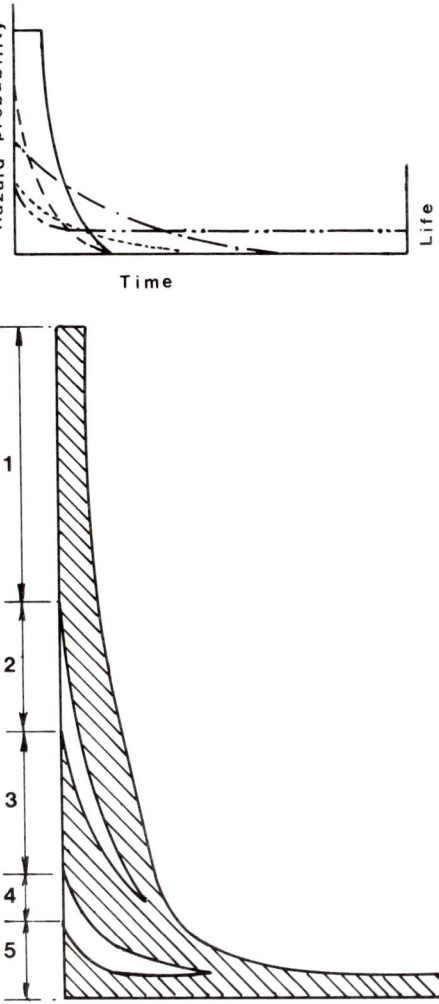

Fig. 8.1. Effect of constituent defects on infant machine mortality failure, and summation of infant machine mortality hazard probabilities.

———— (1) y1 Design defects.
– – – (2) y2 Manufacturing defects.
—·— (3) y3 Installation defects.
······ (4) y4 Commissioning defects.
—··— (5) y5 Operating defects.

(Reproduced by permission of the Council of the Institution of Mechanical Engineers.)

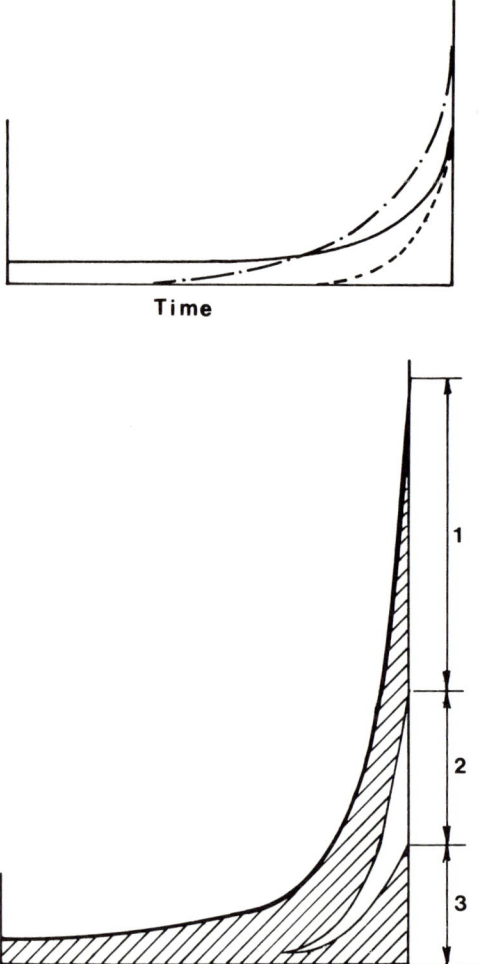

Fig. 8.2. Effect of individual mortality failures and summation of terminal failure probabilities.

——— (1) Overload.
- - - - (2) Looseness.
—·— (3) Metallurgical deterioration.

(Reproduced by permission of the Council of the Institution of Mechanical Engineers.)

ageing. These processes affect the accuracy and performance of a machine over a longer period of time. They are generally random in nature, at least quantitatively so.

8. MACHINE FAILURE

8.4 FAILURE DIAGNOSIS

When failures occur it is generally the symptoms that are observed and not the causes, which are often obscure and usually difficult and, at times, impossible to ascertain. Even when explanations can be given many still have elements of hypothesis and uncertainty; they may be more circumstantial than factual. A few examples, some rather familiar to car owners, will help to illustrate the above propositions:

Example 1. Car steering defective.
Symptom. Car is stiff and heavy to steer.
Possible source of trouble. Tyre pressures are too low. No grease on king-pin. No oil in steering gear. No grease on steering and suspension ball joints. Front wheel toe-in is excessive. Suspension geometry is incorrect. Steering gear is incorrectly adjusted or is too tight. Steering column is badly misaligned or bent.
Basic cause. Lack of proper maintenance. Accidental damage.

Example 2. Starter motor faulty operation.
Symptom. Starter motor operates, but does not turn engine.
Possible source of trouble. Starter motor pinion is sticking on screwed sleeve (Bendix coupling). Pinion or flywheel gear teeth are broken or badly worn.
Basic cause. Dirt or oil on Bendix drive. Mechanical damage.

Example 3. Steam turbo-generator power fall off.
Symptoms. The load capacity of a generator set fell with associated changes in the efficiency and pressure level in the first stage of the turbine.

Thus:

load change $= -4.9\%$

first-stage pressure change $= -4.2\%$

first-stage efficiency change $= -1.8\%$

feed-water flow-rate change $= -3.6\%$

intercept valve pressure change $= -4.4\%$

intermediate stage efficiency change $= -0.4\%$

Deductions. The drop in the first stage and intercept valve pressures suggests a restriction in the flow-passing capacity through the turbine, substantiated by the reduction in feed-water flow-rate. Loss in flow-passing capacity relates to changes in stage pressures. The stages, other than the first and last, operate at constant pressure ratios; change in pressure up-stream will be reflected in an equal percentage change in pressure down-stream. There was a similar percentage change in pressure at the first stage and at the intercept valve.

Conclusions. The loss of flow arises from some restriction at the front end of the turbine, either in the first stage or at the control valve.

Inspection. When the turbine was shut down and examined, it was found that the stem of one of the four control valves was broken, interrupting the sequence of opening.

Example 4. Faulty aircraft propeller performance

Symptoms. A reduction gear of an aircraft propeller engine disintegrated in flight and some debris penetrated the pressurized cabins. Fortunately no major casualties were experienced.

Observations. The engine had been stopped because of a rise in oil temperature, and the propeller was feathered. The flight continued for about 2 hours at 20 000 feet altitude. On descending, the propeller suddenly started to unfeather and rev up; corrective action was ineffective.

Cause. After landing, the aircraft was examined and it was found that the controls did not function properly. Basic cause was faulty maintenance and inspection checks of the control panel.

Example 5. Break-down of compressor in an air-conditioning plant

An air-conditioning plant in a four-storey building incorporated a chiller unit with multi-cylinder compressors working on a gas refrigerant. The units were located in an isolated room on the roof of the building with no

Fig. 8.3. A collection of damaged parts.

monitoring instruments in any of the service areas. A standard service contract had to be signed with the suppliers of the equipment.

Symptoms. During a routine maintenance inspection the service engineer noticed that one of the compressors was running, but no pressure was shown on the output gauge. The oil level in the crank-case appeared to be normal.

Observations. When the compressor was opened up, most of the moving parts were found broken and smashed up and lying in the oil sump of the machine. The crank-shaft, pistons, con-rods, cylinder liners and bearings had suffered extensive damage (Figs 8.3 and 8.4). The top of the cylinders and the valves were in good condition. On closer examination severe score marks were

Fig. 8.4. Damaged cylinder liner.

noted on Nos 1, 2, 3 and 4 pistons, the liner bores and the crank-shaft. All piston-rings of these four pistons were either broken or seized in their grooves.

Lubrication failure was suspected, and on inspection it was found that the oil pressure safety switch on the machine had operated (responding to too low an oil pressure in the system) but the relay switch on the control panel had jammed, allowing the electric motor to continue driving the compressor, despite inadequate lubrication.

Piston No. 5 with its rings were in good condition (Fig. 8.5) as were its small- and big-end bearings and the mating crank-pin area. However, the two big-end cap bolts were broken, and so was the con-rod. The fractured surfaces of

Fig. 8.5. No. 5 piston, connecting rod, big end, bearings, crank shaft, and broken cap bolts.

8. MACHINE FAILURE

the bolts were examined under a scanning electron microscope, and it was revealed that these bolts had failed due to fatigue:

Figure 8.6 shows fatigue marks at the surface of the bolt which failed at the centre. Figure 8.7 is evidence of fatigue striations growing larger as fracture progressed (i.e. as stress intensity decreased). There was only a minute area of final ductile fracture.

Fig. 8.6. Fatigue marks at broken bolts.

These observations suggested that both bolts had failed by fatigue over many stress cycles with low fluctuating loads superimposed upon an initially high but continually reducing mean tensile load.

Causes. The primary cause of failure of the compressor was ascribed to the big-end cap bolts of No. 5 con-rod. These bolts had failed in fatigue due to initial overtightening which caused a high initial mean load. The fluctuating stress due to piston and con-rod masses and frictional resistance would generally be low on this type of machine. Once cracks were initiated their propagation and opening would tend to relieve the initial tightening load, gradually reducing the mean stress level and consequently slowing down the rate of crack growth.

The sequence of events following the breaking of the above big-end bolts were reasoned to have been as follows:
  (1) The opening-up of the big-end bearing of No. 5 con-rod allowed oil to gush out through the hole in the crank-pin.

(2) The consequent loss of oil pressure should have caused the electric drive motor to be switched off, but this did not occur due to the jamming of the control switch.
(3) Continued running caused the No. 5 big-end to separate from the crank-pin and then, hanging loosely from the now stationary piston, the con-rod was broken by the rotating crank.

Fig. 8.7. Fatigue striations increasing in pitch.

(4) The other four pistons continued working with no lubrication; they heated and seized up within a short time. The crank-shaft, still being turned by the electric motor, would then cause the severe damage to the four con-rods, bearings, etc.
(5) With all obstructions broken away, the crank-shaft would be relatively free to rotate.

Conclusions. Three primary deficiencies caused the extensive damage to the compressor:
   (1) The lack of proper monitoring equipment to indicate the operational condition of the unit.
   (2) Initial overtightening of the No. 5 big-end cap-bolts.
   (3) Failure of the pressure-controlled relay switch to cut out the electric motor when the oil pressure fell off.

8.4.1 Diagnostic procedures

Service failures Most failures, or malfunctions of equipment, naturally occur in service. To ensure proper feedback of such failures it is necessary to:

(1) Establish an appropriate field and service data retrieval system (see, for example, Fig. 8.8);

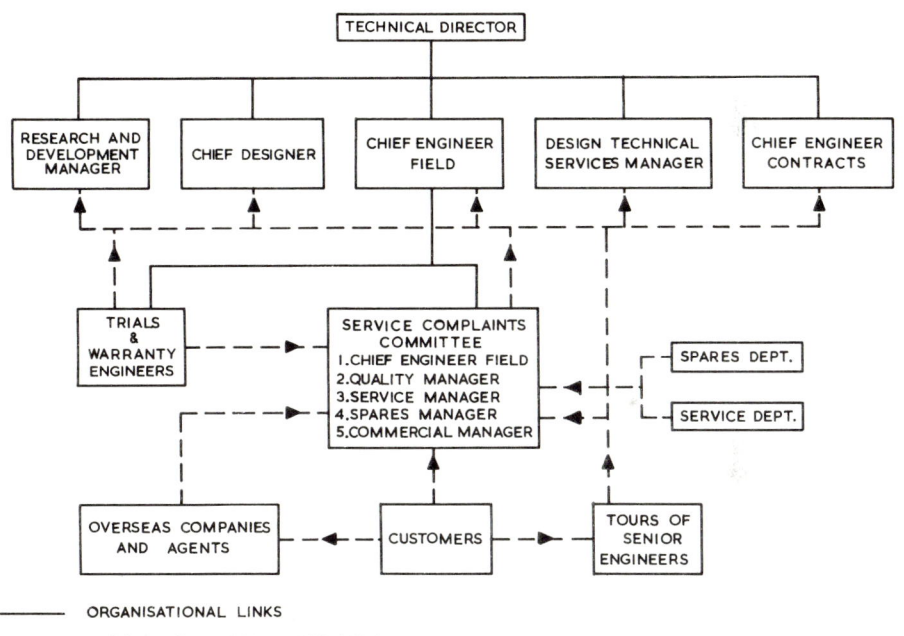

Fig. 8.8. Organization of field information grouping (by courtesy of Mirlees Blackstone Ltd.).

(2) Conduct careful and detailed examination of every failed and associated part;
(3) Critically review designs, including relevant theoretical analysis;
(4) Carry out statistical surveys of service experience of both failed and surviving components and parts, noting common factors such as batch numbers, manufacturing unit or supplier, operating conditions;
(5) Be able to introduce emergency operating restrictions on surviving equipment or parts, where safety is at risk;
(6) Simulate, as exactly as possible, service failures on test rigs to confirm or negate any diagnosis and to provide integrity for any modifications;

(7) Establish an effective modification section which will,
   (a) ensure that all modifications are as simple as possible and, whilst aiming to overcome existing defects, will not introduce new unknown secondary effects;
   (b) plan and organize the introduction of approved modifications on existing and new models and ensure that the effect on spares is appropriately dealt with.

Design defect analysis. Two intensive fault-analysis procedures during the design phases have been described by Green (1969–70). The first is called Potential Fault Analysis and considers each component and its relation to the system as a whole. The second is called Systematic Design Review and aims to consider every aspect and function of a system and its components to ensure that no factor has been overlooked. These procedures bear a strong resemblance to value analysis and other "check-list" techniques. One must note with some amazement how many questions managers, organizers and operational researchers can think up for *others* to answer. The question that never seems to be put is whether the designer will have any time left to do his actual designing.

Having said that, there certainly is a need for "devil's advocate" sessions during the design stages. The effects on the safety and efficient operation of the equipment of any possible errors in manufacture, assembly and usage should be critically examined, and where possible either made impossible (fool-proof) or minimized.

Defect surveys. Organizations and companies do find it necessary to carry out defect surveys of equipment and machinery of concern to them. Two typical surveys, one by an engineering insurance organization and one by the National College of Agricultural Engineering are illuminating.

Insurance companies usually have many years experience in inspecting and examining failures of machinery and plant. The respondent in this instant wished to emphasize that the true causes of many failures are frequently obscure. In practice, failures are often due to a combination of a number of contributory causes: mal-practice, mal-operation, deliberate or incidental misuse, overloading, carelessness, etc. Typical failure rates of various plant and equipment are listed in Table 8.1. These are comparative failure frequencies as between the range of equipment dealt with, not true statistical rates of failure for each item. The relative cost-losses are not indicated.

If this one-year sample is representative, steam-plant, IC engines, air-compressors, pumps, turbines, electric generators and converters, motors and transformers appear to have fair levels of operational reliability.

A more detailed analysis of a survey of farm equipment was reported by

Table 8.1. Failure rates based on 12-month period (1969–70).

Class of plant	Description	Failure rate (percentage of total)
Lifting and handling machinery	Lifts	1·9
	Electric overhead travelling cranes	7·6
	Mobile cranes	8·6
	Tower cranes	4·9
	Rail-mounted jib cranes	19·8
	Other jib cranes	3·9
	Fork lift trucks	5·2
	Excavators	10·5
	Tractors, etc.	11·4
Steam plant	Water tube boilers	2·5
	Shell boilers	0·7
	Heating and DHW boilers	1·7
	Economizers	1·0
	Ovens, etc.	1·0
	Piping installations	0·25
Engines and general plant	Internal combustion engines	2·1
	Air compressors	1·7
	Pumps	0·7
	Refrigerators (small)	10·0
	Refrigerators (large)	13·5
	Turbines	3·2
	Printing machinery	6·7
Electrical plant	Generators and convertors	2·1
	Motors	3·3
	Portable tools	10·5
	Submersible pumps	7·0
	Transformers	1·2
	Capacitors	8·1
	Oil burners	5·1
Rectifiers		1·7
Furnaces		22·0
Lifting magnets		5·5

Richardson (1967; Table 8.2), who classified the failed parts according to the nature of their defects. This enabled him to focus on the weak parts in farm machinery and to examine the possible causes of the weaknesses.

Table 8.2. A pilot survey of the durability of farm machinery (from Richardson, 1967, reprinted by permission of the Institute of Agricultural Engineers).

Summary of machine "failures" (April–December 1966)

Incident	Type of incident	Degree of anticipation	Failure mechanism	Function	Availability of failed parts
Total cases 1251	Normal work 1185	Routine replacements 390	Wear 420	Drive components 282	Failed parts examined 576
(3305) (items)	Accident 52	Casualty 861	Fracture 536	Parts engaging the worked material 515	Most failure information available 327
	Mis-use 14		Others 191	Tyres, wheel and axles 37	Some failure information available 310
	(safety 16)		Not known 38	Structure 171	Not known 38
				Others 175	
				Power Units (elec.) 5	

Cases of failure classed according to function

Drive components		Parts engaging the worked material		Structure		Others	
Shafts	35	Shares	111	Main frames and components	111	Shafts	3
Bearings	38	Discs	13	Major components (casings, castings, etc.)	13	Bearings	14
Gears	16	Mouldboards	20	Details	51	Gears	1
Springs	2	Other soil wearing parts	109	Not known	5	Hydraulics	26
Universal joints	11	Leaf spring tines	61	Other	18	Cams	4
Chains/sprockets	80	Wire coil tines	37			Springs	15
Belts/pulleys	52	Legs and stems	11			Guards	17
Others	64	Cutter bar parts	84			others	102
		Flails and hammers	10				
		Others	74	Total	198	Total	182

	Tyres, wheels and axles		Power units	
	Wheels	24	Electric motors	5
	Shafts and axles	6		
	Bearings	37		
	Total	37	Total	5

| Total | 298 | | 531 | | 235 | | 187 |

Proposed failure classification

FRACTURE

Fatigue Black spring
Wire spring
Soil wearing part
Other
{ Near weld
At thread or tooth root
At shaft step
At spline or keyway
At hole
At plain section
At bend or change of section
Other }

Impact Crop cutting part
Structure
Other

Ductile

Brittle Grey cast iron
Other

Shearbolt

WEAR

Metal–soil Cutting consolidated soil
Other

Metal–metal rubbing (including chains) Plain bearings { Soil present
Other }

Metal–metal rolling (gears) { Open
Encased } { Lubrication failure
Contaminant
Other }

Metal–metal rolling (bearing) { Open
Encased
Sealed }

Table 8.2. (cont.)

Proposed failure classification

OTHER			
Corrosion			
Distortion			
Drive belt	Near weld	Metal–crop or plant matter	Rubbing (e.g. augers)
Hydraulic	Other		Cutting (e.g. knives and flails)
Instrument			Grinding (e.g. mills)
Other		Other	Crop only (perhaps with contaminant)
			Soil wear as well (e.g. mower flails)
			Other

8.5 MECHANISMS OF COMPONENT FAILURES

Six basic failure mechanisms can be distinguished:
(1) *Failures due to static load* may be
   (a) plastic deformation caused by stressing beyond the elastic limit of the material,
   (b) fracture, which may be preceded by some or no plastic deformation.
(2) *Fatigue failures.* Materials fail in fatigue at fluctuating stress-levels much below their static strength. Fatigue is associated with the initiation and growth of cracks; ductility of a material alleviates to some extent the rate of such growth. Fluctuating stresses may be due to mechanical and/or thermal effects.
(3) *Creep* is a delayed non-reversible or partly reversible deformation. Reversibility is usually rather slow after release of load. Excessive creep leads to rupture.
(4) *Wear.* This describes processes which effect the removal, or relocation of particles from a solid surface by mechanical means. The action is often aided by corrosive attack.
(5) *Degradation.* There are some time-dependent non-reversible surface changes caused by external environments. Such processes are corrosion and erosion, which may result from thermal, chemical, electrical and mechanical actions.
(6) *Ageing.* Most materials and products exhibit slow gradual changes with time. They occur independently of the methods of operation or usage of the product. Examples are: embrittlement of rubber and plastics; growth and distortion of castings; embrittlement of soldered joints; stress relaxation, radiation effects. Accretion can be viewed as a special kind of ageing resulting from depositions and furring-up of valves and pipes.

Extensive specialized findings have been published dealing with all of these processes, both qualitatively and quantitatively. Some of the processes are briefly discussed in this book to provide designers with a guide for initial selection of materials and approximate estimation of performance.

8.5.1 Failure due to static loading

Most qualified engineers have a fair grounding in the principles of strength of materials. Suffice it here to emphasize a few principal points.

Ductile materials (elongation $>5\%$) pass through a slow plastic deformation phase before final rupture. Failure is usually in the shear mode. Where no distinct yield point exists, the 0.2% proof stress is often used.

Impacting loads induce stresses well above those caused by the same stationary load, because kinetic energy has to be absorbed in the form of strain energy.

We know from basic theory that

$$dU = \tfrac{1}{2} \cdot \frac{\sigma^2}{E} \cdot dV \qquad (8.1)$$

in simple tension or compression.

Writing equation (8.1) as

$$dU = \tfrac{1}{2} \cdot \frac{\sigma}{E} \cdot \sigma \cdot dV$$

we note that

$$\sigma \propto \frac{1}{A} \quad \text{and} \quad dV \propto A$$

$$\therefore \sigma \cdot dV = \text{constant}$$

Thus, increasing the area of the section will reduce σ and result in a lowering of capacity to absorb energy.

Similarly, for bending, we have

$$\begin{aligned}dU &= \frac{1}{2} \cdot \frac{(\sigma_{mean})^2}{E} \cdot dV \\ &= \frac{1}{2} \cdot \frac{(\sigma_{mean})^2}{E} \cdot ds \cdot A \\ &= \frac{1}{2} \cdot \frac{(M \cdot k)^2}{E \cdot I} \cdot ds \cdot \left(\frac{I}{k^2}\right)\end{aligned}$$

or

$$dU = \frac{1}{2} \cdot \frac{M^2}{EI} \cdot ds \qquad (8.2)$$

Again we see that energy absorption capacity is reduced if the I value of the section is increased. Thus generally for high impact energy absorption we should use the highest safe working stress throughout the member and increase its volume by increasing its length. This will, of course, be restricted by the maximum deflection acceptable.

Real materials are not perfectly homogeneous. They invariably contain flaws and weaker areas, or even cracks and cavities. Brittle fracture is generally associated with some flaws and a tensile stress beyond a critical value. Fracture usually occurs at high speeds (H. F. Zalud and A. H. Zaludova, personal communication) given by the relationship

$$v = \frac{\sigma_L}{E} \cdot \frac{a}{s} \cdot v_0 \qquad (8.3)$$

8. MACHINE FAILURE

where
v = velocity of fracture
E = Young's modulus
σ_L = critical stress level
a = crack length
s = length of plastically deformed material at root of crack
v_0 = velocity of sound in the material

Modern fracture mechanics considers crack propagation as being a function of a stress intensity factor,

$$K_I = \sigma . \sqrt{\pi . a} . \alpha \tag{8.4}$$

where
K_I = stress intensity factor (in notch or crack)
σ = gross tensile stress perpendicularly across crack
a = crack length
α = a factor depending on the geometry

K_I-values have been established experimentally for various situations and are by now fairly well disseminated in the literature. Begley (1973) reports on "a standard ASTM method for K_{IC} (= critical intensity factor) tests. This K_{IC} measurement point essentially represents 2% crack growth in a fatigue pre-cracked bend specimen. The specimen must be large enough to fail in the essentially elastic range."

$$\left. \begin{array}{c} \text{Crack size, remaining ligament} > 2.5\left(\dfrac{K_{IC}}{\sigma_y}\right)^2 \\[1em] \text{and thickness} > 2.5 . \left(\dfrac{K_{IC}}{\sigma_y}\right)^2 \end{array} \right\} \tag{8.5}$$

Here, σ_y is taken as the 0.2% proof strength.

To design against brittle fracture, flaw size and loading must be controlled so that

$$K_I < K_{IC}$$

Tests yielded results which are shown in Fig. 8.9 and indicate the approximate and probabilistic nature of the values obtained (Begley, 1973). If crack length is plotted against stress levels (Fig. 8.10), it is possible to estimate a point representing the transition from ductile to brittle fracture. This critical length is denoted by

$$a_t = \frac{1}{\pi} \frac{(K_{IC})^2}{\sigma_y} \tag{8.6}$$

and values for K_{IC}, σ_y and a_t for a number of materials are given by Begley (1973) in Table 8.3. Some K_I values are presented in Table 8.4.

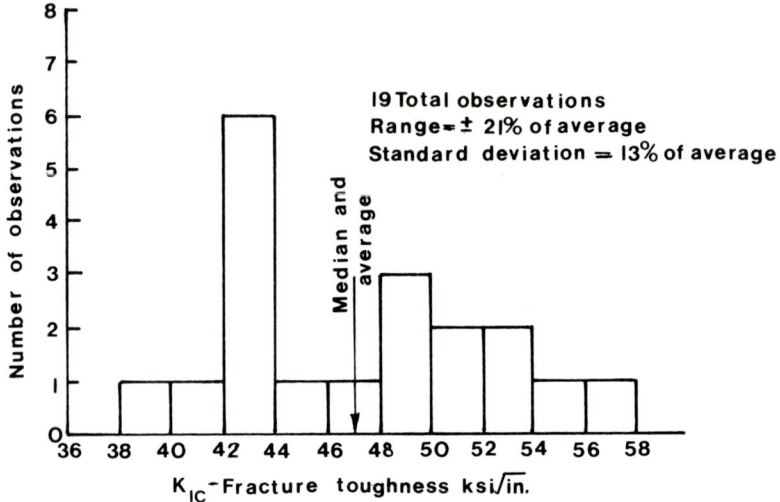

Fig. 8.9. K_{IC} distribution curve for CrMoV steel (124k406) at 0 °F.

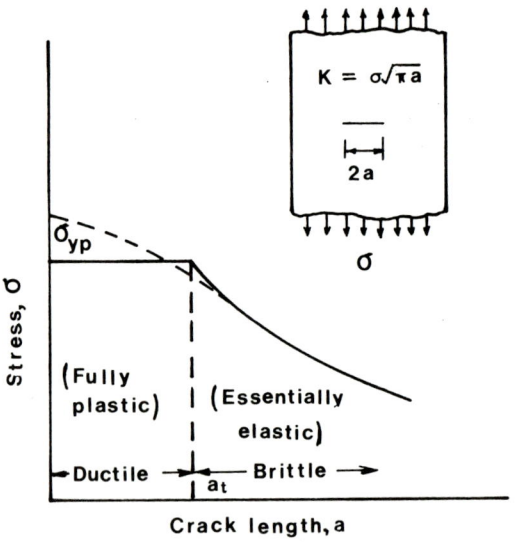

Fig. 8.10. Failure stress v. crack length for an infinite centre cracked panel.

While high-strength steels, aluminium and titanium alloys show relatively less effect of temperature (as measured by Charpy tests), low- and medium-strength steels, most metals, rubbers, plastics, etc., all exhibit marked reduction in fracture toughness at low temperatures. The change from ductile to

Table 8.3. Typical room temperature mechanical properties (after Begley, 1973).

Material	Toughness fraction K_{IC}(k.s.i. $\sqrt{\text{in.}}$)	0·2% yield strength σ_{yp}(k.s.i.)	Transition crack a_t (inches)
Tungsten	10	210	0·0001
Beryllium	10	35	0·026
7075-T651	25	75	0·035
4340 (800 °F temper)	80	200	0·051
18 Ni maraging	60	250	0·018
CrMoV forgings	60	90	0·141
NiMoV forgings	120	90	0·570
NiCrMoV forgings	200	120	0·884
A535 Grade B Class II, heavy section plate	200	70	2·600

Table 8.4. Stress intensification factor (K_I): plain strain.

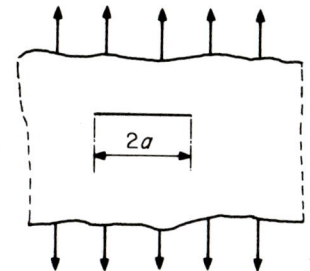

Stress intensification factor. $K_I = \sigma \sqrt{(\pi . a)}$
Validity. Infinite plane
Reference. Pook (1973)

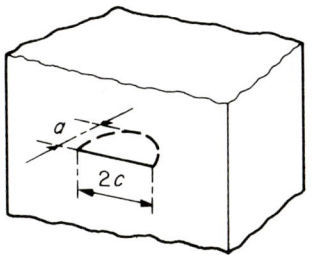

Stress intensification factor. $K_I = 0·7 \sigma \sqrt{(\pi . a)}$; c = a
$K_I = 1·0 \sigma \sqrt{(\pi . a)}$; c = 3a
$K_I = 1·1 \sigma \sqrt{(\pi . a)}$; c ⩾ a

Validity. Face crack in block or thick plate
Reference. Pook (1973)

Table 8.4 (*cont.*)

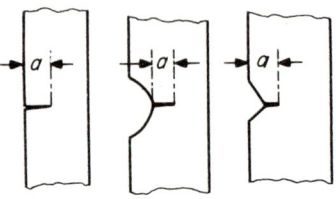

Stress intensification factor. $K_I = 1 \cdot 1 \sqrt{(\pi \cdot a)}$

Validity. $a \ll$ other plate dimensions

Reference. Pook (1973)

Geometry. Embedded circular flaw

Stress intensification factor. $K_I = \dfrac{2}{\pi} \sigma \sqrt{(\pi \cdot a)}$

Reference. Begley (1973)

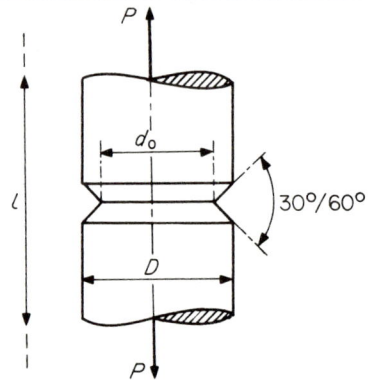

Stress intensification factor. $K_I = \dfrac{PY}{D^{3/2}}$

$$Y = 1 \cdot 72 \dfrac{D}{d} - 1 \cdot 27$$

$$d = d_0 - \dfrac{K_I^2}{3\pi \sigma_y^2}$$

Validity. Round notched bar: $l > 8D$; $0 \cdot 5 < \dfrac{d}{D} < 0 \cdot 8$

Recommend: $\dfrac{d_0}{D} = 0 \cdot 5$

Reference. Osgood (1969)

8. MACHINE FAILURE

Table 8.4 (*cont.*)

(thickness = B)

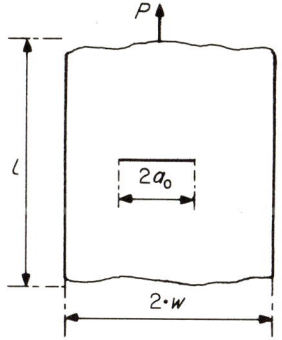

Stress intensification factor. $K_I = \dfrac{PY}{2Bw^{1/2}}$

$$Y = 1 \cdot 77 \left(\dfrac{a}{w}\right)^{1/2} + 0 \cdot 227 \left(\dfrac{a}{w}\right)^{3/2} - 0 \cdot 51 \left(\dfrac{a}{w}\right)^{5/2} + 2 \cdot 7 \left(\dfrac{a}{w}\right)^{7/2}$$

$$a = a_0 + \dfrac{K_I^2}{6\pi\sigma_y^2}$$

Validity. Centre cracked plate; $l \geqslant 8w$; $0 < \dfrac{a}{w} < 0 \cdot 70$; $\dfrac{w}{5} < B < \dfrac{w}{2}$

Recommend: $a_0 = \dfrac{w}{3}$

Reference. P. F. Thomason (personal communication)

(thickness = B)

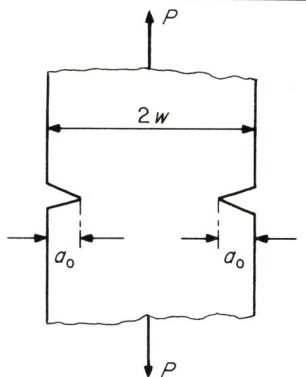

Stress intensification factor. $K_I = \dfrac{PY}{2Bw^{1/2}}$

$$Y = 1 \cdot 98 \left(\dfrac{a}{w}\right)^{1/2} + 0 \cdot 36 \left(\dfrac{a}{w}\right)^{3/2} - 2 \cdot 12 \left(\dfrac{a}{w}\right)^{5/2} + 3 \cdot 42 \left(\dfrac{a}{w}\right)^{7/2}$$

$$a = a_0 + \dfrac{K_I^2}{6\pi\sigma_y^2}$$

Table 8.4 (cont.)

Validity. Double-edge cracked plate; $0 < \frac{a}{w} < 0.70$; $\frac{w}{5} < B < \frac{w}{2}$

Recommend: $a_0 = \frac{w}{3}$

Reference. P. F. Thomason (personal communication)

(thickness = B)

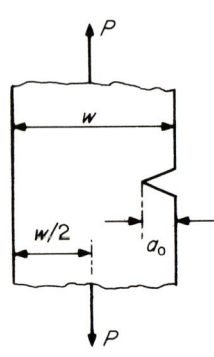

Stress intensification factor. $K_I = \frac{PY}{Bw^{1/2}}$

$$Y = 1.99\left(\frac{a}{w}\right)^{1/2} - 0.41\left(\frac{a}{w}\right)^{3/2} + 18.7\left(\frac{a}{w}\right)^{5/2} - 38.48\left(\frac{a}{w}\right)^{7/2} + 53.85\left(\frac{a}{w}\right)^{9/2}$$

$$a = a_0 + \frac{K_I^2}{6\pi\sigma_y^2}$$

Validity. Single-edge cracked plate; $0 < \frac{a}{w} < 0.6$; $\frac{w}{3} < B < \frac{w}{8}$

Recommend: $a_0 = \frac{w}{3}$

(thickness = B)

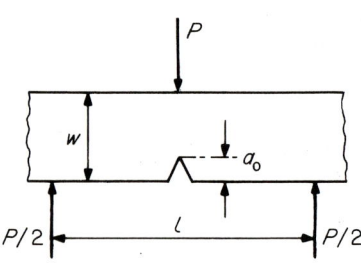

Table 8.4 (cont.)

Stress intensification factor. $K_I = \dfrac{6PY}{Bw^{1/2}}$

$$Y = 1\cdot 93\left(\dfrac{a}{w}\right)^{1/2} - 3\cdot 07\left(\dfrac{a}{w}\right)^{3/2} + 14\cdot 53\left(\dfrac{a}{w}\right)^{5/2} - 25\cdot 11\left(\dfrac{a}{w}\right)^{7/2} + 25\cdot 8\left(\dfrac{a}{w}\right)^{9/2}$$

$$a = a_0 + \dfrac{K_I^2}{6\pi\sigma_y^2}$$

Validity. Single-edge crack, 3-point bending; $l = 4w$; $0 < \dfrac{a}{w} < 0\cdot 6$; $\dfrac{w}{8} < B < w$

Recommend: $a_0 = \dfrac{w}{3}$; $B = \dfrac{w}{2}$

Reference. P. F. Thomason (personal communication)

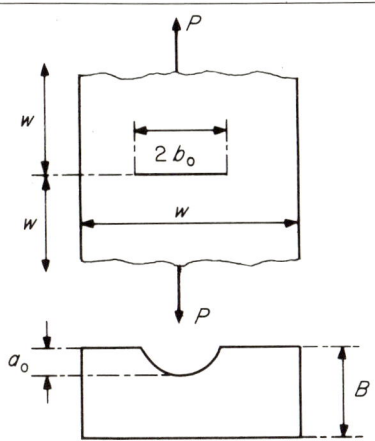

Stress intensification factor (approximate solution).

$$K = \dfrac{1\cdot 2P^2 a_0}{wB^2}\left[\dfrac{1}{\phi^2 - 0\cdot 2\left(\dfrac{P^2}{w^2 B^2 \sigma_y^2}\right)}\right]$$

$$\phi = \int_0^{\pi/2}\sqrt{\left(1 - \dfrac{b_0^2 - a_0^2}{b_0^2}\cdot\sin^2\theta\, d\theta\right)}$$

Note. $\sigma = 0\cdot 2\%$ proof stress where definite yield point does not exist

Validity. Semi-elliptical-surface cracked plate; $\dfrac{w}{B} \geqslant 6$; $a_0 < \dfrac{B}{2}$; $2b_0 \leqslant \dfrac{w}{3}$

Reference. P. F. Thomason (personal communication)

brittle characteristics is termed the transition point, and may range from 0 °C to −120 °C. A typical transition plot is shown in Fig. 8.11 (Begley, 1973), and similar curves (with Charpy impact values) can be obtained from material manufacturers.

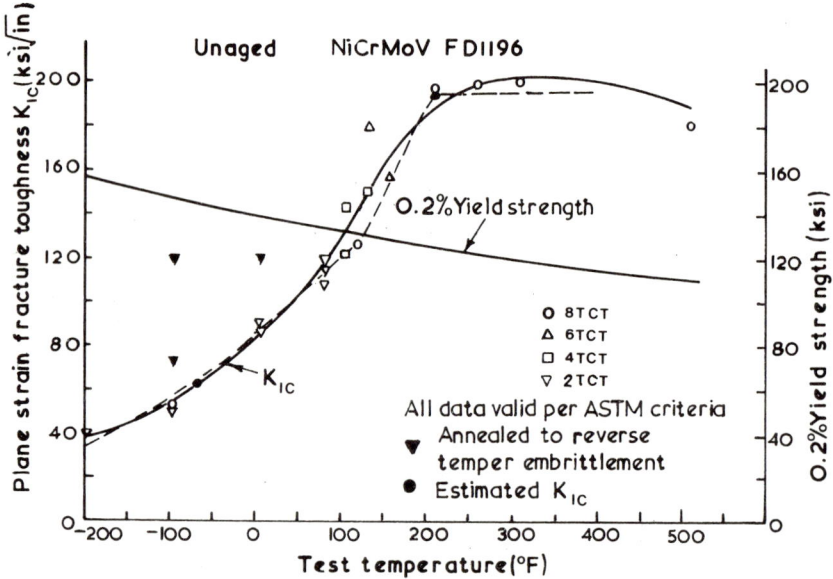

Fig. 8.11. Temperature dependence of the plane-strain fracture toughness (K_{IC}) of a NiCrMoV steel (196).

8.5.2 Fatigue Failures

(A) Fatigue strength concept. Components fail in fatigue due to the initiation and progressive growth of cracks (or flaws) resulting from fluctuating applications of load. Traditional test and analysis of fatigue characteristics of materials concentrated on establishing maximum nominal stress levels required to initiate cracks for any given number of stress cycles.

In standard tests polished cylindrical specimens, 0·30 in. diameter, were subjected to rotary bending (cantilever fashion), and the number of revolutions (cycles of stress reversals) to failure was noted for various stress levels. If the maximum reversed stress is plotted against the log of the number of cycles, one obtains the well-known S/N curves. For steels, there appeared to exist a limiting stress level below which the material could be cycled indefinitely without failure. This was denoted the "endurance limit" which lay somewhere between 10^6 and 10^7 stress cycles. No such definite "knee" is observable with other materials, and the endurance limit for these simply indicates the stress

8. MACHINE FAILURE

levels the material can tolerate for a specified number of stress reversals. A few typical endurance limit values are given in Table 8.5. A more extensive range of materials is listed by Lessels (1954), and Fig. 8.12 from the *Fulmer Materials Optimiser* (1974) gives a useful review.

Table 8.5. Endurance limit values for some materials at $N = 2 \times 10^6$ cycles of reverse bending

Material	$\sigma_{ult\,tensile}$ (ton in.$^{-2}$)	σ_e (ton in.$^{-2}$)
Cast iron	25–50	6–18
Cast steel	60–80	24–32
Cast steel	60–150	25–75
Nitralloy	125	80
Brasses	25–75	7–20
Copper	32–50	12–17
Phosphor-bronze	55	12
Aluminium alloy—cast	18–40	6–11
Aluminium alloy—wrought	25–70	8–18
Magnesium alloy	20–45	7–17
Molybdenum	98	80

The fatigue strength (endurance limit) of a machine component is not the same as the numerical values listed in the tables. These represent only the test specimens under standard test procedures.

Component endurance limits may be derived from standard σ_e values by the relationship,

$$\sigma_{e_{comp}} = \sigma_{e_{matl}} \cdot A \cdot B \cdot C \cdot D \tag{8.7}$$

where

- A = factor for stress mode different from reversed bending;
  - $A = 0.85$ for reversed direct stress
  - $A = 0.6$ for reversed shear stress
- B = factor allowing for size different from 0·3 in. dia. (see Fig. 8.13)
- C = factor for surface finish effects (see Fig. 8.14)
- D = factor covering various influences such as interference fits, corrosion, residual stresses and adjustment for reliability (see Section 3.3)

Most fluctuating stress situations may be represented by a mean (static) stress upon which is superimposed an alternating (fully reversed) stress. For such stress fluctuations it is necessary to establish an "equivalent" stress which will represent the same damage effect as the combined stress system.

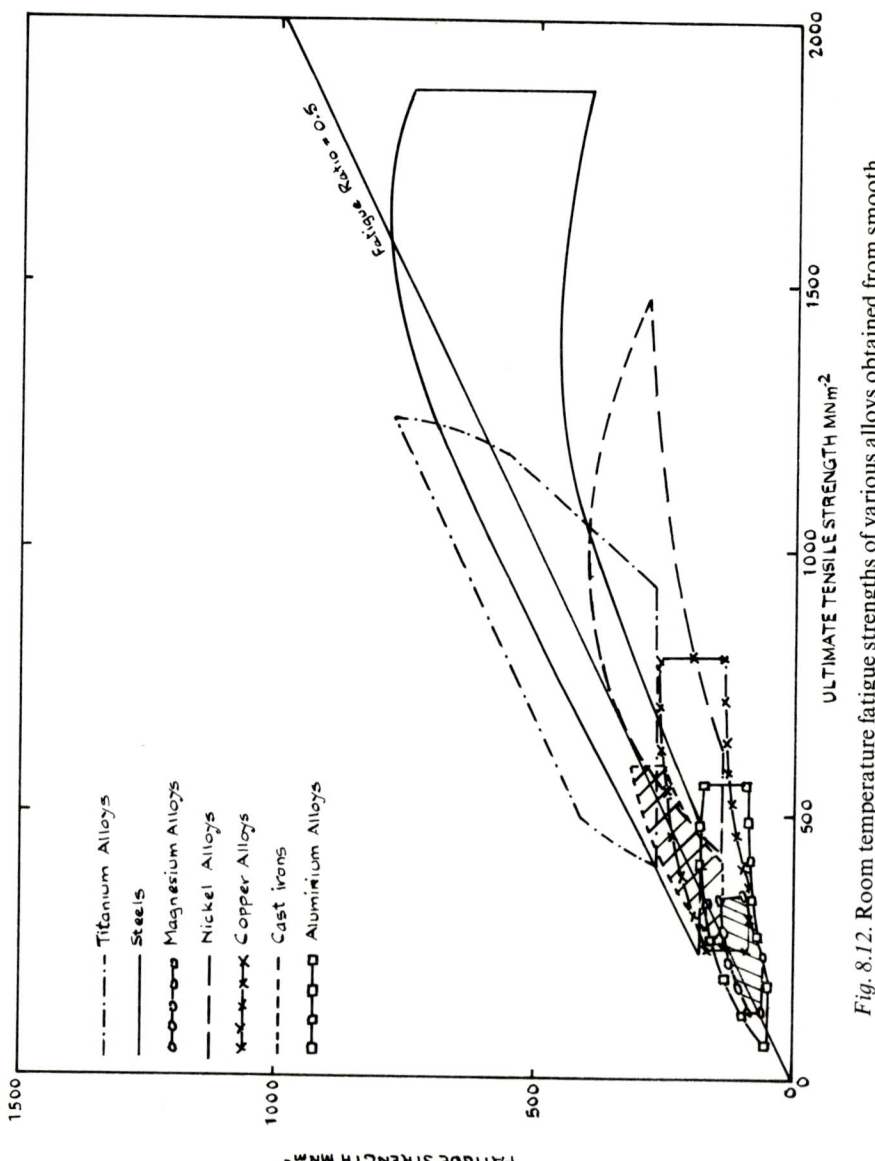

Fig. 8.12. Room temperature fatigue strengths of various alloys obtained from smooth test pieces. At low ultimate tensile stresses the fatigue ratio is at least 0·5; however, at higher ultimate stresses the fatigue ratio decreases for most alloy systems (*Fulmer Materials Optimiser*, 1974)

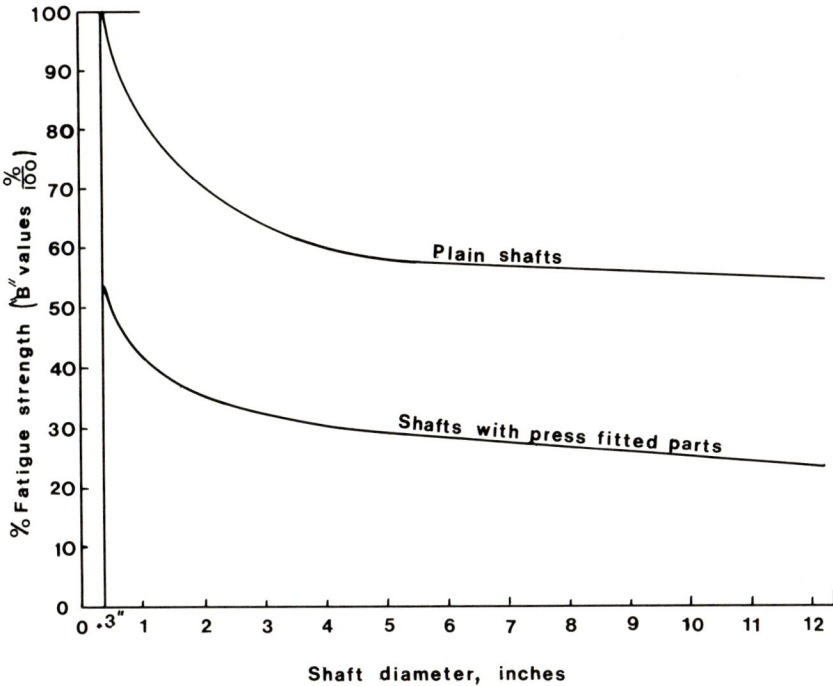

Fig. 8.13. Size effect on fatigue strength of a plain, medium-carbon steel shaft and the additional reduction in strength caused by pressing components along it, e.g. gear, flywheel. Graph shows intrinsic material fatigue strength for material with UTS range 30–50 tons in^{-2} (e.g. medium-carbon steels). Standard test specimen 0·3 inches in diameter. (From lecture notes prepared by H. N. Green, University of Salford.)

(*i*) Static + alternating stresses of the same kind:
Correlations available:

$$\left.\begin{array}{ll} \text{(i) Goodman} & \pm \sigma_a = \sigma_{e\text{comp}}\left(1 - \dfrac{\sigma_m}{\sigma_{\text{ult}}}\right) \\\\ \text{(ii) Gerber} & \pm \sigma_a = \sigma_{e\text{comp}}\left[1 - \left(\dfrac{\sigma_m}{\sigma_{\text{ult}}}\right)^2\right] \\\\ \text{(iii) Soderberg} & \pm \sigma_a = \sigma_{e\text{comp}}\left(1 - \dfrac{\sigma_m}{\sigma_y}\right) \end{array}\right\} \quad (8.8)$$

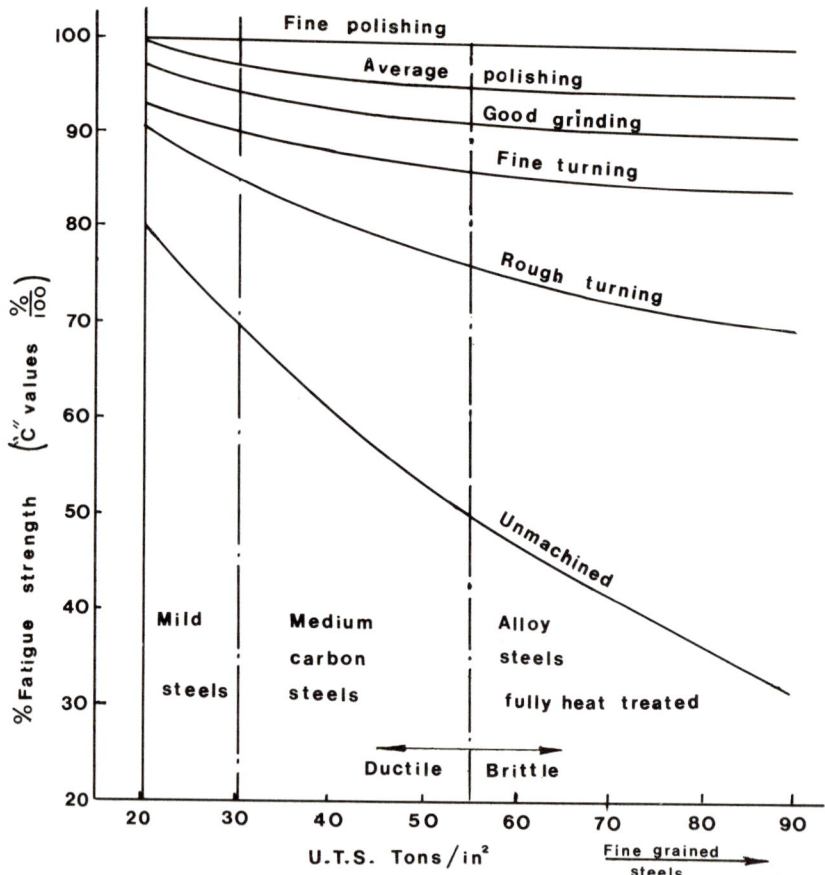

Fig. 8.14. Effect of surface finish on fatigue strength of steels indexed by their ultimate tensile strength. Classifications of steels are approximate. (From lecture notes prepared by H. N. Green, University of Salford.)

Equation (8.8) can be written in a general form

$$1 = \left(\frac{\sigma_a}{\sigma_{e_{\text{comp}}}}\right) + \left(\frac{\sigma_m}{\sigma_{\text{ult}}}\right)^m$$

or (8.9)

$$1 = \left(\frac{\sigma_a}{\sigma_{e_{\text{comp}}}}\right) + \left(\frac{\sigma_m}{\sigma_y}\right)^m$$

$m = 2$, for Gerber correlation

$m = 1$, for Goodman and Soderberg's equations

8. MACHINE FAILURE

Using Soderberg's (most pessimistic) correlation, we can also write it in the following form

$$\left(\frac{\sigma_y}{n}\right)_{equiv} = \sigma_m + \left(\frac{\sigma_a}{\sigma_{ecomp}}\right) \qquad (8.10)$$

where n is a factor of safety.

In the above equations the values of σ_m and σ_a must be those that actually arise in the section or region one is considering. Whereas components of uniform shapes have nominal stress-distributions as for direct, bending, torsion and shear loading, discontinuities in component create distortions in such distributions, increasing considerably the stress levels at the discontinuities. These distortions, or stress concentrations, are operative in brittle materials and in all materials experiencing alternating loads. Stress concentration factors have been established theoretically and experimentally for a large number of discontinuities in engineering components (e.g. Peterson, 1965).

Denoting in general, stress concentration factors for static stress = k_m (applicable only for brittle materials) and for alternating stress = k_a (valid for all materials, attenuated by ductility), equations (8.9) and (8.10) may be written as

$$1 = \left(\frac{k_a \cdot \sigma_a}{\sigma_{ecomp}}\right) + \left(\frac{k_m \cdot \sigma_m}{\sigma_y}\right)^m \qquad (8.11)$$

and

$$\left(\frac{\sigma_y}{n}\right)_{equiv} = k_m \cdot \sigma_m + \frac{k_a \cdot \sigma_a \cdot \sigma_y}{\sigma_{ecomp}} \qquad (8.12)$$

Similarly for shear stress conditions,

$$1 = \left(\frac{k_{as} \cdot \tau_a}{\tau_{ecomp}}\right) + \left(\frac{k_{ms} \cdot \tau_m}{\tau_y}\right)^m \qquad (8.13)$$

and

$$\left(\frac{\tau_y}{n}\right)_{equiv} = k_{ms} \cdot \tau_m + \frac{k_{as} \cdot \tau_a \cdot \tau_y}{\tau_{ecomp}} \qquad (8.14)$$

The left-hand side of equations (8.12) and (8.14) can be treated as equivalent static stresses, representing the total fluctuating stress system, and they can be combined in the usual way to obtain the equivalent principal stresses, e.g.

$$\left(\frac{\tau_y}{n}\right)_{principal} = \tfrac{1}{2}\left[\left(\frac{\sigma_y}{n}\right)^2 + 4\left(\frac{\tau_y}{n}\right)^2\right]^{\frac{1}{2}} \qquad (8.15)$$

Both equations (8.11) and (8.12) can be used to determine the maximum peak stress acceptable by a given component.

Let

$$\text{maximum peak stress} = \sigma_{max}$$

and

$$\text{minimum peak stress} = \sigma_{min}$$

then

$$\left. \begin{array}{c} \sigma_m = \dfrac{\sigma_{max} + \sigma_{min}}{2} \\ \\ \sigma_a = \dfrac{\sigma_{max} - \sigma_{min}}{2} \end{array} \right\} \quad (8.16)$$

and

We note that $\sigma_{max} = \sigma_m + \sigma_a$, and if

$$\sigma_a = a \cdot \sigma_{max} \quad \text{and} \quad \sigma_m = b \cdot \sigma_{max}$$

$$[(a+b) = 1]$$

We can rewrite equation (8.11) thus

$$1 = \frac{k_a \cdot a \cdot \sigma_{max}}{\sigma_{e_{comp}}} + \left(\frac{k_m \cdot b \cdot \sigma_{max}}{\sigma_y} \right)^m \quad (8.17)$$

which for $m = 1$ gives

$$\sigma_{max} = 1 \bigg/ \left[\left(\frac{a \cdot k_a}{\sigma_{e_{comp}}} \right) + \left(\frac{b \cdot k_m}{\sigma_y} \right) \right] \quad (8.18)$$

Introducing a safety factor $= n$, equation (8.18) becomes

$$\sigma_{max} = \frac{1}{n} \bigg/ \left[\left(\frac{a \cdot k_a}{\sigma_{e_{comp}}} \right) + \left(\frac{b \cdot k_m}{\sigma_y} \right) \right] \quad (8.19)$$

The same result is obtained from equation (8.12).

The effect of corrosion on the fatigue strength of various materials is given in Table 8.6, and the influence of protecting surface coatings is referred to in Section 8.5.5. For convenience we supplement this information with the data in Table 8.6.

(B) Fracture mechanics concept. The traditional approach to fatigue of materials gives no information on the relative contribution of crack initiation and crack growth to the total fatigue life of a component. Fracture mechanics

8. MACHINE FAILURE

Table 8.6. Surface coatings on steel; effect on fatigue strength (*Fulmer Materials Optimiser*, 1974).

Coating	σ_e for 10^7 cycles (lbf/in^2)	
	In air	In salt spray
None	36 700	9 000
Enamel	38 400	25 000
Galvanized	33 000	37 000
Sheradized	33 000	34 000
Zinc plated	36 000	33 000
Cadmium plated	34 000	30 800
Cadmium + enamel	35 400	30 000
Cadmium + oil	35 400	30 000
Phosphated	39 600	29 000
Nitriding	Complete protection	

introduces the concept of a "flawed component" which possesses both "strength" and "toughness". Pook (1973) claims that "most fatigue failures originate from pre-existing flaws, so that life is entirely occupied by crack growth". Fatigue crack growth is essentially a strain-controlled process.

Fracture toughness is a measure of the ability of a material to resist crack growth. It is defined by a stress intensity factor, viz.

$$K_I = \sigma\sqrt{M.a} \qquad (8.20)$$

where
- a = flaw size (= widest opening to tip of flaw)
- σ = nett nominal tensile stress perpendicular across the crack
- M = flaw parameter depending on the geometry and loading conditions (Note: $M \equiv \alpha^2 . \pi$ was used in equation (8.4).)

K_I values for a number of geometries are given in Table 8.4, from which the M parameter can be calculated. In Section 8.5.1 we discussed the criterion for brittle fracture under steady or impact loads. The critical stress intensification factor for brittle fracture was denoted by K_{IC}. It can also be considered the critical factor when rapid crack growth can be expected to occur. Values of K_{IC} are generally closely related to the yield strength, or 0·2% proof strength, of a material:

$$K_{IC} = f.\sigma_y\sqrt{M.a} \qquad (8.21)$$

If K_{IC} is known, a combination of (f. σ_y) and "a" can be chosen so that $K_I \leqslant K_{IC}$ to avoid rapid fatigue crack growth. This of course implies that for a given working stress any cracks likely to exist in the component or structure must somehow be measurable. Values of K_{IC} for various materials have a considerable range. A few values are given in Table 8.7, and Fig. 8.15 is a panoramic chart.

Table 8.7. Strength and Toughness data for some Al, Ti and Fe alloys.

Material	σ_y (MN m^{-2})	K_{IC} (MN m$^{-3/2}$)	C	B_{min} (mm)
Steel Fe–9Ni–4Co–1Mo.2C	1350	160	1.10^{-12}	35
Maraging Steel Fe–18Ni–5Mo–7Co	2300	45	1.10^{-12}	0·96
Titanium Ti–6Al–6V–2.5Sn	1320	60	2.10^{-12}	5·2
Aluminium Al–2Cu–2.7Mg–6.8Zn	620	20	4.10^{-11}	2·60

Crack propagation. The rate of fatigue crack growth can be estimated from the equation

$$\frac{da}{dN} = C(\Delta K)^m \qquad (8.22)$$

where

$\dfrac{da}{dN}$ = rate of crack growth per one stress cycle

m = exponent determined experimentally; varies between 2 and 6 and is usually about 4

$$\Delta K = K_{I\max} - K_{I\min}; \quad \text{with } K_{I\min} \geq 0 \qquad (8.23)$$

(negative, i.e. compressive stresses are not considered contributing to crack growth)

C = a material constant influenced partially by σ_{mean}, the cycle frequency, the environment, etc., but is mainly a function of K_I (see Table 8.7). C for a number of materials may be estimated from Table 8.8 (Pook, 1973) using equation (8.22) above. R in the table is the ratio $\sigma_{\min}/\sigma_{\max}$, with $\sigma_{\min} \geq 0$, and ΔK values are for $da/dN = 10^{-6}$ mm/cycle.

Equation (8.22) suggests that a crack will grow, however small ΔK is. This is not so. Converting S/N fatigue plots into ΔK v. N plots shows the even clearer existence of a knee: a threshold value of ΔK_c (corresponding to σ_e), below which there appears to be infinite life, i.e. no crack growth. Some values of ΔK_c are listed in Table 8.9 (Pook, 1973). Pook and Frost (1973) discuss this

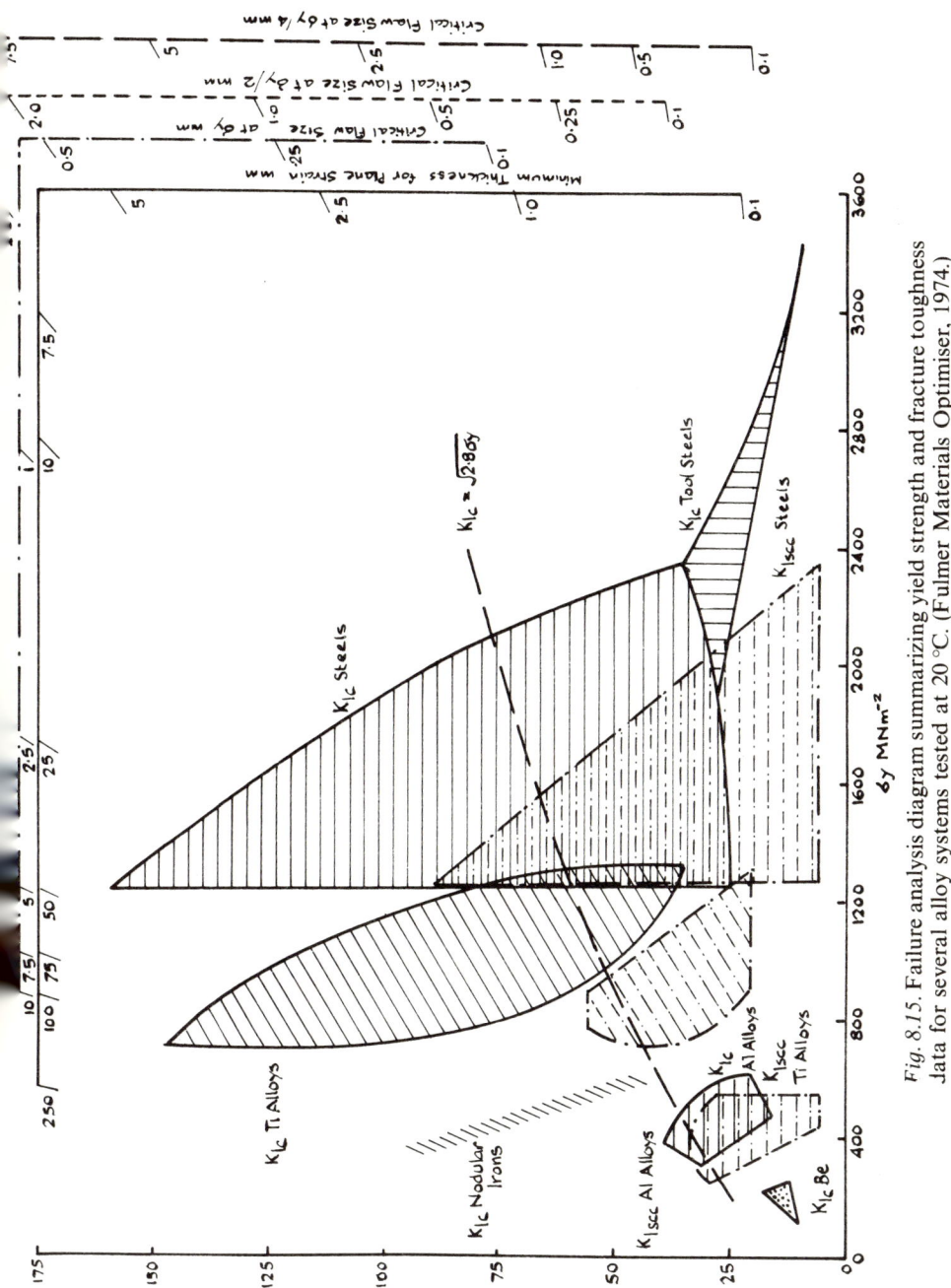

Fig. 8.15. Failure analysis diagram summarizing yield strength and fracture toughness data for several alloy systems tested at 20 °C. (Fulmer Materials Optimiser, 1974.)

Table 8.8. Fatigue crack growth data for various materials (after Pook, 1973, reprinted by permission of the Director, National Engineering Laboratory).

Material	Tensile strength (MN m^{-1})	0·1 or 0·2% proof stress (MN m^{-1})	R	m	ΔK for $da/dN = 10^{-6}$ mm/c^{-1} (MN m$^{-3/2}$)
Mild steel	325	230	0·06–0·74	3·3	6·2
Mild steel in brine[a]	435	—	0·64	3·3	6·2
Cold rolled mild steel	695	655	0·07–0·43	4·2	7·2
			0·54–0·76	5·5	6·4
			0·75–0·92	6·4	5·2
Low alloy steel[a]	680		0–0·75	3·3	5·1
Maraging steel[a]	2010		0·67	3	3·5
18/8 Austenitic steel	665	195–255	0·33–0·43	3·1	6·3
Aluminium	125–155	95–125	0·14–0·87	2·9	2·9
5% Mg-aluminium alloy	310	180	0·20–0·69	2·7	1·6
HS30W aluminium alloy (1% Mg, 1% Si, 0·7% Mn)	265	180	0·20–0·71	2·6	1·9
HS30WP aluminium alloy (1% Mg, 1% Si, 0·7% Mn)	310	245–280	0·25–0·43	3·9	2·6
			0·50–0·78	4·1	2·15
L71 aluminium alloy (4·5% Cu)	480	415	0·14–0·46	3·7	2·4
L73 aluminium alloy (4·5% Cu)	435	370	0·50–0·88	4·4	2·1
DTD 687A aluminium alloy (5·5% Zn)	540	495	0·20–0·45	3·7	1·75
			0·50–0·78	4·2	1·8
			0·82–0·94	4·8	1·45
ZW1 magnesium alloy (0·5% Zr)	250	165	0	3·35	0·94
AM503 magnesium alloy (1·6% Mn)	200	107	0·5	3·35	0·69
			0·67	3·35	0·65
			0·78	3·35	0·57
Copper	215–310	26–513	0·07–0·82	3·9	4·3

Phosphor bronze[a]	370		0·33–0·74	3·9	4·3
60/40 brass[a]	325		0–0·33	4	6·3
Titanium	555	440	0·51–0·72	3·9	4·3
5% Al titanium alloy	835	735	0·08–0·94	4·4	3·1
15% Mo titanium alloy	1160	995	0·17–0·86	3·8	3·4
			0·28–0·71	3·5	3·0
			0·81–0·94	4·4	2·75
Nickel[a]	430		0–0·71	4	8·8
Monel[a]	525		0–0·67	4	6·2
Inconel[a]	650		0–0·71	4	8·2

[a] Data of limited accuracy obtained by an indirect method.

Table 8.9. Values of ΔK_c for various materials (after Pook, 1973, reprinted by permission of the Director, National Engineering Laboratory).

Material	Tensile strength (MN m^{-2})	R	ΔK_c (MN m$^{-3/2}$)	a_c (mm)
Mild steel	430	−1	6.4	0.06
		0.13	6.6	
		0.35	5.2	
		0.49	4.3	
		0.64	3.2	
		0.75	3.8	
Mild steel in brine	430	−1	⩽1.5	
		0.64	1.15	
Mild steel in tap water	430	−1	7.3	
or SAE30 oil	835	−1	6.3	0.005
Low alloy steel	680	0	6.6	
		0.33	5.1	
		0.50	4.4	
		0.64	3.3	
		0.75	2.5	
Maraging steel	2010	0.67	2.7	
18/8 Austenitic steel	685	−1	6.0	0.012
	665	0	6.0	
		0.33	5.9	
		0.62	4.6	
		0.74	4.1	
Aluminium	77	−1	1.0	0.17
		0	1.7	
		0.33	1.4	
		0.53	1.2	
L65 Aluminium alloy	450	−1	2.1	0.007
(4.5% Cu)	495	0	2.1	
		0.33	1.7	
		0.50	1.5	
		0.67	1.2	
L65 Aluminium alloy (4.5% Cu) in brine[a]		0.50	1.15	
ZW1 Magnesium alloy	250	0	0.83	
(0.6% Zr)		0.67	0.66	
AM503 Magnesium	165	0	0.99	
alloy (1.6% Mn)		0.67	0.77	
Copper	225	−1	2.7	0.24
	215	0	2.5	
		0.33	1.8	
		0.59	1.5	
		0.69	1.4	
		0.80	1.3	
Phosphor bronze	325	−1	3.7	0.07

Table 8.9 (*cont.*)

Material	Tensile strength (MN m^{-2})	R	ΔK_c (MN m$^{-3/2}$)	a_c (mm)
	370	0·33	4·1	
		0·50	3·2	
		0·74	2·4	
60/40 brass	330	−1	3·1	0·08
	325	0	3·5	
		0·33	3·1	
		0·51	2·6	
		0·72	2·6	
Titanium	540	0·60	2·2	
Nickel	455	−1	5·9	0·25
	430	0	7·9	
		0·33	6·5	
		0·57	5·2	
		0·71	3·6	
Monel	525	−1	5·6	0·025
		0	7·0	
		0·33	6·5	
		0·50	5·2	
		0·67	3·6	
Inconel	655	−1	6·4	0·07
	650	0	7·1	
		0·57	4·7	
		0·71	4·0	

a Unpublished data.

criterion in greater depth and suggest that ΔK_c is associated with the minimum crack size of one lattice spacing which equals about $3·6 \text{ Å} = 3·6 \times 10^{-10}$ m. The critical value of ΔK_c is then derived to be

$$\Delta K_c = 1·12 \times 10^{-5} \cdot E \quad \text{MN m}^{-3/2} \tag{8.24}$$

(E = Young's modulus of elasticity in GN m^{-2} units)

Fatigue life. The fatigue life of a component or structure is primarily dependent on crack growth, i.e. its fracture toughness and very much less so on its tensile strength.

If a component has an initial crack, a_o, and the final critical crack length is a_c, how many cycles of stress is the component able to endure before rapid fracture occurs?

Equation (8.20) can be differentiated to give

$$\Delta K = \Delta \sigma \cdot \sqrt{M \cdot a}$$

and substituting this into equation (8.22) we obtain

$$\frac{da}{dN} = C\left[\Delta\sigma\cdot\sqrt{M\cdot a}\right]^m \quad (8.25)$$

where $\Delta\sigma = \sigma_{max} - \sigma_{min}$; $\sigma_{min} \geqslant 0$. Integrating between $a = a_o$ and $a = a_c$, we get

$$N = \frac{(1/a_o^{(\frac{1}{2}m-1)}) - (1/a_c^{(\frac{1}{2}m-1)})}{C\cdot(\Delta\sigma)^m\cdot M^{m/2}\cdot(\frac{1}{2}m-1)} \quad (8.26)$$

Since $a_o \ll a_c$, equation (8.26) can be simplified to

$$N = \frac{1}{a_o^{(\frac{1}{2}m-1)}\cdot C\cdot(\Delta\sigma)^m\cdot M^{m/2}\cdot(\frac{1}{2}m-1)}$$

or

$$N = \frac{C_1}{a_o^{(\frac{1}{2}m-1)}\cdot(\Delta\sigma)^m} \quad (8.27)$$

where

$$C_1 = \frac{1}{(\frac{1}{2}m-1)\cdot C\cdot M^{m/2}}$$

Critical material thickness. Experiments have shown that K_I values tend to reduce as the thickness of the test specimen increases up to a certain value, after which K_I remains constant relative to plate thickness. The limiting thickness is given by (*Fulmer Materials Optimiser*, 1974):

$$B = 2.5\left[\frac{K_{IC}}{\sigma_y}\right]^2 \quad (8.28)$$

B values are listed in Table 8.7 and they can also be obtained from Fig. 8.15.

Fracture safe design. It is clear from equation (8.20) that if rapid fatigue fracture is to be avoided the working stress levels must be chosen so that

$$\sigma_w < \frac{K_{IC}}{\sqrt{M\cdot a_c}}$$

and, as noted before, this implies careful control and inspection to ensure that no crack sizes exist which are greater than a_c.

An alternative approach, applicable to some structures such as pipes and pressure vessels, is to design for "leak before break". The thickness of the pressure vessel wall may be chosen so that leakage will occur before rupture due to critical crack size. Consider, for example, a pipe with 0·10 m radius and a working pressure of 7 MN m^{-2} ($\simeq$ 1000 p.s.i.). Assume $M = 4$. If possible we would wish to select a material so that the thickness needed to withstand the

8. MACHINE FAILURE

operating pressure would be equal or less than the critical length of crack, i.e. we would experience leakage, before sudden bursting.

Now, hoop stress $= f \cdot \sigma_y = (p \cdot R)/t$ and critical fracture toughness $K_{IC} = f \cdot \sigma_y \sqrt{M \cdot a_c}$. Thus,

$$t = \frac{p \cdot R}{f \cdot \sigma_y} \tag{i}$$

and

$$a_c = \left(\frac{K_{IC}}{f \cdot \sigma_y}\right)^2 \cdot \frac{1}{M} \qquad (f \leqslant 1) \tag{ii}$$

We want to make $t \leqslant a_c$. Combining (i) and (ii),

$$\frac{(K_{IC})^2}{\sigma_y} \geqslant f \cdot M \cdot p \cdot R \tag{iii}$$

Taking $f = 1$ (maximum value) and the values given for $M = 4$; $R = 0 \cdot 1$ and $p = 7$, we get

$$\frac{(K_{IC})^2}{\sigma_y} \geqslant 2 \cdot 8$$

or $\tag{iv}$

$$K_{IC} \geqslant \sqrt{2 \cdot 8 \cdot \sigma_y}$$

This curve is shown in Fig. 8.15. Any material which lies above the line will have adequate toughness for a working stress equal to its yield (or 0·2% proof) strength. Table 8.10 gives a number of acceptable solutions.

Steel will require the thinnest plate, but titanium will give the lightest pipe. Cost criteria could also be added to establish which material would be the cheapest to use.

Critical flaw size. Consider the case of a ship's propeller cast in bronze, which experiences fluctuating stresses of 124 ± 60 MN·m^{-2}. Normal fatigue calculation forecast infinite fatigue life. Early failures were attributed to casting flaws.

The working stress was $\sigma_w = 124 \pm 60$ MN m^{-2}

$$\left. \begin{array}{l} \therefore \sigma_{max} = 184 \text{ MN m}^{-2} \\ \sigma_{min} = 64 \text{ MN m}^{-2} \end{array} \right\} R = \tfrac{64}{184} = 0 \cdot 348 \, (\approx \tfrac{1}{3}).$$

$$\Delta \sigma = 120 \text{ MN m}^{-2}$$

On inspection of the broken propeller, flaws ranging from 0·3 to 3·0 mm in size were found at the cracked surface. What life could have been expected from such a casting?

Table 8.10. Acceptable solutions; "leak before break" (*Fulmer Materials Optimiser*, 1974).

Alloy	$(K_{IC})^2 \sigma_y$	f	σ_y	Wall thickness $t \leqslant a_c$	Mass per unit area
Steel	2·8	1	2120	0·33	2·6
Aluminium	2·8	1	463	1·51	4·2
Titanium	2·8	1	1330	0·565	2·36

From Table 8.8 for phosphor bronze:

$$\Delta K = 4·3 \quad \text{for} \quad \frac{da}{dN} = 10^{-9}\,\text{m/cycle} \quad \text{and} \quad m = 3·9 \qquad \text{(i)}$$

From Table 8.9, we have

$$\Delta K_c = 4·1 \quad \text{for} \quad R = \tfrac{1}{3} \qquad \text{(ii)}$$

From (i) we can estimate the value of C, thus:

$$\frac{da}{dN} = 10^{-9} = C(4·3)^{3·9} \therefore C = 3·38 \times 10^{-12} \quad \text{(in MN and m units) (iii)}$$

Using equation (8.27) and inserting the following numerical values,

$$C = 3·38 \times 10^{-12}; \quad m = 3·9; \quad \Delta\sigma = 120; \quad M = \pi$$

we get for $a_o = 0·30$ mm,

$$N = \frac{1}{(0·0003)^{(0·95)} \times (3·38 \cdot 10^{-12}) \times (120)^{3·9} \times \pi^{(0·95)} \times (0·95)}$$

$$\therefore N \simeq 1·82 \times 10^6 \text{ cycles of life}$$

and for $a_o = 3$ mm, similarly,

$$N = 2·0 \times 10^5 \text{ cycles of life}$$

Note: For consideration of the effects of temperature, corrosion, random loading, etc., the reader is referred to the rapidly growing specialized literature.

Damage-tolerant design. It seems appropriate to conclude this section with a practical designer's viewpoint.

Osgood (1969), accepting the reality of engineering life, says, "it is not likely that a completely fatigue resistant structure would be economically feasible, or that total protection can be obtained against damage from any source" therefore "some service failures must be expected". The failure, or survival of engineering products is consequently a reliability problem.

Fail-safe, or damage-tolerant design accepts:
(a) the probability of failures in practice,
(b) reduced capacity and/or efficiency of operation, and aims
(c) to avoid complete (dangerous or costly) collapse.

To achieve these aims, fail-safe design philosophy should include:
(1) Provision for multiple load paths (redundancy);
(2) Plastic deformation so as to re-distribute load sharing;
(3) Specifying reduced load capacity acceptable—i.e. residual strength required;
(4) Provision of crack-stoppers (reinforcements or span-wise discontinuities);
(5) Ability to inspect visually critical areas in service, or after test;
(6) Ability to repair damaged elements by simple replacements or the use of doublers or straps.

A number of design changes on aircraft are described by Osgood (1969).

8.5.3 Creep and Stress Relaxation

Creep is a time- and temperature-dependent deformation (strain) which occurs under sustained loading of a material. External loads cause change in shape (dimensions) which can lead to rupture. In internally or self-loaded members, such as tightened bolts and interference fits, creep is apparent as a stress relaxation and is thus unlikely to cause rupture.

Some materials can undergo slow continuous deformation at quite low stress levels. This phenomenon is particularly prevalent with thermo-plastics, even at low ambient temperatures. Thermo-sets and "glassy" plastics are fairly stable. Metals exhibit significant creep levels only at more elevated temperatures. As a general guide for metals, creep is not significant if the ratio,

$$\frac{\text{absolute working temperature (K)}}{\text{absolute melting temperature (K)}} < 0.48$$

Critical creep temperatures for some metals are:
Titanium, $t \geqslant 720\,°C$
Ferrous materials, $t \geqslant 628\,°C$
Aluminium, $t \geqslant 250\,°C$
Zinc, $t \geqslant 60\,°C$
Lead, $t \geqslant 20\,°C$

Conventional long-duration tests of metals at constant loads and temperatures give families of creep curves, ε v. t, as in Fig. 8.16. Three phases of creep rates can generally be distinguished:
(a) Primary, or transitional creep, where the rate of strain decreases with time,

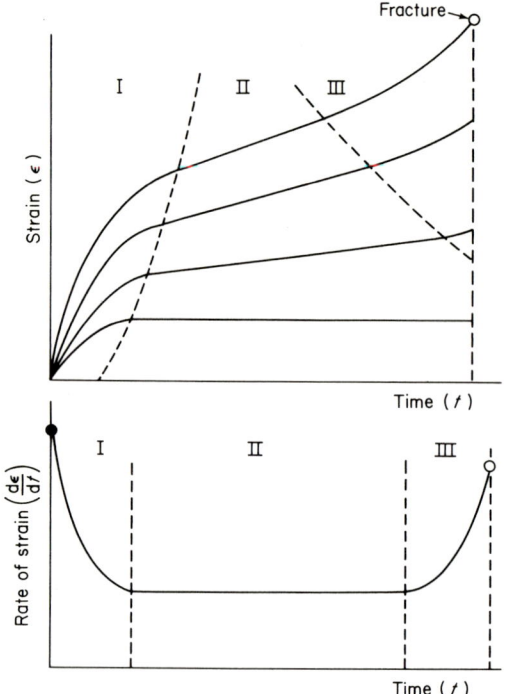

Fig. 8.16. Creep curves at constant temperature.

(b) Secondary, or steady creep, with a steady strain rate,
(c) Tertiary, or accelerated creep, where the rate of strain increases leading rapidly to rupture.

It has also been observed that creep deformation contains two elements;
  (i) reversible strain,
  (ii) non-reversible strain.

This is noted when an applied load is relieved for a time and re-applied (see Fig. 8.17). Steady non-reversible strain is present during the primary phase, represented by the dotted lines in the diagram. This feature is particularly pronounced with plastic materials. It is believed that "transitional" creep in steels is principally associated with yielding of ferrite grains, whereas in "steady" creep plastic deformation predominates at the boundaries of the grains.

Stress relaxation describes a process taking place in a body stressed to a definite constant strain (or total deformation) produced within a short time after loading. This total deformation may be partly elastic and partly plastic. In the course of time a part of the reversible (elastic) deformation can change into a non-reversible (plastic) deformation due to creep in the material.

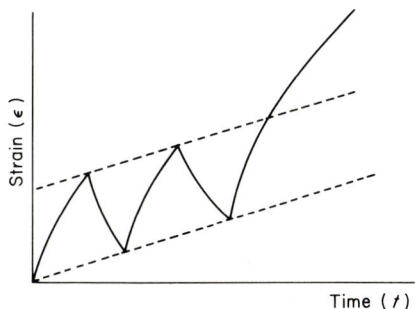

Fig. 8.17. Strain and recovery (steel).

Heating to an elevated temperature accelerates this process, which results in a relaxation of the initially induced stresses. This is, of course, what engineers in effect do when they stress-relieve materials by annealing.

Design criteria for creep

When designing a component which may experience creep due to loading, temperature and time, two criteria must be borne in mind:
  (a) dimensional stability, i.e. maximum acceptable level of deformation at a given operating temperature and in a specified period of time,
  (b) limiting stress capacity, i.e. rupture should not occur at the operating temperature before the completion of a specified period of service

Considerable creep data from extensive research and test programmes on a large range of materials are available in the literature. A most useful source is the *Fulmer Materials Optimiser* (1974). A few illustrative extracts are included in this chapter.

It can be seen from Table 8.11 that:
$a2 < a4$; i.e. 0.2% proof strength is safer to use as a criteria for the working stress,

Table 8.11. Creep data for plain carbon steels (from *Fulmer Materials Optimiser* 1974).

	Temperature (°C)	0.2% Proof str. (MN m^{-2})	UTS (MN m^{-2})	Stress at rupture		Creep strength 1% strain
				100 000 h (MN m^{-2})	10 000 h (MN m^{-2})	100 000 h (MN m^{-2})
	1	2	3	4	5	6
a	400	150	360	165	220	120
b	450	130	320	85	130	48

$a6 < a2$; i.e. 1% creep strain after 100 000 and 120 MN m^{-2} stress is less than the 0·2% proof strength,

$b2 > b4$; i.e. rupture stress is critical if component operating life required is to be greater than 100 000 h.

The implications from these figures are:
 (i) at temperatures below 400 °C the design stress should be based on the 0·2% proof strength,

Table 8.12. Comparison of Tensile and 1000 h creep rupture strengths[a] (from *Fulmer Optimiser Handbook*, 1974).

Alloy type	Stress (MN m^{-2})	Temperature (°C)								
		0	100	200	300	400	500	600	700	800
Plain carbon	0·2% PS	200	195	185	160	150	125	—	—	—
	UTS	400	360	400	420	360	280	—	—	—
	1000 h	—	—	—	—	290	110	—	—	—
Plain carbon quenched and tempered	0·2% PS	390	345	330	280	260	230	—	—	—
	UTS	550	450	490	525	465	310	—	—	—
	1000 h	—	—	—	—	—	—	—	—	—
Cr–Mo	0·2% PS	200	190	175	160	150	130	100	65	—
	UTS	430	425	420	410	370	320	230	—	—
	1000 h	—	—	—	—	—	170	60	—	—
Super 12 Cr Class I	0·2% PS	470	465	445	405	350	270	170	50	—
	UTS	565	550	530	480	415	325	225	—	—
	1000 h	—	—	—	—	—	240	70	20	—
Super 12 Cr Class II and III	0·2% PS	825	800	775	735	670	570	400	200	—
	UTS	940	925	880	810	705	575	415	150	—
II	1000 h	—	—	—	—	600	370	135	—	—
III	1000 h	—	—	—	—	—	620	270	—	—
Austenitic stainless	0·2% PS	165	125	100	85	75	72	70	70	—
	UTS	500	450	415	395	380	360	315	230	—
	1000 h	—	—	—	—	—	—	190	80	25
25 Cr–20 Ni	0·2% PS	220	205	190	175	165	145	130	105	90
	UTS	600	575	550	530	510	470	410	305	175
	1000 h	—	—	—	—	—	—	—	—	80
15 Cr–25 Ni precipitation hardening	0·2% PS	680	650	645	650	660	655	640	570	—
	UTS	1070	1010	970	955	945	915	850	740	—
	1000 h	—	—	—	—	—	—	—	160	90

[a] The values quoted for the 0·2% proof stress, ultimate tensile stress and 1000 h rupture stress are taken from the lower limits of the scatter bands in Figs II.A—5.b-1, 11.A—5.b-3 and II.A—5.c-1 respectively of the *Fulmer Materials Optimiser* (1974).

8. MACHINE FAILURE

(ii) at temperatures above 400 °C the design stress should be selected relative to creep criteria.

Tables 8.12 and 8.13 give some idea of the relative values of tensile and rupture strengths of a number of steels. Figures 8.18 and 8.19 relate rupture life, rupture stress and temperature for variously treated plain carbon steels.

Both short-term tensile strength and creep strength change with temperature, but the latter is time dependent and the former is not. This is clearly illustrated in Fig. 8.20a. By using the Larsen–Miller parameter,

$$P = T(20 + \log t) \times 10^{-3}, \qquad (8.29)$$

where

T = absolute temperature (K)
t = time in hours

common curves (Fig. 8.20b) can be obtained for
 (i) the UTS and rupture stress,
 (ii) the 0·2% proof stress and the 0·2% creep stress.

The designer will appreciate the convenience of these plots which allow direct comparison between conventional static (short-term) stress situations and the effects of creep. The *Fulmer Materials Optimiser* (1974) has further extended the convenience of using these plots for a range of materials by adding a P v. temperature/time nomogram on the graphs. A worked example is shown as Fig. 8.21 for cast iron, plain carbon steel, Cr–Mo steel and a 12% Cr steel. It will be noted that at $P < 15$, both UTS and the rupture stresses are constant, i.e. they are both independent of temperature and time, but for $P > 15$ the strength of the materials reduces rapidly. In this latter case design should, therefore, be based on creep criteria.

Tables 8.14 and 8.15 present two further sets of numerical data on creep.

8.5.4 Wear

Wear refers to undesirable changes in surface conditions caused by the relative movement between:
 two solids;
 a solid and a bulk material;
 a solid and a fluid.

The actions are essentially mechanical and the effects on the surface are changes in:
 roughness;
 hardness;
 geometry (shape and dimensions).

Table 8.13. Comparison of tensile properties, 100 000 h and 10 000 h rupture stresses and 1% creep strengths in 1 000 000 h for several types of steel (from *Fulmer Materials Optimiser*, 1974).

Alloy type	Temperature (°C)	0.2% PS (MN m^{-2})	UTS (MN m^{-2})	100 000 h (MN m^{-2})	10 000 h (MN m^{-2})	1% in 100 000 h (MN m^{-2})
Carbon steels	400	150	360	165	220	120
	450	130	320	85	130	48
½Cr–½Mo–¼V	460	196	385	290	360	265
	500	180	360	173	260	155
	550	170	320	88	145	85
Super 12 Cr Class III	500	680	710	360	560	290
10 Cr–7 Co+Ni, Mo, V, Nb, W	550	600	640	175	300	140
	600	380	440	60	100	—
Austenitic steels	550	72	335	140	215	—
	600	71	310	88	148	—
	650	70	270	45	90	—
	700	70	230	20	55	—
15 Cr–25 Ni precipitation hardening austenitic	550	650	895	410	500	—
	600	640	850	250	345	215
	650	620	800	155	220	130
	700	570	740	60	130	60

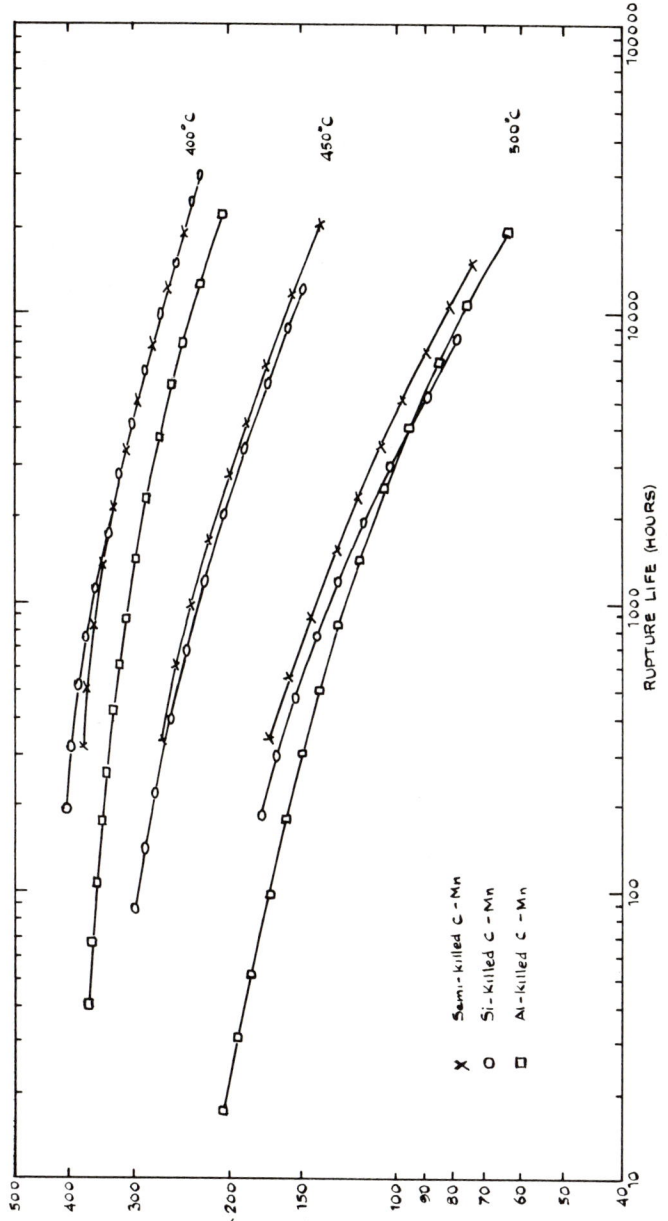

Fig. 8.18. Stress plotted against rupture life for plain carbon steels.

	%C	%Mn
Semi-killed C-Mn	0.14–0.19	1.08–1.5
Si-killed C-Mn	0.12–0.24	1.02–1.44
Al-killed C-Mn	0.13–0.2	1.02–1.54

(*Fulmer Materials Optimiser*, 1974.)

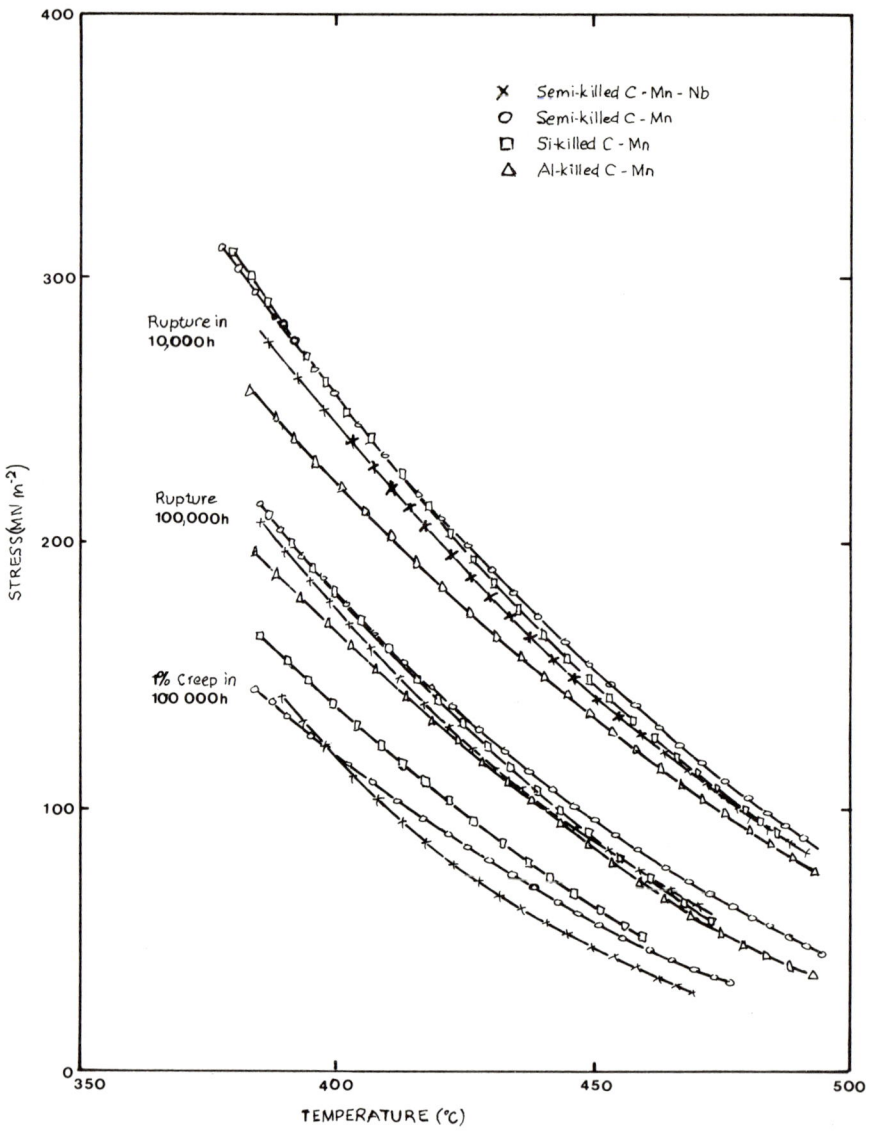

Fig. 8.19. 10 000 hours and 100 000 hours rupture stress and 1% creep strength for 100 000 hours for plain carbon steels. (*Fulmer Materials Optimiser*, 1974.)

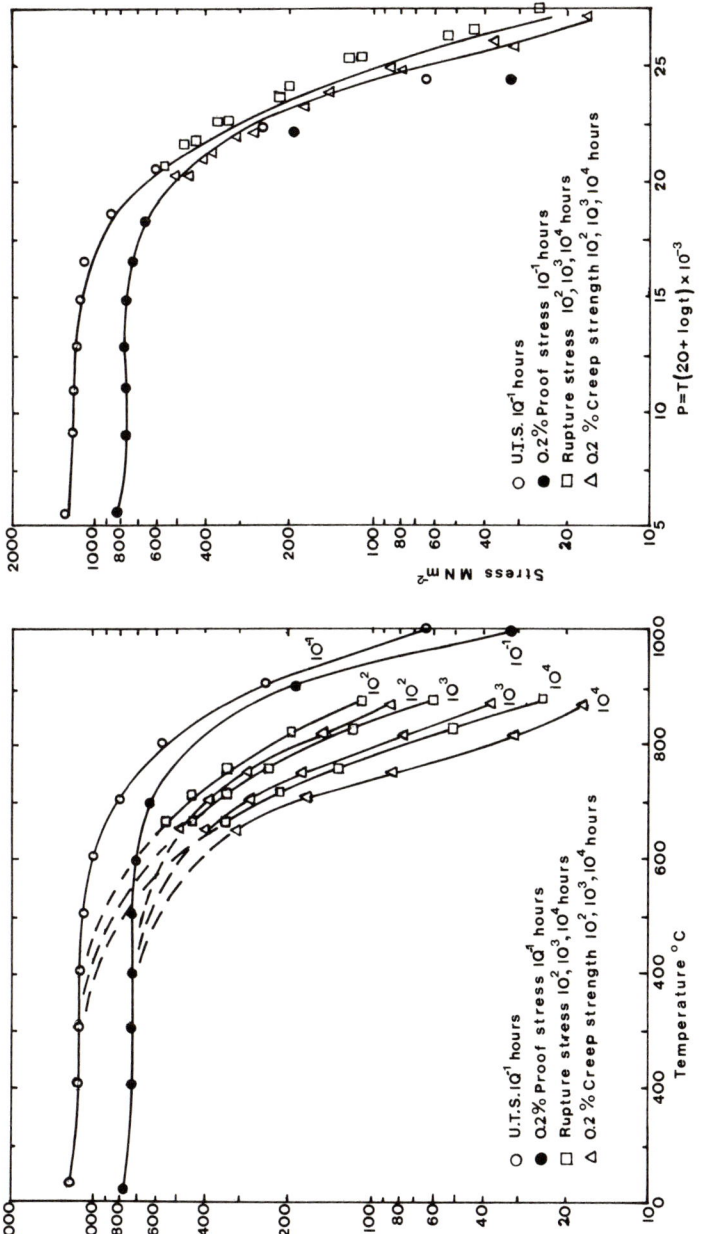

Fig. 8.20. (*Left*) Tensile and creep data for Nimonic 90 as a function of temperature. (*Right*) Same data plotted as a function of the Larson–Miller parameters, P. (*Fulmer Materials Optimiser*, 1974).

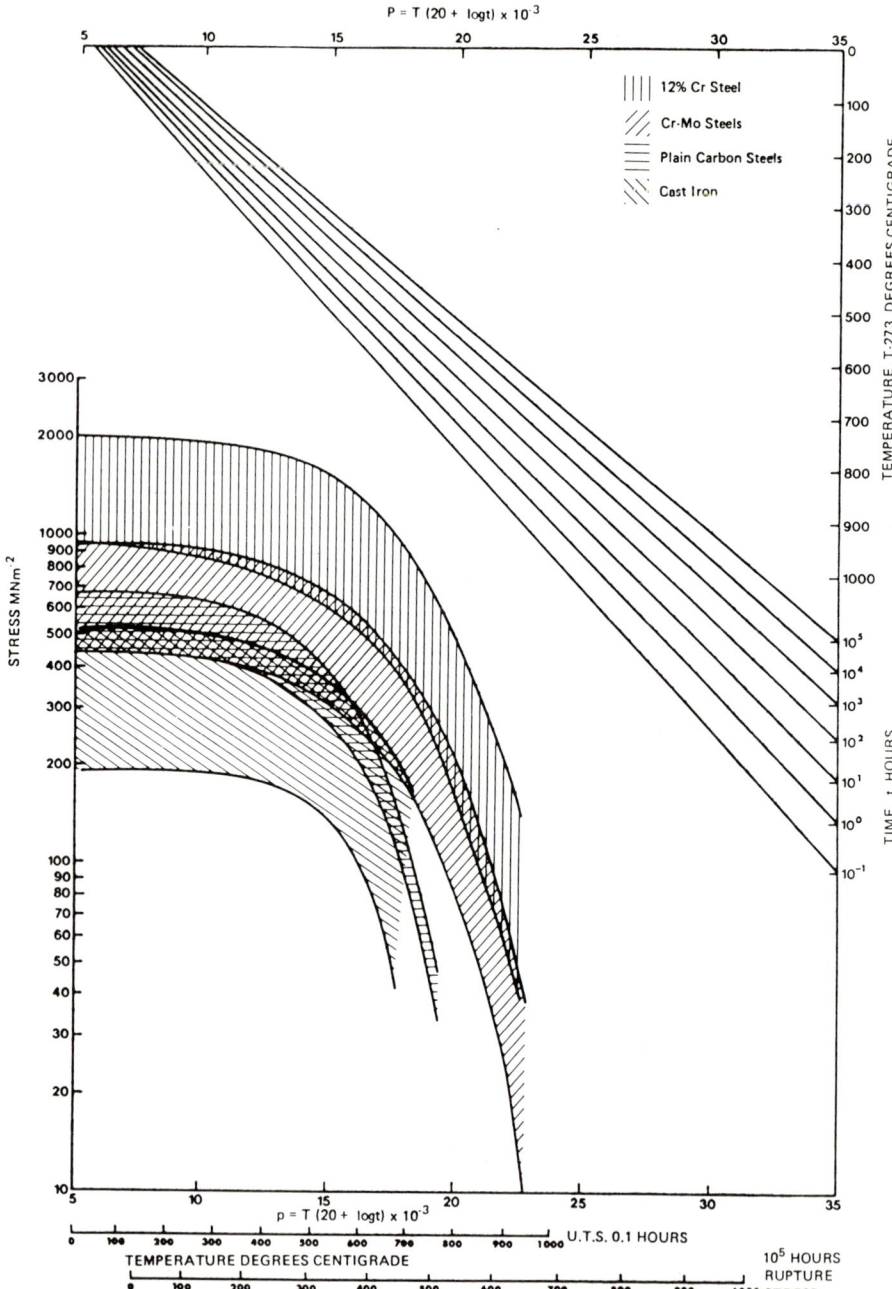

Fig. 8.21. Ultimate tensile and rupture stress for various alloys systems plotted against the Larson–Miller parameter, *P*. (*Fulmer Materials Optimiser*, 1974.)

Table 8.14. Creep properties of materials in pressure vessel design.[a]

(i) BS.1501—845 Ti steel
($\sigma_{ult} = 35$ tons in.$^{-2}$; $\sigma_{yield} = 13 \cdot 5$ tons in.$^{-2}$ at room temperature)

Rupture stress (tons in.$^{-2}$)	Rupture time (hours)	Temperature (°C)	Rate of strain ((in/in)·h^{-1})	Stress level (tons in.$^{-2}$)
13	10 000	600	10^{-7}	7·0
5	10 000	700	10^{-7}	2·5

(ii) Rupture stress at 1000 h (tons in.$^{-2}$)

Steels	450 °C	500 °C	550 °C	600 °C
1% Cr + ½% Mo	25	15	7	—
2½% Cr + 1% Mo	—	16	8	4
3% Cr + ½% Mo	—	14	6	—

(iii) Nickel-based alloys

Material	UTS (tons in.$^{-2}$)		0·2% proof stress (tons in.$^{-2}$)		Rupture stress in 1000 h (tons in.$^{-2}$)		
	20 °C	600 °C	20 °C	600 °C	600 °C	800 °C	1000 °C
Nickel	30	—	8	—	—	—	—
Monel 400	31	15·8	11	7·7	—	—	—
Inconel	38	33·5	15·5	10·0	6·5	1·5	0·8
Inconel X	72	55	41	23	—	8	1·5

[a] From Bickell and Ruiz (1967), reprinted by permission of Macmillan and Co. Ltd.

The satisfactory operation of machine element pairs—or mechanical joints—depends on the correct relationship between the surfaces with respect to all three conditions.

Mechanism of wear

In broad terms it is sufficient to classify wear mechanisms into six principal categories:

1. *Adhesive wear* is usually experienced with two surface materials which readily weld or fuse together. Under sliding conditions local pressures at surface asperities can be quite high, and instantaneous friction welds are formed and sheared. To avoid adhesive wear weld-incompatible material

Table 8.15. Typical creep rates for metals at various temperatures.[a] (Figures in tables represent stress levels 1000 lb in.$^{-2}$ producing 0·1% elongation for 1000 h application of load.)

Material	413 °C	468 °C	550 °C	612 °C	669 °C
Wrought steel 0·2 C + 0·5 Mo	26–33	18–25	9–16	2–6	1–2
Commercial iron 0·2 C + 4·6 Cr	30	10–20	7–10	1	—
0·16 C + 1·2 Cr	—	18	10–15	3	1
CI	20	8	4	—	—

	1100 °C	1200 °C	1300 °C	1400 °	1500 °C
Wrought 18–8 Cr Ni steel	10–18	5–11	3–10	2·5	2·5

Material	350 °C	400 °C	550 °C	700 °C	800 °C	1000 °C
60/40 Brass	8	2	—	—	—	—
Phosphor bronze	—	15	6	4	1	—
Nickel	—	—	—	—	20	10

Material	300 °C	500 °C	600 °C	750 °C
Duralumin	22	5	1·5	—
70 Cr + 30 Ni	—	—	28	13·18

[a] From Baumeister and Marks (1952), reprinted by permission of McGraw-Hill Book Company.

pairs should be used (see Fig. 8.22). Oxides, lubricants and polymers are especially suitable for metal surfaces.

2. *Abrasive wear* is normally the result of mechanical action by hard sharp particles which act as abrasives, scouring the surface, cutting into softer parts, and even gouging out harder crystals. Abrasive wear is the most common form experienced in machine parts.

3. *Fatigue wear.* When counter-conformal surfaces experience loading and unloading conditions (both in pure sliding and rolling), high fluctuating shear stresses are produced just below the surfaces. These fluctuating shear stresses

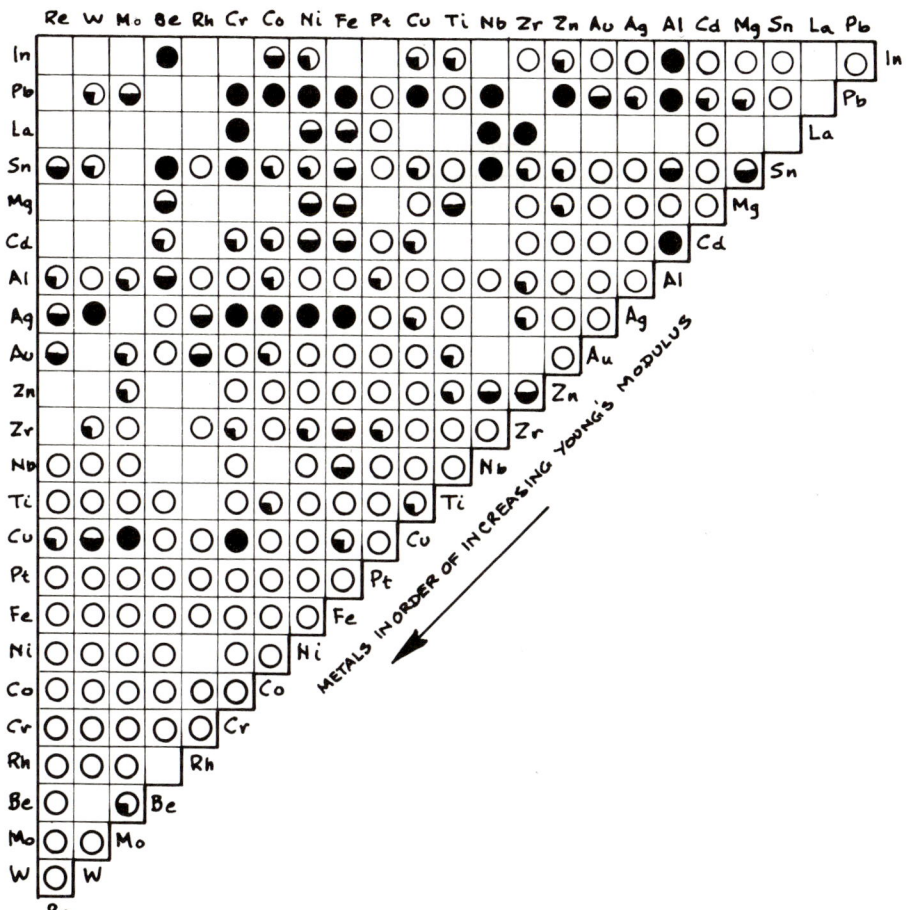

Fig. 8. 22. Compatibilities of metal pairs. The blacker the square, the less compatible—and therefore the more suitable—that combination is for sliding applications.

● Two liquid phases, solid solution less than 0·1%.
◐ Two liquid phases, solid solution less than 0·1%, or one liquid phase solid solution less than 0·1%.
◑ One liquid phase solid solution between 0·1% and 1%.
○ One liquid phase, solid solution over 1%.
(Blank boxes indicate insufficient information.)

Ti	Titanium	Cr	Chromium	Cd	Cadmium	Re	Rhenium
Cu	Copper	In	Indium	Al	Aluminium	Ag	Silver
Pt	Platinum	Pb	Lead	Rh	Rhodium	Au	Gold
Fe	Iron	La	Lanthanum	Be	Beryllium	Zn	Zinc
Ni	Nickel	Sn	Tin	Mo	Molybdenum	Zr	Zirconium
Co	Cobalt	Mg	Magnesium	W	Tungsten	Nb	Niobium

ultimately cause flaking, or pitting after stress cycles ranging between 10^5 and 10^9 fluctuations. Fatigue wear is typically experienced with:
  (i) rolling element bearings at contact pressures of 1–$4\,\text{GN m}^{-2}$;
  (ii) gears at contact pressures of $0{\cdot}6$–$2{\cdot}5\,\text{GN m}^{-2}$;
  (iii) cams and tappets at contact pressures of $0{\cdot}2$–$1{\cdot}9\,\text{GN m}^{-2}$;
  (iv) railway wheels, steel mill rollers and plain bearings.

4. *Erosive wear.* Disperse systems often contain hard, sharp particles which may impact the container surfaces. Wear rates are generally proportional to flow rates, particle concentration, hardness and sharpness. The loss in kinetic energy can cause:
  (a) deformation wear, mainly due to perpendicular impact;
  (b) cutting wear, due to flow at acute angles relative to the solid surface;
  (c) combination of (a) and (b).
Resistance to (a) is best by ductile surface materials which can absorb kinetic energy; for (b) hard surfaces with high shear strength is required.

5. *Cavitation wear.* Continuous bombardment by imploding cavities can lead to erosion of surfaces. This is often experienced on equipment handling liquids, such as pumps and pipes. Good resistance to this type of erosion wear is offered by materials which have high resilience, i.e.

$$\text{high } R = \frac{1}{2}\frac{(\sigma_{\text{yield}})^2}{E}$$

6. *Fretting wear.* A number of stationary (but detachable) mechanical joints are made by press fitting, rivetting, bolting, keying, splines and serrations. Under conditions of vibrations such joints may experience microsliding, and at amplitudes between 10^{-1} and $10^{-2}\,\mu\text{m}$ wear will occur at the contact faces. Once started this type of wear develops rapidly as no oxidation or lubrication can enter to inhibit the deterioration. If and when the fine-powdered dry debris can escape, loss of the fit results.

Fretting wear can be reduced by the use of:
unlike materials in contact,
phosphate and sulphide surface coating on steel,
anodized coatings on aluminium,
oil impregnation,
solid MoS_2 coatings,
electro-plated coatings,
also by design changes, such as
  elimination of vibrations,
  use of interlayers (rubber, PTFE) to absorb vibrations,
  increased contact pressure to eliminate micro-movements.

Types of sliding joints

Pronikov (1973) classifies joints according to wear conditions into two principal types and four groups (see Fig. 8.23):

Type I are joints in which the displacement of one member relative to the other is unidirectionally controlled.

Type II are joints which are free to align themselves according to the shape of the worn surface.

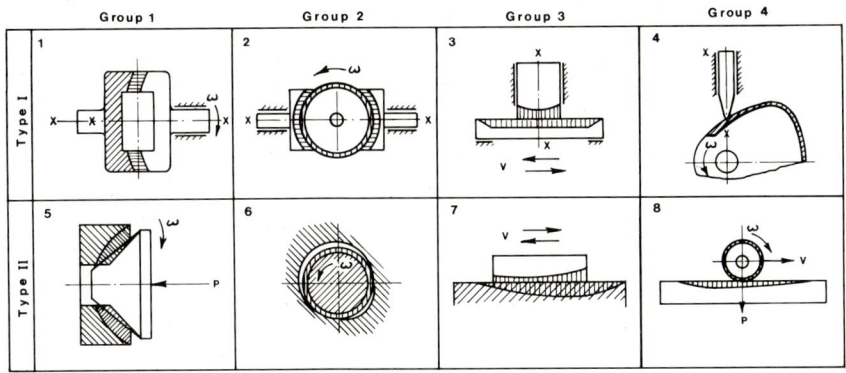

Fig. 8.23. Classification of joints according to conditions of wear.

The four groups are:

Group 1. Similar wear characteristics for both members along the displacement. Conformity of the contact surfaces is maintained. (Examples: axial clutches and thrust bearings.)

Group 2. Uniform wear conditions on the contact face of one of the members. Conformity of the contact surfaces is maintained. (Examples: drum brakes, journal bearings.)

Group 3. Joints with "lower" pairs. Initial conformity of contact surfaces is lost. (Examples: guides and crank mechanisms.)

Group 4. Joints with "higher" pairs. No conformity of contact faces. (Examples: rolling element bearings, cam mechanisms.)

Estimation of joint wear

When designing a joint, the designer will wish to make some estimate of the probable life expectancy of his mechanism.

Exact quantitative methods for calculating wear profiles or total wear volumes can hardly be expected to be available, since almost every wear situation is in some respect unique. However, attempts have been made to develop approximate correlations which involve some basic parameters and experimental data. These correlations relate primarily to abrasive wear situations.

Pronikov (1973) presents Khrushchov's law for abrasive wear:

$$U \propto (p \text{ and } S)$$

i.e.
$$U = k \cdot p \cdot S = k \cdot p \cdot (v \cdot t) \tag{8.30}$$

and
$$\gamma = \frac{U}{t} = k \cdot p \cdot v \tag{8.31}$$

where

U = total volume of wear
p = pressure intensity normal to the sliding surfaces
S = sliding distance
v = velocity of sliding
t = time of sliding
γ = time rate of wear
k = specific wear resistance for the particular materials and operating conditions

In a more generalized form, for various surfaces and lubrication conditions, equation (8.31) may be written as

$$\gamma = k \cdot p^m \cdot v^n \tag{8.32}$$

It is noted that equation (8.30) makes the total wear independent of velocity. This is in line with Coulomb's law of friction.

Equations (8.30) and (8.31) can be applied to both members of a joint, by using the appropriate k-value for each surface. Thus

$$\left. \begin{aligned} U_1 &= k_1 \cdot p \cdot S = \gamma_1 \cdot t \\ U_2 &= k_2 \cdot p \cdot S = \gamma_2 \cdot t \end{aligned} \right\} \tag{8.33}$$

or
$$\psi = \frac{U_1}{U_2} = \frac{k_1}{k_2} = \frac{\gamma_1}{\gamma_2} \tag{8.34}$$

The combined wear for the joint is thus

$$U_{\text{Total}} = U_1 + U_2 \tag{8.35}$$

If k, p, v, t are known, the above equations allow an estimation of the total wear in the joint with time, but they do not enable an assessment of the wear profiles (pattern) of the contacting faces. Wear profiles can be established by reference to the type of joint being considered. Two examples examined by Pronikov (1973) are presented here.

Example 1: Group 1, Type I class joint (see Fig. 8.24). The linear sliding velocity at the contacting surfaces will vary with the radius.

$$v = 2\pi \cdot n \cdot \rho \tag{8.36}$$

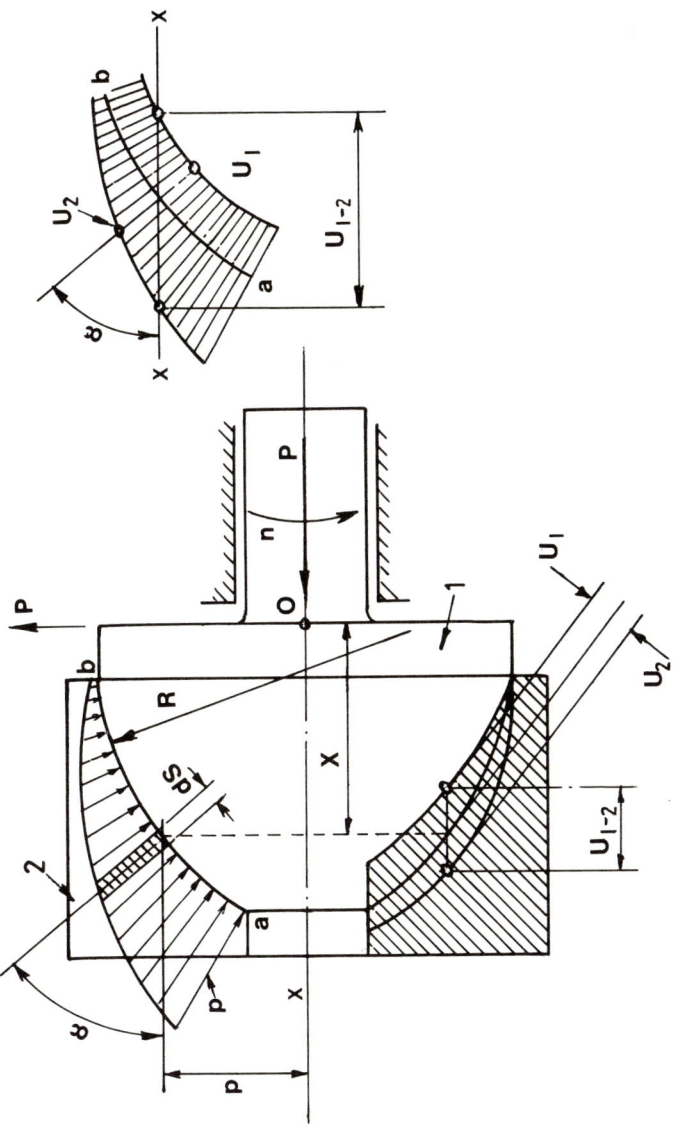

Fig. 8.24. Diagram for calculating the wear of Group 1 joints.

258 APPROXIMATE METHODS IN ENGINEERING DESIGN

For abrasive wear, at any radial position, ρ,

$$\gamma_1 = 2\pi \cdot n \cdot k_1 \cdot p \cdot \rho$$
$$\gamma_2 = 2\pi \cdot n \cdot k_2 \cdot p \cdot \rho \tag{8.37}$$

γ_1, γ_2 and p are parameters along the normal to the contacting surfaces. The combined rate of wear along the normal is

$$(\gamma_1 + \gamma_2) = 2\pi n \cdot p \cdot \rho (k_1 + k_2) \tag{8.38}$$

and the corresponding axial rate of wear is

$$\gamma_{1-2} = 2\pi n \cdot p \cdot \rho \frac{k_1 + k_2}{\cos \alpha} \tag{8.39}$$

Since the axial displacement is constant for this type of joint, i.e. γ_{1-2} = const, then

$$p = \frac{\gamma_{1-2} \cdot \cos \alpha}{2\pi n \cdot (k_1 + k_2) \cdot \rho} = C \cdot \left(\frac{\cos \alpha}{\rho}\right) \tag{8.40}$$

which indicates a non-uniform pressure distribution over the contacting faces of the sliding members. It is also noted that though conformity will be maintained, the contacting profiles will change with wear. To estimate the axial wear rate of the joint, we have to relate this to the externally applied axial force (P).

Now

$$\left. \begin{array}{c} P = \int_0^s p \cdot \cos \alpha \cdot ds \\ \\ P = \int_0^s c \cdot \dfrac{\cos^2 \alpha}{\rho} \cdot ds \end{array} \right\} \tag{8.41}$$

or

If the initial geometry is known it will be possible to express ds in terms of α, e.g.

$$ds = R \cdot d\alpha$$

and the limits of integration will then be $\alpha = \alpha_1$ to $\alpha = \alpha_2$.

Example 2. For the flat disc (Fig. 8.25) equation (8.41) yields

$$P = \int_r^R 2\pi \cdot p \cdot \rho \cdot d\rho$$

and substituting

$$P = \frac{\gamma_{1-2} \cdot 1}{2\pi n(k_1 + k_2) \cdot \rho}$$

we get

$$\gamma_{1-2} = \frac{(k_1 + k_2) \cdot P \cdot n}{(R - r)} \tag{8.42}$$

for a circular flat disc.

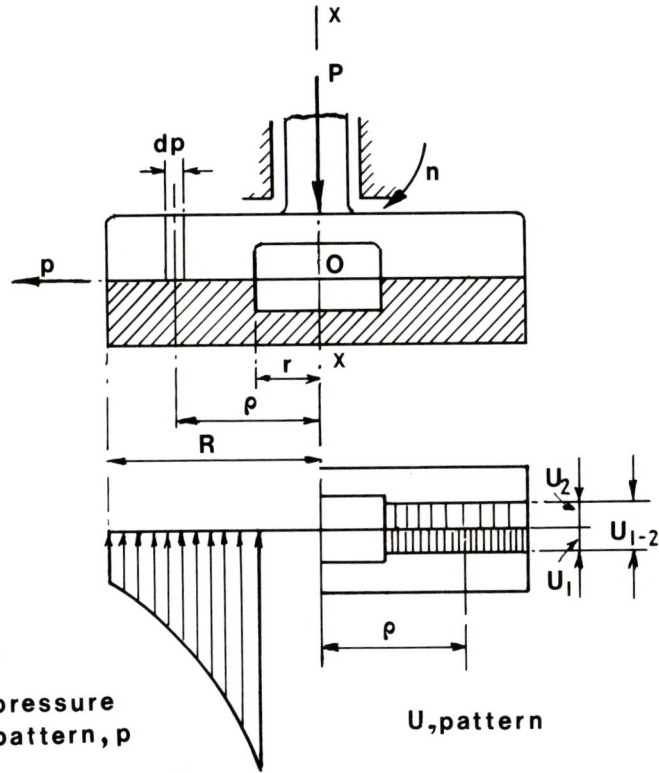

Fig. 8.25. Diagram for determining wear between plain circular discs.

The shape of the worn surfaces is obtained from equations (8.33) and (8.34). Thus,

$$\left. \begin{array}{l} U_1 = \gamma_{1-2} \cdot \cos\alpha \dfrac{\psi}{1+\psi} \cdot t \\[2mm] U_2 = \gamma_{1-2} \cdot \cos\alpha \dfrac{1}{1+\psi} \cdot t \end{array} \right\} \quad (8.43)$$

For flat discs, therefore,

$$\left. \begin{array}{l} U_1 = k_1 \cdot \dfrac{P \cdot n}{R-r} \cdot t \\[2mm] U_2 = k_2 \cdot \dfrac{P \cdot n}{R-r} \cdot t \end{array} \right\} \quad (8.44)$$

Thus, if α is constant (flat or cone), the axial wear will be constant and the members retain their initial shape. Curved clutches, however, will tend to change their shapes towards flatness. The reader is referred to Pronikov (1973) for further examples, including the more "complex analysis" of rectilinear slide-ways.

Bayer and Ku (1964), dealing primarily with small mechanisms, introduce some novel design concepts for wear. First, they distinguish between "zero" and "non-zero" wear. Zero wear they define as being a wear scar depth of not more than half the peak-to-peak value of the unworn surface finish. Secondly, they define a type A wear as one where material transfer predominates (unlubricated faces), and a type B wear as one where little or no material transfer occurs and the scars are relatively smooth. Thirdly, to assess the wear life of each member in a joint, they introduce a "unit of operation" and "number of passes" as relevant parameters.

Apart from Pronikov's group 1 joints (see Fig. 8.23) it may be seen that rubbing pairs have one "loaded" and one "unloaded" member. In the loaded member the area of contact experiences no unloading during a unit of operation, whereas the unloaded member does experience this. For example, when a shaft rotates in a journal bearing with a undirectional radial load, the shaft is the unloaded member and the bearing the loaded one. Similarly, in a A/C motor, the brush pieces are the loaded members and the slip-ring the unloaded ones:

Number of passes may be defined as

$$n = \frac{S}{W} \tag{8.45}$$

where S = sliding distance per unit of operation and W = length of surface contact in direction of sliding.

For the two examples noted above, we can see that in the case of the shaft and journal bearing, any surface element on the shaft (unloaded member) experiences one loading and one unloading per revolution, i.e.

$$n_{shaft} = 1 \text{ pass per revolution}$$

For the loaded bearing,

$$n_{brg} = \frac{\text{sliding distance}}{\text{contact length}} = \frac{\pi D}{\frac{1}{2}\pi D}$$

$$\therefore n_{brg} = 2 \text{ passes per rev.}$$

For the A/C motor,

$$n_{slip\text{-}ring} = 1 \text{ pass per rev.}$$

$$n_{brush} = \frac{\pi \cdot d_{mean}}{W_{brush}}$$

If, for example, brush width = ¼ in., and $d_{mean} = 4$, then $n_{brush} = (\pi \cdot 4)/\tfrac{1}{4} = 50\cdot 25$ passes per rev.

If the number of unit operations for a life period is L, then the total number of passes during this period is

$$N = n \cdot L \tag{8.46}$$

Calculation of zero wear. Bayer and Ku (1964) state that zero wear conditions are satisfied if the maximum shear stress generated by the contact pressure and friction

$$\tau_{max} \leqslant \gamma \cdot \tau_{yield} \tag{8.47}$$

and

$$\gamma = \left(\frac{2000}{N}\right)^{1/9} \cdot \gamma_R \tag{8.48}$$

or combined if

$$\tau_{max} \leqslant \left(\frac{2000}{N}\right)^{1/9} \cdot \gamma_R \cdot \tau_y \tag{8.49}$$

The symbols used are defined as

γ = critical fraction of the shear yield strength of the surface material for a required life of N passes

γ_R = fraction of shear yield strength for 2000 passes (see Table III.3 of Bayer and Ku, 1964)

which have been established experimentally for various combinations of materials and lubricating conditions. For combinations not listed in their book, Bayer and Ku recommend

$\gamma_R = 0\cdot 20$ for situations of high susceptibility of material transfer

$\gamma_R = 0\cdot 54$ for low material transfer

γ_R = value in Table III.3 (of Bayer and Ku, 1964) for similar surface combinations

τ_y = surface shear yield strength

The shear strength of the material only a few thousandths of an inch below the surface is not the same as that of the bulk solid material. Where this is not known, the authors have established a relationship between τ_y and H_m, the micro-hardness of a surface (see Fig. 8.26). A conversion chart of various hardness values is appended for convenience (Fig. 8.27). Layered (non-homogeneous) materials may be treated as homogeneous if the outer surface layer thickness is

(a) $t \geqslant 0\cdot 005$ in. for conforming surfaces,
(b) $t \geqslant 2 \cdot a$ for Hertzian geometries (a = half major axis of contact area),
(c) $t \geqslant 2 \cdot b$ for parallel cylinders.

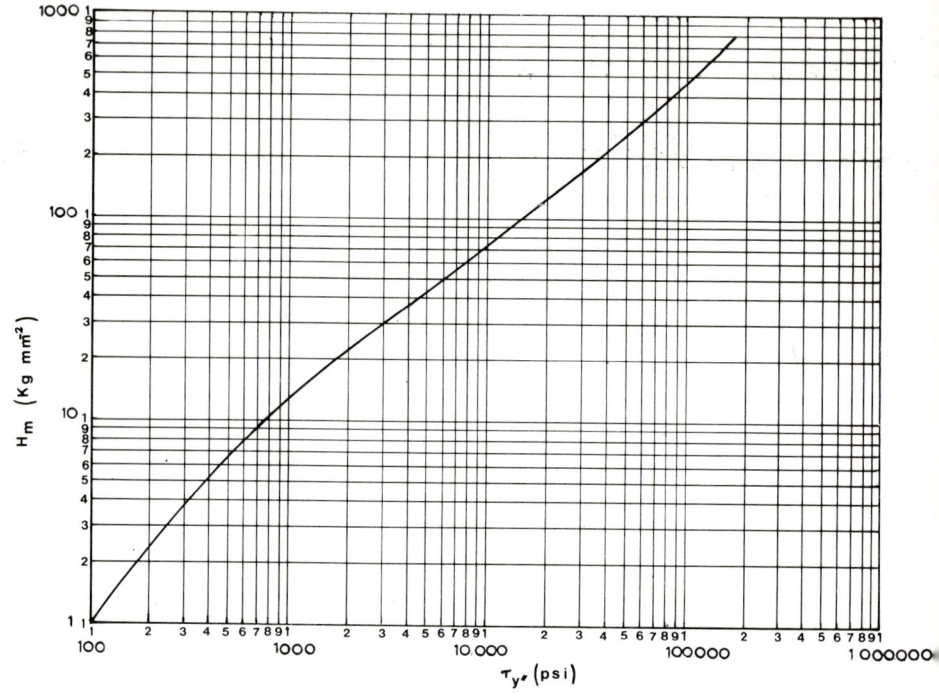

Fig. 8.26.

Fig. 8.27 (opposite). Types of tests: (1) indentation—Brinell, Rockwell, Knoop, Durometer; (2) rebound or dynamic—Scleroscope; and (3) scratch—Mohs. These 10 standards together with the materials to which they apply are as follows:

1. Brinell: ferrous and nonferrous metals, carbon and graphite
2. Rockwell B: nonferrous metals or sheet metals
3. Rockwell C: ferrous metals
4. Rockwell M: thermoplastic and thermosetting polymers
5. Rockwell R: thermoplastics and thermosetting polymers
6. Knoop: hard materials produced in thin sections or small parts
7. Vickers diamond pyramid hardness: all metals
8. Durometer: rubber and rubberlike metals
9. Scleroscope: primarily used for ferrous alloys
10. Mohs: minerals

(From Niebel and Draper (1974) "Product Design and Process Engineering" Copyright McGraw-Hill Book Co. 1974 and used with permission of McGraw-Hill Book Company.)

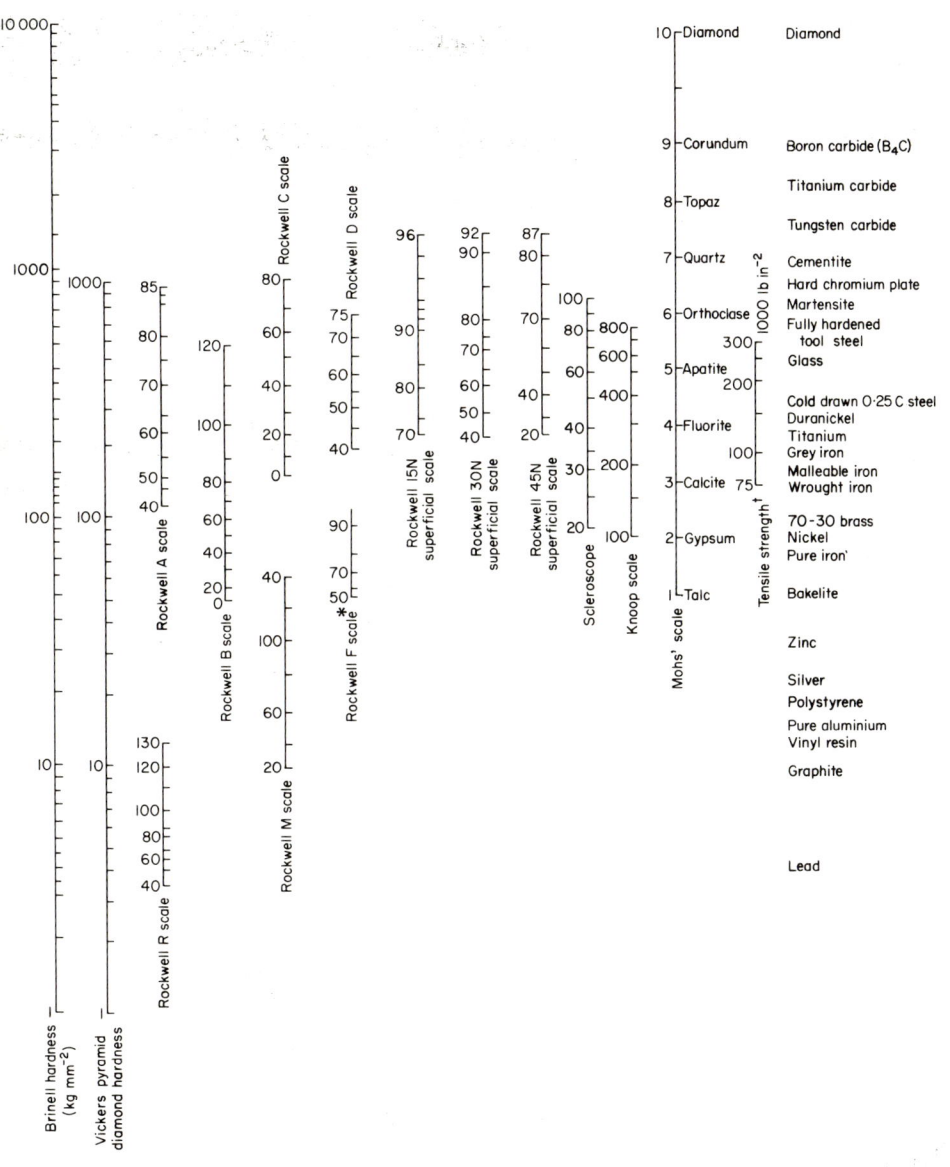

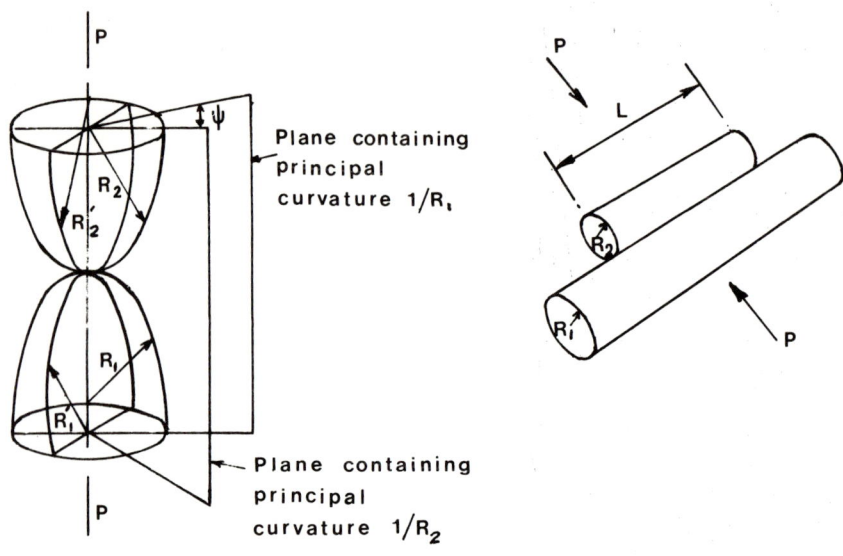

Fig. 8.28. Expression for τ_{max}.

(μ = Coefficient of friction; v = Poisson's ratio; E = Young's modulus; K = Stress concentration factor.)

If sliding is parallel to axis of cylinder

$$\tau_{max} = Kq_0 \sqrt{(\tfrac{1}{2})^2 + \mu^2}$$

If sliding is perpendicular to axis of cylinder

$$\tau_{max} = K\left(\frac{1+\mu}{2}\right)q_0$$

where

$$q_0 = \frac{2}{\pi}\frac{P}{Lb} \qquad b = 2\sqrt{\frac{P}{L}\frac{R_1 R_2 (k_1 + k_2)}{R_1 + R_2}} \qquad k = \frac{1-v^2}{\pi E}$$

Conforming Geometries

$$\tau_{max} = Kq_0 \sqrt{\tfrac{1}{2}^2 + \mu^2} \quad \text{where} \quad q_0 = P/A$$

A is the area of the contact projected onto a plane perpendicular to the line of action of P.

If sliding is parallel to a-axis, τ_{max} is given by largest of

$$\tau = q_0\left(\frac{1-2v}{3}\right)\frac{b}{a}$$

$$\tau = q_0 \sqrt{\tfrac{1}{4}\left[(1-2v)\frac{a}{a+b}\right]^2 + \mu^2}$$

$$\tau = 0{\cdot}31 q_0$$

8. MACHINE FAILURE

It is now possible to determine the critical right-hand value of equation (8.49). It is next necessary to evaluate the τ_{max} actually experienced by the working joint. Bayer and Ku (1964) have established correlations for a number of standard contact geometries and these are summarized in Fig. 8.28.

A number of examples of zero wear calculations are presented by Bayer and Ku; one of these is presented below to demonstrate the procedure.

Example. A cylindrical rod of 0·060 in. diameter oscillates (sliding axially) across a flat plate 0·10 in. wide. The two members are to be made of stainless steel; the rod of grade 440C with micro-hardness $H_m = 296$, and the plate of grade 302, $H_m = 270$. An oil grade "A" is to be used (see Table III.5 of Bayer and Ku, 1964). The stroke of the oscillations is $\frac{1}{16}$ in. under a contact load of $\frac{2}{3}$ lbf. It is required to maintain zero wear for up to 10^{10} strokes. We start with the critical equation (8.49),

$$\tau_{max} \leqslant \left(\frac{2000}{N}\right)^{1/9} \cdot \gamma_R \cdot \tau_y \qquad (i)$$

If sliding is parallel to b-axis, τ_{max} is given by largest of

$$\tau = q_0\left(\frac{1-2v}{3}\right)\frac{b}{a}$$

$$\tau = q_0 \sqrt{\frac{1}{4}\left[(1-2v)\frac{b}{(a+b)}\right]^2 + \mu^2}$$

$$\tau = 0.31 q_0$$

where

$$q_0 = \frac{3}{2}\frac{P}{\pi ab}$$

$$a = m \sqrt[3]{\frac{3\pi}{4}\frac{P(k_1+k_2)}{(B+A)}} \quad b = n \sqrt[3]{\frac{3\pi}{4}\frac{P(k_1+k_2)}{(B+A)}}$$

m and n are found from $\cos \theta$ by means of Table III-1 (Bayer and Ku, 1964).

$$\cos \theta = (B-A)/(B+A)$$

$$B-A = \frac{1}{2}\left[\left(\frac{1}{R_1}-\frac{1}{R_1'}\right)^2 + \left(\frac{1}{R_2}-\frac{1}{R_2'}\right)^2 + 2\left(\frac{1}{R_1}-\frac{1}{R_1'}\right)\left(\frac{1}{R_2}-\frac{1}{R_2'}\right)\cos 2\phi\right]^{1/2}$$

$$B+A = \frac{1}{2}\left(\frac{1}{R_1}+\frac{1}{R_1'}+\frac{1}{R_2}+\frac{1}{R_2'}\right)$$

$$k = \frac{1-v^2}{\pi E}$$

266   APPROXIMATE METHODS IN ENGINEERING DESIGN

For the particular combination of materials and lubricant specified, Bayer and Ku (1964) give $\gamma_R = 0.54$, and from Fig. 8.26 we get for

$$H_m = 296, \quad \tau_y = 630 \times 10^3 \text{ p.s.i.}$$
$$H_m = 270, \quad \tau_y = 50 \times 10^3 \text{ p.s.i.}$$

Thus for the two components we have

$$\tau_{\max \atop \text{rod}} \leqslant \left(\frac{2000}{N_{\text{rod}}}\right)^{1/9} \cdot (34 \times 10^3) \text{ p.s.i.} \tag{ii}$$

$$\tau_{\max \atop \text{plate}} \leqslant \left(\frac{2000}{N_{\text{plate}}}\right)^{1/9} \cdot (31.4 \times 10^3) \text{ p.s.i.} \tag{iii}$$

The calculations then follow a trial and error procedure because the equations to be applied depend on the contact geometry selected.

1. *Try rod on plain flat plate*

The appropriate equations in this case are those for two parallel cylinders, sliding parallel to their axes (the plate is a cylinder of infinite radius). Thus,

$$\tau_{\max} = K \cdot q_0 \cdot \sqrt{[(\tfrac{1}{2})^2 + \mu^2]} \tag{iv}$$

where, in this case,

K = stress concentration factor (for rounded edges) = 3

q_0 = Hertzian contact pressure = $\dfrac{2}{\pi} \cdot \dfrac{P'}{b}$

P' = load per unit length of contact = $\dfrac{0.6666}{0.10}$ = 6.67 lbf/in.

b = contact width = $2 \cdot \left[\dfrac{P \cdot R_1 \cdot R_2 (k_1 + k_2)}{L \cdot (R_1 + R_2)}\right]^{1/2}$

$R_1 = 0.030$

$R_2 = \infty$

$\dfrac{P}{L} = 6.67$

$k_1 = k_2 = \dfrac{1 - v^2}{\pi E} = \dfrac{1 - (0.30)^2}{\pi \cdot (27 \cdot 10^6)} = 1.073 \times 10^{-8} \text{ (p.s.i.)}^{-1}$

$\therefore b = 1.31 \times 10^{-4}$ in.

and

$q_0 = 3.24 \times 10^4$ p.s.i.

$\mu = 0.15$ (from Bayer and Ku, 1964, Table III.3)

Therefore,
$$\tau_{max} = 3(3.24 \times 10^4 \cdot \sqrt{[(\tfrac{1}{2})2+(0.15)^2]}$$
$$\tau_{max} = 5.07 \times 10^4 \text{ p.s.i.} \qquad (v)$$

Number of passes. In this mechanism the rod is the unloaded member and the plate the loaded one. But the rod is never completely unloaded since the contact length of 0·10 in. is greater than the stroke ($= 0.0625$ in.). For unit operation, therefore, we take two strokes, i.e. one complete oscillation, Thus,

$$n_{rod} = \frac{S}{W} = \frac{2 \times 0.0625}{0.1} = 1.25 \text{ passes/oscillation}$$

$$n_{plate} = 2 \text{ passes/oscillation}$$

Now the total number of passes for 10^{10} strokes is

$$N_{rod} = 1.25 \times \tfrac{1}{2} \times 10^{10} = 0.625 \times 10^{10} \text{ passes} \qquad (vi)$$
$$N_{plate} = 2 \times \tfrac{1}{2} \times 10^{10} = 1.0 \times 10^{10} \text{ passes} \qquad (vii)$$

Substituting (vi) and (vii) into (ii) and (iii) and comparing with equation (v), we get

$$\tau_{max \atop rod} = \left(\frac{2000}{0.625 \times 10^{10}}\right)^{1/9} \cdot (3.4 \times 10^4) = 0.645 \times^4$$

i.e. 5.07×10^4 p.s.i. $\not\leqslant 0.645 \times 10^4$

and

$$\tau_{max \atop plate} = \left(\frac{2000}{10^{10}}\right)^{1/9} \cdot (3.14 \times 10^4) = 0.565 \times 10^4$$

i.e. $5.07 \times 10^4 \not\leqslant 0.565 \times 10^4$

The results indicate that the contact stress levels are too high for both members, and neither could sustain 10^{10} strokes with zero wear.

2. Try a flat plate with a conforming groove

The groove must clearly be $\geqslant 0.030$ in. radius. The sensible thing is to make the groove radius $=$ the cylinder radius 0·030 in. But how wide should the groove be? Let groove width $= x \cdot (\text{diameter of rod}) = (0.06) \cdot x$, clearly $x \leqslant 1$.

Using the equation for conforming diameters (Fig. 8.28) we have

$$\tau_{max} = K \cdot q_0 \sqrt{(\tfrac{1}{2})^2 + (\mu)^2} \qquad (viii)$$

$K =$ stress concentration factor $= 3$ as before

$$q_0 = \frac{P}{A} = \frac{\tfrac{2}{3}}{0.01 \cdot x \cdot 0.06} = \frac{100}{0.9 \cdot x} \quad \text{or} \quad q_0 = 111.11\left(\frac{1}{x}\right)$$

$$\therefore \tau_{max} = 3 \cdot \frac{1}{x} \cdot 111.11 \cdot \sqrt{[(\tfrac{1}{2})^2 + (0.15)^2]}$$

i.e.

$$\tau_{max} = \frac{1}{x} \cdot 580 \text{ p.s.i.} \tag{ix}$$

The number of passes remain the same as for trial one, therefore,

$$\text{for rod, } 0{\cdot}645 \times 10^4 \leqslant \frac{1}{x} \cdot 580 \text{ p.s.i.}$$

$$\text{for plate, } 0{\cdot}565 \times 10^4 \leqslant \frac{1}{x} \cdot 580 \text{ p.s.i.}$$

x will have to be larger for the plate, i.e.

$$x \geqslant \frac{580}{0{\cdot}565 \times 10^4}$$

$$x \geqslant 0{\cdot}1025$$

The groove width needs to be just greater than $\frac{1}{10}$ of the (rod diameter), say $b = 0{\cdot}0062$ in., to give the mechanism a zero wear life of 10^{10} strokes.

Design procedure for non-zero wear

Bayer and Ku (1964) postulate, but do not present derivation or proof, differential equations representing the two mechanisms of wear:

(a) $dQ = C' \cdot dN$ for wear with predominance of material transfer (unlubricated surfaces),

(b) $dQ = C''(\tau_{max} \cdot W)^{9/2} \cdot dN + \frac{9}{2} \cdot \frac{Q}{(\tau_{max} \cdot W)} \cdot d(\tau_{max} \cdot W)$

for wear with little or no material transfer (lubricated surfaces).

Here,

Q = cross-section area of the wear scar perpendicular to the direction of sliding
W = length of contact in direction of sliding
N = number of passes
τ_{max} = maximum shear stress in the contact area.

The problem usually is to determine the wear progress after the conditions of zero wear have been violated. A number of examples are worked through by Bayer and Ku (1964), and interested readers are referred to their book.

It is of interest to compare the two approaches discussed: Pronikov's methods are generally used by most authors; his analysis of the wear profiles of the two contacting surfaces is a notable development of these techniques. The differential degree of wear between the surfaces is taken care of by the use of individual specific wear resistances. However, his method neglects the effects of the different rubbing frequencies each member experiences within the working period of the joint. This is neatly taken care of by Bayer and Ku (1964), who introduce the concept of "number of passes". They, on their part, dealing mainly with small-scale mechanisms, neglect to consider the wear profiles of the individual members. There is clearly a need for some cross-fertilization between the two methods.

Design data for wear
Considerable experimental data have already been accumulated for a variety of wear situations, material combinations and environmental conditions. The most comprehensive sources of design informaion are: The *Fulmer Materials Optimiser* (1974) (Vol. 1, Section c); Neale (1973); and a number of ESDU Data Sheets (Institution of Mechanical Engineers, London).

Quantitative and, in some cases only qualitative, information is given for five basic wear processes:
Abrasive wear,
Fatigue wear,
Erosive wear,
Cavitation wear,
Fretting wear.
The data can be divided into:
Selection of materials and material combinations,
Specific wear rates,
Design factors, such as maxim load capacities, max. velocities,
Coefficient of friction, etc.,
Surface treatments,
Surface coatings.
Tables 8.16 and 8.17 and Figs 8.29 to 8.34 from the *Fulmer Materials Optimiser* (1974) are included to illustrate the type of information available.

Points to note in design
(1) A low coefficient of friction is more often than not associated with high wear rates (dry lubrication).
(2) The cost of materials for longer wear life must be compared with cheaper materials and the cost of more frequent service and replacements.
(3) Heat dissipation must always be borne in mind.
(4) Dimensional changes can occur due to:
thermal effects,
moisture (or oil) absorption,
stress relaxation.
(5) Practical machining tolerances differ for polymers, rubbers, ceramics, carbons and metals.
(6) Avoid weld-compatible pairs even for lubricated surfaces.

Table 8.15. Relative costs of typical rubbing bearing materials (from *Fulmer Optimiser Handbook*, 1974).

Material	Common types	Typical trade names	Cost in bearing form relative to nylon[a]

Thermoplastics	Nylon/polyamide
	—monocast
	—extruded: plain,
	  reinforced,
	  solid lubricant
	  filled (graphite/MoS_2)
		Nylon 66
		Nylatron
		Zytel
		Plaslubes
		Maranyl
		Ertalon
		Nylastic
		Nylasint
			1

	Polyacetal: plain,
	  solid lubricant
	  filled (PTFE)
		Delrin
		Glacetal
		Railko Pv80
		Fulton
		Kemetal
			1

Metal-backed thermoplastic	Nylon on steel Polyacetal + additive – impregnated in porous bronze on steel	Nyliner Glacier DX	2–2.5
Reinforced thermosetting plastics	Fabric reinforced phenolic resins (some with solid lubricant additions) Fabric reinforced polyester resins (some with solid lubricant additions)	Tufnol Ferobestos Railko Xylok Orkot	2–3.5
PTEF impregnated metal	PTFE-lead—impregnated in porous bronze on steel	Glacier DU	3–5

Filled PTFE	Glass filled	Crane CF2	
	Mica filled	Fluon	
	Bronze–graphite filled	Fluorosint	
	Graphite filled	Teflon	
	Bronze–lead oxide filled	Glacier DQ	3–8
	Ceramic type filled	Graflon	
		Nobrac M7	
		Permaflon	
		Rulon	
		Klingerflon	
Woven PTFE fibre	PTFE-cotton weave—thermoset reinforced + graphite	Fiberglide	
		Fiberslip	
	PTFE-glass weave—thermoset reinforced	Uniflon	
		Unimesh	6–20
	PTFE-wire weave—thermoset reinforced	Fabroid	
		Fabrilube	
		Pydane	
Polyimides	Glass filled	Vespel	
	Solid lubricant filled (graphite/PTFE/MoS$_2$)	Feuralon	6–12
		Kerimid	
		Kinel	
Graphites	High carbon	Morganite carbon	
	Low carbon/electrographite with copper and lead with white metal impregnated with thermoset	Nobrac carbon	10–15
Graphite-impregnated metal	Graphite filled irons + MoS$_2$	Deva metal	12
	Graphite filled bronzes + MoS$_2$		

[a] Based on 1000 components.

Table 8.17. A general guide to selection of materials for practical types of abrasive wear (from *Fulmer Optimiser Handbook*, 1974)

Operating condition	Relation to Basic abrasive wear mechanisms	Physical description of situation	Surface finish after wear	Typical practical situations and depth wear rates in μm/h	Material properties required	Typical materials
High stress abrasion	Involves large amounts of deformation and cutting. Rubbing wear less prevalent at high stresses as abrasive particles fracture and so remain angular	High specific contact pressures resulting from abrasion at high overall loads and impact of large particles.	Considerably roughened with a combination of large gouges and/or pits.	Hammers in impact pulverizers 1–25 Shovel dipper teeth 0·1–13 Wearing blades in coarse ore scrapers 0·1–2·5 Ball mill scoop lips 0·08–0·4 Chrusher liners for crushing silicaceous ores 0·05–0·5 Chute liners handling coarse silicaceous ores 0·003–0·25	High toughness rather than hardness. Ability to work-harden.	Austenitic manganese steel. Hardfacings. Thick resilient rubber. Cast-iron—avoid large particle impact.
Medium stress abrasion	Considerable amounts of cutting wear prevails with smaller contributions made by the rubbing and deformation processes.	Medium specific contact pressures arising from physical situations which are intermediate between high and low stress abrasion. Most common form of abrasive wear.	Surfaces have a scratched and abraded appearance with some small surface pitting.	Rod and ball mill liners in silicaceous ores 0·01–0·1 Grinding balls in wet grinding silicaceous ores 0·004–0·01 Grinding balls in wet grinding raw cement slurries. 0·001–0·004 Grinding balls in dry grinding cement clinkers. 0·0001–0·0003	High hardness. Toughness less important.	Hardfacings. Hardened and/or heat-treated metals, e.g. cast iron, ceramics, quarry tiles, concretes. Cements, e.g. tungsten carbide for maximum wear resistance if cost is justified.

Low stress abrasion	Considerable amounts of rubbing wear as well as cutting wear. Wear by deformation does not occur to any appreciable degree.	Low specific contact pressures which are caused by a suitable combination of low overall loads and rounded and smooth particles, or by small abrasive grains.	Smooth polished surfaces with slightly scratched appearance. No large scale pits or gouges.	Sandblast nozzles 2·5–25 Sandslinger liners 1·3–6·4 Pump runner 0·003–0·1 vanes pumping abrasive mineral slurries. Agitator and 0·001–0·03 flotation impellers in abrasive mineral slurries. Screw type 0·001–0·005 classifier wear shoes in sand slurries	High hardness—becoming less essential at lower stresses. Toughness unimportant at low stresses. Low coefficient of friction.	Tungsten carbide for maximum wear resistance if cost justified. Ceramics, smooth metal surfaces. **Rubbers**, polyurethanes and plastics (especially at lowest stresses).

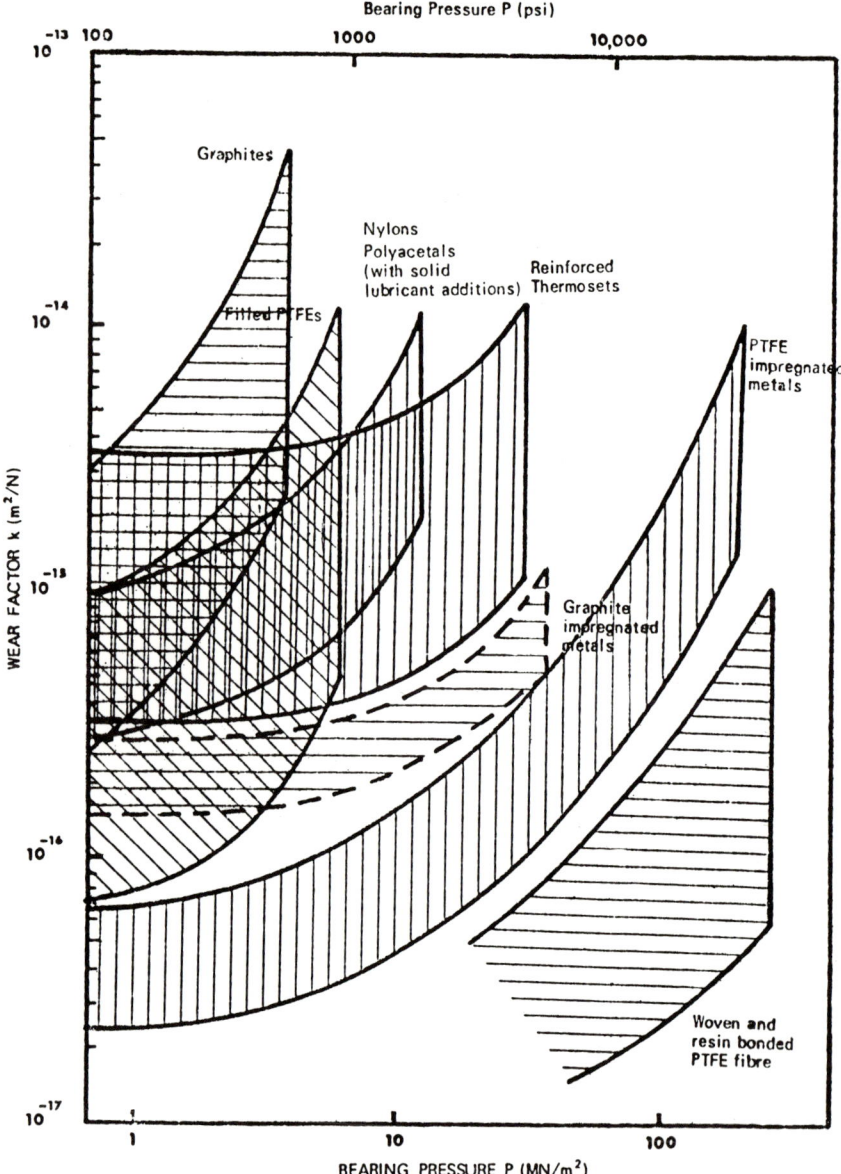

Fig. 8.29. Guide to the wear rates at room temperature of typical rubbing bearing materials at various bearing pressures against steel. (*Fulmer Materials Optimiser*, 1976.)

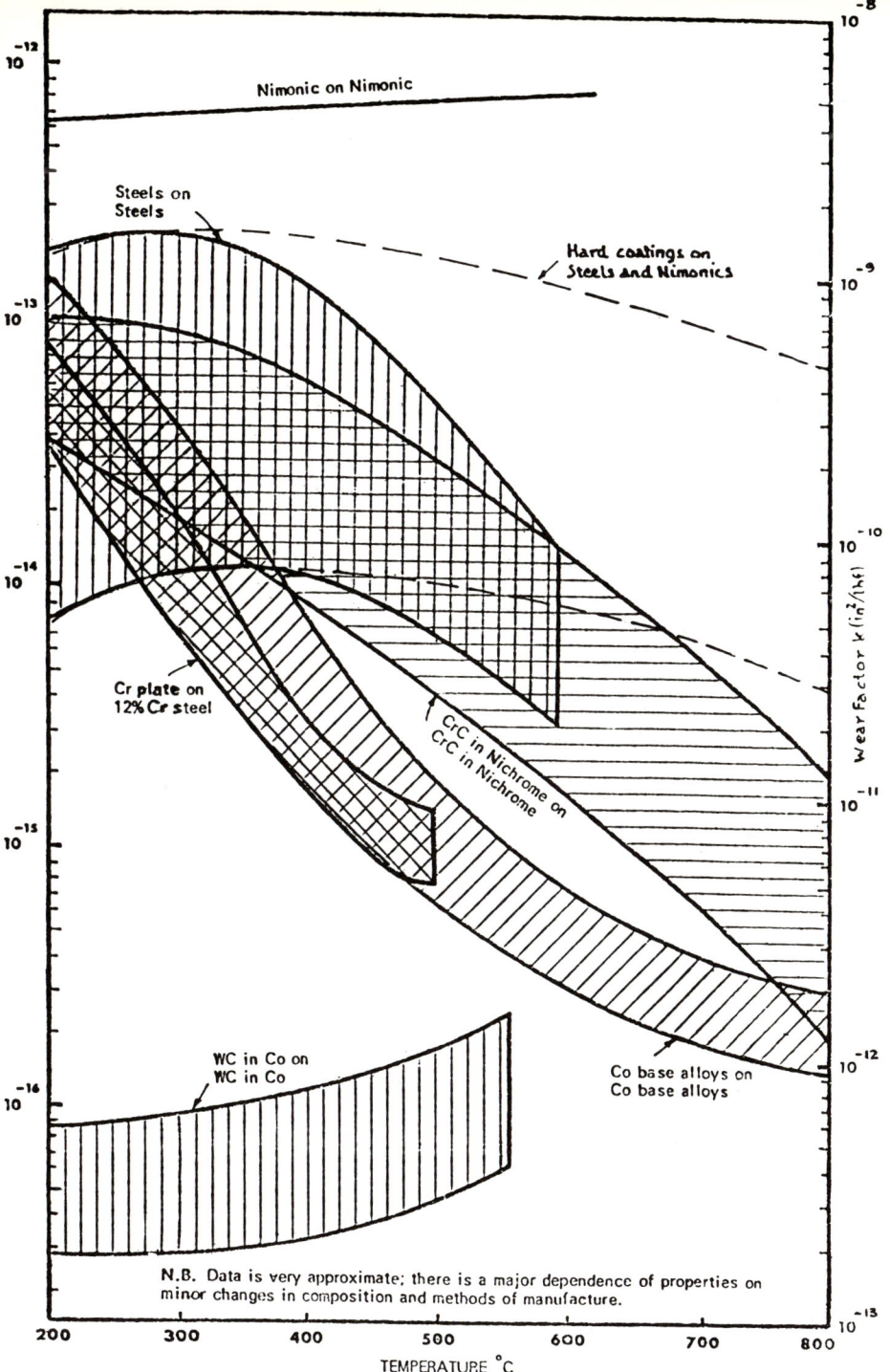

Fig. 8.30. Wear factors for hard materials. (*Fulmer Materials Optimiser*, 1976.)

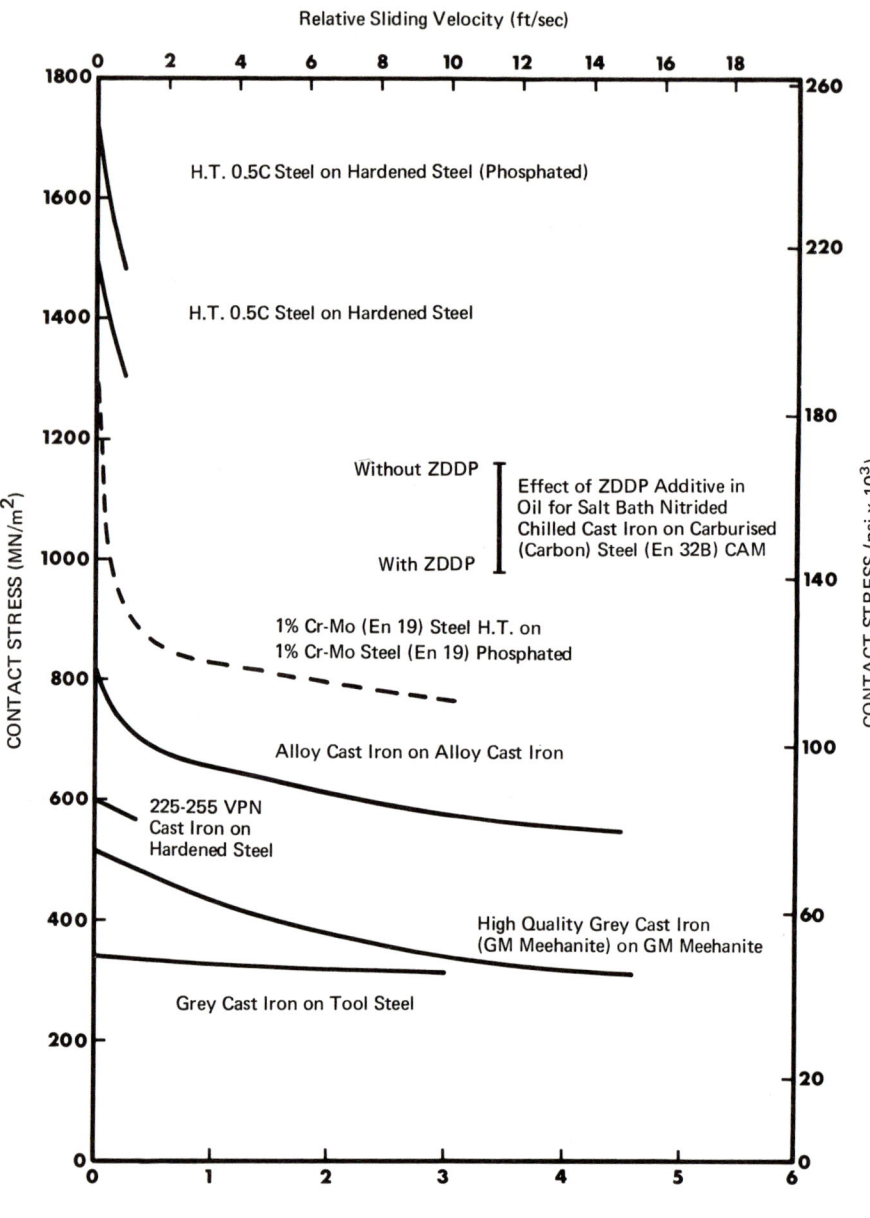

Fig. 8.31. Variation of contact stress with relative sliding velocity for 10^8 cycles pitting life (*Fulmer Materials Optimiser*, 1974).

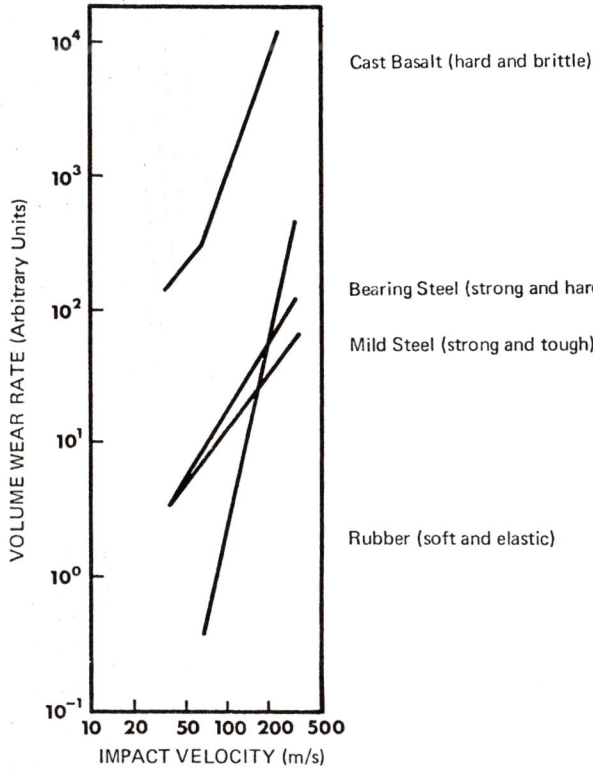

Fig. 8.32. Factors in materials selection for erosive wear resistance (*Fulmer Materials Optimiser*, 1974).

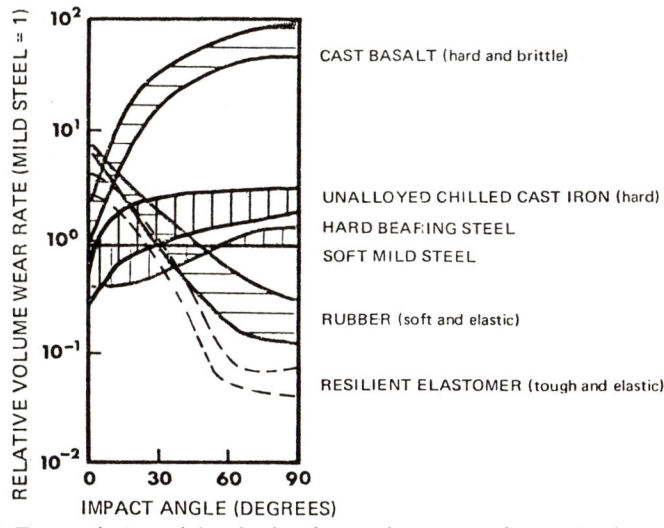

Fig. 8.33. Factors in materials selection for erosive wear resistance (*Fulmer Materials Optimiser*, 1974).

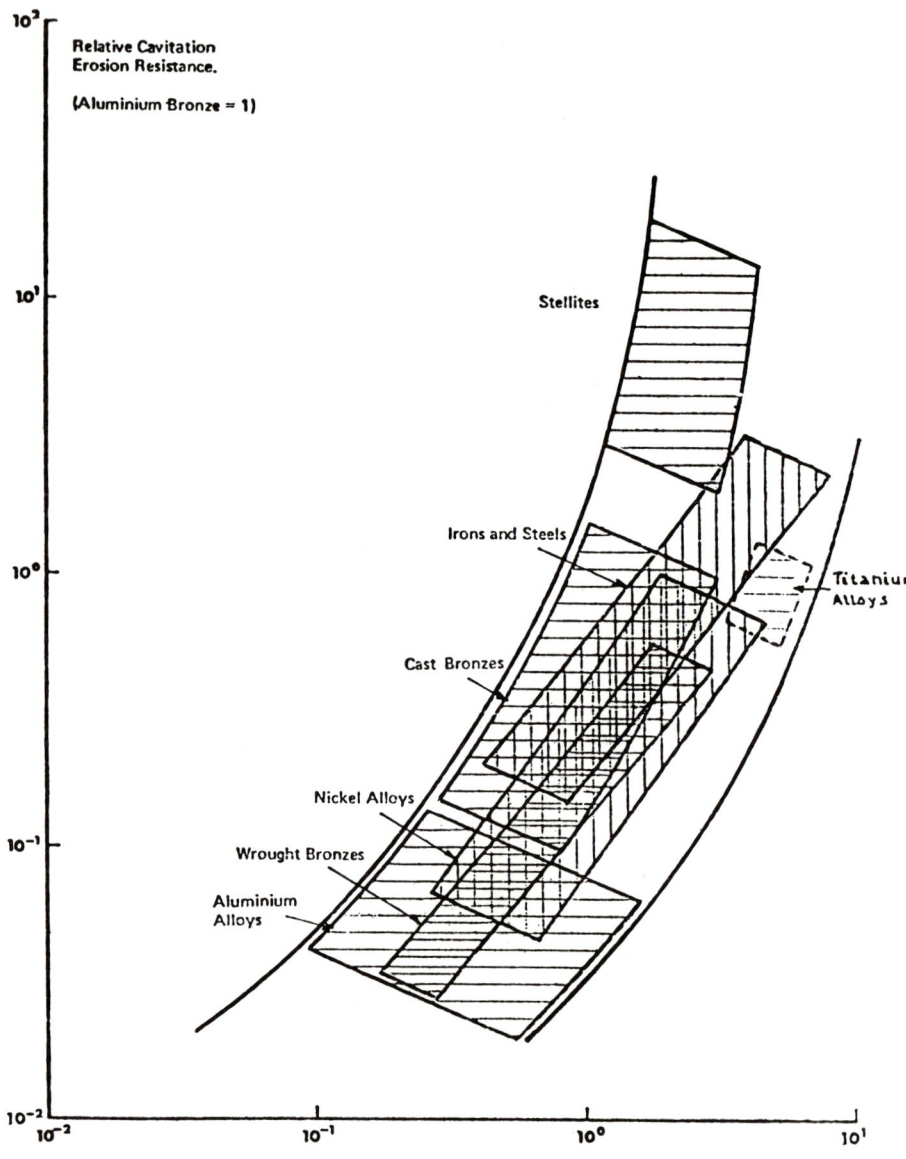

Fig. 8.34. General guide to the selection of metals for resistance to cavitation erosion (*Fulmer Materials Optimiser*, 1974).

(7) The more expensive part, or the member experiencing the greater number of "passes", should be made with the harder surface. Surface finishes for shafts are recommended to be 0·1 to 0·2 μm cla for dry journals.

(8) Wear is essentially a surface phenomenon; surface treatment should always be considered (see Table 8.18).

8.5.5 Corrosion

Corrosion of a material takes place as a result of chemical or electrochemical action in the presence of humidity. It implies a deterioration in its physical properties and/or its appearance. Some corrosion processes are quite complex and are influenced by many factors.

Several classifications of corrosive phenomena have been put forward. The *Fulmer Materials Optimiser* (1974) lists thirteen categories, but these can be condensed into seven basic types:
galvanic corrosion,
caustic embrittlement,
crevice corrosion,
corrosion fatigue,
direct chemical attack,
fretting corrosion,
stress corrosion.

(1) *Galvanic corrosion* occurs principally between two dissimilar metals and the presence of an electrolytic fluid (see Table 8.19). The further apart the metals are on the galvanic table of nobility, the stronger the galvanic action. The electrochemical reaction causes material to be removed from the anode.
"Cathodic" protection may be introduced by the use of
(a) a sacrificial material of lower anodic level (e.g. zinc to protect the mild steel of a ship's hull with a bronze propeller),
(b) a reverse electric potential to counter the galvanic current.

(2) *Caustic embrittlement* is experienced mainly with mild steel. It is a highly selective form of attack in which the material between the grains of the metal is corroded, or weakened, to the extent that the metal crystals can separate and micro-cracks are initiated. To reduce caustic embrittlement, residual stresses should be relieved and working stress levels kept well below the material yield point.

(3) *Crevice corrosion.* The difference in chemical composition between the environment inside a crevice and that outside can stimulate pitting. The simplest type is shown in Fig. 8.35a. Here the heaviest attack has occurred at the root of the crevice, where the oxygen concentration is lowest. When moisture entering the crevice contains salts (e.g. splashed from the road) an

Table 8.18. Some surface treatments available to reduce friction and wear (from *Fulmer Optimiser Handbook*, 1974).

Process	Base materials	Size effects	Advantages	Costs (relative) pence kg^{-1} (1974)
Phosphating	All ferrous metals	Adds 5 μm to surface	Resist scuffing	inexperience
Tufftriding	Cast irons, C-steels, alloy steels	Adds 5–10 μm to surface; depth up to 1 mm	Outer 10–16 μm most effective to resist wear, fatigue, corrosion	18–30
Sulfinuz	All ferrous metals, also stainless steels and Nimonics	May add 3 μm and loose 5 μm at surface. Depth up to 0·5 mm	Low coefficient of friction. Outer 4–25 μm best resist wear, fatigue, scuffing	16–30
Noskuff	Case hardening and direct hardening steels	Adds 5–10 μm to surface. Depth = 45–50 μm	Antiscuffing. Addition of N and S at surface	25
Sulf B.T.	All ferrous metals —not stainless	Removes 2–5 μm from surface. Addition of sulphur	Low temperature process; low distortion. Anti-scuffing	24–55
Tin plating; electro, or hot dip.	All ferrous metals	—	Low distortion. Aids conformability	
Silver plating electro-	Best on top of copper underlay	—	Corrosion resistance, except against oil sulphur additives	Expensive
Copper plating; electro-	All ferrous metals	—	Reduces scuffing on gears and piston rings	Inexpensive

Process	Material	Depth/Addition	Properties	Cost
Boriding	Low and medium C steels. Low alloy steels	Adds ≃ 20 μm to surface	1800–2200 VH. Abrasive resistance	Expensive
Carburizing (case hardening)	Low and medium C steels, alloy steels and Nimonics	Some growth; case depth up to 1·5 mm	Thermal distortion. Resist abrasive wear and shocks. Poor adhesive wear resistance	10–15
Carbonitriding	Low to medium C steels. Alloy steels and Nimonics	Depth 0·075–0·5 mm	Lower distortion. Resist scuffing	10–15
Nitriding NH$_4$ salt bath. Ion	Special steels	Adds up to 20 μm to surface. Depth up to 0·74 mm	VH up to 1250. Low distortion. No after grinding	Expensive
Flame and induction hardening	Medium alloy, steels, cast irons	Depth 0·25–3·2 mm	VH up to 740. Gear teeth, cams, tappets, etc.	50 per m^2

Table 8.19. Galvanic series of metals.

CORRODED END
(*anodic, or least noble*)

Magnesium
Magnesium alloys

Zinc

Aluminium 1100

Cadmium

Aluminium 2017

Steel or iron
Cast iron

Chromium–iron (active)

Ni-resist irons

18–8 Chromium–nickel–iron (active)
18–8–3 Cr–Ni–Mo–Fe (active)

Lead–tin solders
Lead
Tin

Nickel (active)
Inconel (active)
Hastelloy (active)

Brasses
Copper
Bronzes
Copper–nickel alloys
Monel

Silver solder

Nickel (passive)
Inconel (passive)

Chromium–iron (passive)
Titanium
18–8 Chromium–nickel–iron (passive)
18–8–3 Cr–Ni–Mo–Fe (passive)

Table 8.19. (*cont.*)

Silver

Graphite
Gold
Platinum

PROTECTED END
(*cathodic, or more noble*)

---

electrolyte is effectively formed. In this case the main attack is at the upper edge of the crevice where the salt concentration is highest (Fig. 8.35*b*).

(4) *Corrosion fatigue*. It is well known that the fatigue strength of materials is drastically reduced in the presence of corrosive action. The mechanism is associated with breaking down of oxide films and the exposure of virgin

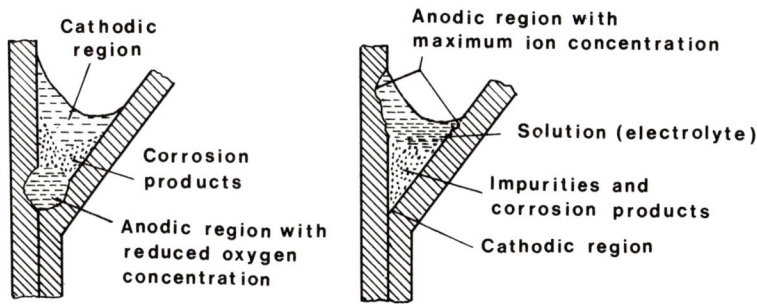

Fig. 8.35. Crevice corrosion. (*Left*) Corrosion in presence of pure condensate. (*Right*) Corrosion in presence of ion concentration.

surfaces during crack propagation to corrosive attack which decreases the fracture toughness at the root of the crack. The combined effects of fatigue stress and corrosion is denoted by the damage ratio =

$$\frac{\text{(fatigue + corrosion) strength}}{\text{normal fatigue strength}}$$

Experimental results of tests in air and in salt water have shown that
$DR = 0.2$ for plain carbon steels
$DR = 0.5$ for stainless steels
$DR = 0.4$ for aluminium alloys
$DR = 1.0$ for copper

Evans (1975) makes the point that "most fatigue cracking observed in ordinary air is really corrosion fatigue—doubtless accelerated by atmospheric contamination". Table 8.20 gives the effects of corrosion on the fatigue endurance limits of some sixteen materials tested in air, fresh water and salt water respectively.

Table 8.20. Effect of corrosion on fatigue-strength (after Evans, 1975).

Material	Ultimate tensile strength (tons in^{-2})	Endurance limit (tons in^{-2}) approx. 5×10^7 cycles		
		Air	Fresh water[a]	Salt water[b]
0·16% Carbon steel (hardened and tempered)	29·3	16·0	8·9	4·0
0·24% Carbon steel	24·8	10·3	7·6	—
Copper steel (0·98 Cu, 0·14 C)	27·5	14·3	8·9	3·5
1·09% Carbon steel	46·2	17·85	9·4	—
Ni steel (3·7 Ni, 0·26 Cr, 0·28 C)	40·5	22·0	10·1	7·1
Cr-V steel (0·88 Cr, 0·14 V)	67·2	30·0	8·0	—
Ni-Cr steel (1·5 Ni, 0·73 Cr, 0·28 C)	62·0	30·3	7·2	6·2
Stainless iron (12·9 Cr, 0·11 C)	40·0	24·5	17·0	13·0
Stainless steel (14·5 Cr, 0·23 Ni, 0·38 C)	42·0	23·2	16·0	16·0
Ni-Si Steel (3·1 Ni, 1·6 Si, 0·5 C)	112·0	48·7	7·5	—
Monel metal (fully annealed)	36·5	16·0	11·6	12·5
Pure nickel	34·0	14·8	10·4	—
Nickel (cold-rolled)	58·7	22·3	13·0	10·7
Duralumin	31·0	7·8	4·5	3·6
Aluminium bronze (7·5 Al)	40·2	14·5	11·2	9·8
Pure copper (annealed)	13·6	4·2	4·5	—

[a] Well water containing calcium carbonate.
[b] River water with a salinity about one-sixth that of sea water.
Speed of tests = 1450 cycles per min.

(5) *Direct chemical attack*. Most metals are attacked by strong acids, natural rubber is dissolved by oil, plastics are softened by plasticizers, and even ceramics can be oxidized at high temperatures or degraded by some liquid metals and strong alkaline solutions. It is well known by jewellers that silver is embrittled by soft-solder.

The problem is simply to ascertain which material is suitable for contact with the substance(s) to be handled at the operating temperature and environmental circumstances. The *Fulmer Materials Optimiser* (1974) gives extensive coverage on materials in hostile environments, and the type of degradation they experience. This can greatly assist the designer in the selection of materials for known duties. The information includes: corrosion susceptibility of metallic materials, corrosion degradation of ceramics and degradation of plastics (*Fulmer Materials Optimiser*, 1974, Section 1B). Another useful review of materials for chemical plants was published in *The Journal of Chemical Processing* (1962).

(6) *Fretting corrosion* is experienced with vibrating equipment at the interfaces of close fits and highly loaded stationary contacting surfaces. The mechanism is micro-slippage at the contact points of surface asperities which cause shear and exposure of virgin surfaces at relatively high localized temperatures under which conditions there is enhanced susceptibility of corrosive attack. The danger is the initiation of fine surface cracks which can develop into fatigue failure. Most metals and some non-metals such as wood and glass are susceptible to fretting corrosion. Among the most sensitive metals is stainless steel, on itself or on other metals. Brass has a fairly good resistance to this form of corrosion.

Steps that may be taken to eliminate, or reduce fretting corrosion, are:
remove vibrations, if possible;
increase surface contact pressure to *stop* micro-slippage;
increase surface hardness;
use lubricants to lower frictional heat and prevent local welding.

(7) *Stress corrosion*. Metals supporting stresses well within their normal capabilities can crack in certain environments. Examples of such degrading situations are:
Nitrates attack iron and steels; presence of aluminium increases resistance to such attack.
Chlorides tend to attack stainless steels; more so austenitic and less so ferritic.
Ammonia compounds attack stressed copper.
Mercury attacks copper.

Brasses with zinc contents of 20% to 40% are highly susceptible to stress corrosion, and so are high strength aluminium alloys, mainly due to residual stresses in thick sections.

Partial remedies possible are:
annealing during processing,
thermal stress relieving after processing,
surface work-hardening (e.g. shot-peening) to counteract tensile stresses,
nitriding, which produces hard nitrides, and because of their expansion also tends to induce compressive stresses in the surface layers.

Corrosion prevention. Since most corrosion attacks are surface phenomena it stands to reason that only the exposed or contacting surfaces need

Table 8.21. Rates of corrosion damage[a] (after Polar, 1961).

	Test conditions[b]							Average corrosion rates (inches per year)				
Corrosion mediums	(1) Industry (process)	(2) Type of test	(3) Average temperature (°F)	(4) Duration (days)	(5) Aeration	(6) Agitation		Type 304	Type 316	Type 317	"20"	Ni-0-NEL
HEXANE												
Hexane vapor, low-boiling vapor from tall oil, sulphur dioxide 0·3%, water 5%	Chemical	F	257	250	—	—		<0·0001ad	<·0001	<0·0001	—	—
HYDRAZINE												
Hydrazine, various concentrations	Research	L	—	—	—	—		OK	(*)c	(*)	(*)	(*)
HYDROBROMIC ACID												
Hydrobromic acid, various concentrations, decomposition products of ethylene dibromide	Chemical (distillation)	F	200–212	37	—	×		0·018cd	0·0035cd	—	—	—
Hydrobromic acid, hydrochloric acid, crude ethylene dibromide	Chemical (distillation)	F	168	55	—	—		0·0274bd	0·0058bd	—	—	—
HYDROCHLORIC ACID												
28%	Metal (plating)	F	124	1·2	—	—		0·18	0·067	—	—	0·036
Hydrochloric-acid fumes from tank containing hydrochloric acid 19% approximately	Metal (pickling)	F	160–180	41	××	×		0·0086c	0·0071c	—	0·0086a	0·0059c
15% (half immersed)	Water Treatment	L	R.T.	3	—	××		0·153 0·179	0·049	—	—	—

Concentration / Description	Industry	L/F	Temp	Days			Rate 1	Rate 2	Rate 3	Rate 4
10%	Research	L	150	1	—	—	—	—	—	0·00368
10%	Research	L	75	1	—	—	—	—	—	0·00166
10%	Plastic	F	75	12	—	—	—	—	0·0006ad	0·0008a
"Dilute"	Rubber	F	75	105	×	×	—	—	—	—
5%	Research	L	95	6	×	×	—	—	—	—
3·52%	Research	L	77	70	—	—	1·4	54·1	—	—
1%	Research	L	95	6	×	× ×	—	0·16	—	—
1%	Research	L	140	6	×	× ×	—	81·4	—	—
0·5%	Research	L	95	6	×	× ×	—	0·14	—	—
0·36%	Research	L	77	70	—	—	1·4	—	—	—
0·036%	Research	L	77	70	—	—	2·0	—	—	—
Water containing small amount of traces of hydrochloric acid	Metal (pickling)	F	160–180	41	—	×	0·0016	0·0007	0·0022	0·001
Moist air containing hydrogen chloride	—	—	R.T.	137	—	—	0·003cd	0·004cd	—	—
HYDROCHLORIC-ACID MIXTURES										
20% sulphuric acid 5%, nitric acid 5%	Chemical (pickling)	F	120	12	—	—	—	15·23	3·28	—
7·5%, non-ionic detergent 0·188%, amine-type inhibitor 0·125%, water remainder	Soap	F	72	90	—	—	0·003c / 0·003*c	0·0033	0·003	0·0029
6% approximately, "Polyrad" 1110a amine-type inhibitor 0·5%, some ferric chloride and cupric chloride	Petroleum	F	100	4	×	×	0·437c / 0·437*c	0·656	0·485c	0·414c
1% boric acid 0·1% chlorine <100 ppm, water	Chemical	F	77	87	× ×	× ×	—	0·0004d	—	0·0003d
River water, chlorides, various organics, intermittent chlorine (barometric condenser tailpit)	—	—	90	169	—	—	nil	nil	—	—
Kelly Lake water	Mining	F	72	35	×	× ×	<0·0001	nil	—	—
Hudson River water contaminated with organic waste from Poughkeepsie City	Gas Manufacture	F	100	60	× ×	× ×	<0·0001	0·0001	—	—

Table 8.21 (cont.)

Corrosion mediums	Test conditions[b]						Average corrosion rates (inches per year)				
	(1) Industry (process) (2) Type of test (3) Average temperature (°F) (4) Duration (days) (5) Aeration (6) Agitation (1)	(2)	(3)	(4)	(5)	(6)	Type 304	Type 316	Type 317	"20"	Nl-0-NEL
WATER, SEA (IMMERSED)											
Sea water	—	F	70–82	212	××	××	—	<0·0001	—	—	—
Sea water at Kure Beach, N.C.	—	F	R.T.	160	×	×	0·0015c 0·0011*c	<0·0001	<0·0001	—	—
Sea water at Kure Beach, N.C.	Research	F	60	185	—	—	0·0005*c <0·0001	0·0001	—	—	—
Sea water at Kure Beach, N.C.	Research	F	R.T.	483	—	××	0·0001*c	0·0002c	—	—	—
Sea water at Kure Beach, N.C.	Research	F	R.T.	1645	—	××	<0·0001*c	<0·0001c	—	—	—
Sea water at Kure Beach, N.C.	Research	F	R.T.	480	—	—	0·0002*c	—	—	—	—
Sea water at Curacao, Netherlands West Indies, chlorides 20 000 ppm, pH 6·6 (pressure end of pump casing)	Petroleum	F	79–88	122	××	××	0·0001	<0·0001	—	—	—
Sea water at Curacao, Netherlands West Indies, chlorides 20 000 ppm, pH 6·6 (suction end of pump casing)	Petroleum	F	79–88	44	××	××	0·0001ad	0·0001ad	—	—	—
Sea water at Curacao, Netherlands West Indies, chlorides 20 000 ppm, pH 6·6 (pressure end of pump casing)	Petroleum	F	79–88	12	××	××	0·0005ad	0·0004ad	—	—	—
Artificial sea water, pH 7·7–8	Research	L	86	84	××	××	<0·0001a	<0·0001	—	—	—

WATER, SEA, AERATED
(IN SPRAY OR TIDAL ZONE)

Sea water at Kure Beach, N.C. (half-tide)	Research	F	R.T.	359	x x x x	<0·0001*	<0·0001	—	—	—	—
Sea water at Kure Beach, N.C. (half-tide in basin)	Research	F	R.T.	193	x x x x	<0·0001	min	—	—	—	—

WATER, SEA, IN HARBOURS
(IMMERSED)

Sea water at Wilmington, N.C.	Research	F	R.T.	360	—	0·0007cd	<0·0001d <0·0001ad	—	—	—	—
Sea water at Duxbury, Mass	Research	F	R.T.	160	—	0·0001 0·0001*	0·0001	0·0001	—	—	—
6·5 Bé concentrated seawater brine pH 2·8, specific gravity 1·0469 (evaporation pond)	Food	F	68	105	x x x	0·003c	0·0009c	—	0·0004cd	0·0006cd	—

WATER, SEA, HEATED

Los Angeles Harbour West Basin sea water, total solids in suspension 2·5%, organic matter in solution 2·0%, chloride ion 18 800 ppm, sodium ion 10478 ppm, sulphate ion 2724 ppm, magnesium ion 1245 ppm, calcium ion 451 ppm, potassium ion 374 ppm, bicarbonate ion 165 ppm, bromine ion 12 ppm, silica 6 ppm, phosphate ion 5 ppm, borate ion 0·3 ppm, iron ion 0·2 ppm pH 7·5 (18 in. below water level)	Petroleum	F	120–140	350	—	x	0·0005cr	0·0001a	—	—	—	—

^a Information is included on the five types of corrosion-resistant alloys most commonly used in the chemical plant:
Type 304 The basic 18% Cr–8% Ni type for relatively mild corrosion resistance.
Type 316 The "18:8" type with 2·0/3·0% Mo for superior resistance to pitting and to most types of corrosion, particularly in reducing and neutral solutions.

Table 8.21 (cont.)

Type 317 The "18:8" type with 3·0/4·0% Mo, which has moderately better resistance than type 316 in some conditions, such as high concentrations of acetic anhydride and hot acetic acid.

"20" A 29·0% Ni–20%/Cr steel with copper and molybdenum developed specifically for resistance so sulphuric acid.

Ni-o-nel A 42%Ni–21·5% Cr alloy with copper and molybdenum, developed to meet more severe corrosion and stress-corrosion conditions than can be handled by the stainless steels but where nickel-base alloys are not needed. (Huntington Alloy Products Division, The International Nickel Company, Inc.)

While the corrosion data are given in terms of inches per year; this does not necessarily indicate that the specimen has corroded uniformly over the surface and at a constant rate. Crevice attack and pitting are represented by special symbols as shown in the Key below.

[b]Type of test
 F   Field or pilot-plant test
 L   Laboratory test

Temperature
 R.T. Room temperature
 B.P. Boiling

Aeration
 —   None
 ×   Slight to moderate
 × × Strong

Agitation
 —   None
 ×   Slight to moderate
 × × Rapid

Type 304 *Type 302 or Type 304 with carbon over the standard maximum
"20"   Wrought material is Carpenter 20; castings correspond to ACI CN-7M

Corrosion rates
 S   Sensitized
 L   Low-carbon grade (0·03% C max)
 C   Cast
 W   Welded

Capital letters in front of corrosion-rate figures refer to condition of the material tested

Lower-case letters after corrosion-rate figures refer to observed mode of corrosion where the attack was not uniform
 s   Slight pitting (maximum depth of pits from incipient to 0·005 in.)
 a   Slight pitting (maximum depth of pits from incipient to 0·005 in.)
 b   Moderate pitting (maximum depth of pits from 0·005 to 0·010 in.)
 c   Severe pitting (maximum depth of pits over 0·010 in.)
 d   Crevice attack (tendency to concentration-cell corrosion)
 r   Stress corrosion cracking

The compositions reported are those given by the co-operating companies and do not always total 100%.
[c](*) – Molybdenum content above 0·5% catalyses decomposition of hydrazine.

modifying. For this reason a number of surface treatments have been developed, such as:

Electro-plating,
Hot dips,
Diffusion coatings (including ion-plating),
Paint systems and organic coatings,
Sprayed coatings,
Cathodic protection,
Vitreous, or enamel surfacing.

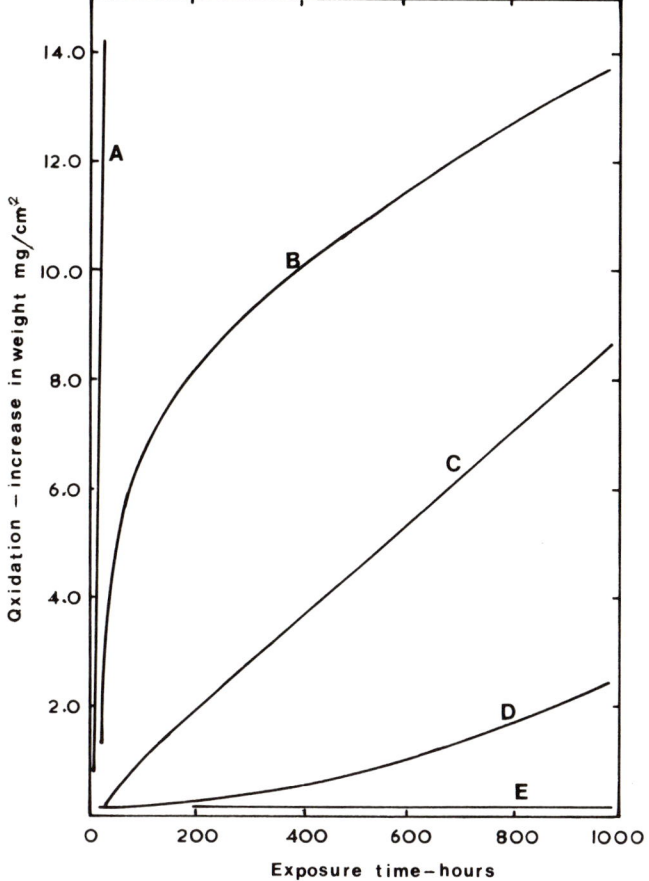

Fig. 8. 36. Oxidation of mild steel at 700 °C in air variously protected. (A) Unprotected. (B) Aluminium spray coating. (C) Commercial vitreous enamel. (D) Special vitreous enamel. (E) Chromized steel (*Fulmer Materials Optimiser*, 1974.)

The methods of applying such surface coatings and their advantages and disadvantages are well covered by the *Fulmer Materials Optimiser* (1974, Vol. I, Section IB—4).

The selection of a suitable material or surface treatment ultimately depends on availability and economics. Expensive materials and/or processes are often cheaper than the cumulative cost of repairs, replacements and shut-down losses. From a reliability point of view the situation could be quite the reverse: cheaper materials would require keener attention and better planned maintenance programmes, whereas expensive materials may induce a sense of false security. In such a case better information would be needed about the likely rates of degradation and damage, and proper monitoring would have to be provided of the progress of such deterioration. Fortunately some work has already been undertaken in this direction, and some quantitative data useful to designers have been published (Table 8.21).

Rates of corrosion damage

Polar (1961) tested five grades of materials under various environmental conditions with a large number of corrosive media. A small sample of his results are presented in Table 8.21 relating materials, test conditions and corrosive media with the corresponding "average corrosion rates" (inches per year = ipy).

At the time of writing the *Fulmer Materials Optimiser* (1974) contained only one page of quantitative data (Fig. 8.36). It gives oxidation rates of mild steel in air at 700 °C, and shows the benefits gained by the use of four different surface coatings.

REFERENCES

Baumeister, T. and Marks, L. S. (1952). "Standard Handbook for Mechanical Engineers." McGraw-Hill Book Co., New York.

Bayer, R. G. and Ku, T. C. (1964). "Handbook of Analytical Design for Wear." Plenum Press, New York.

Begley, J. A. (1973). "Fracture Mechanics in Materials Selection and Design." Proc. ASME.

Bickell, M. B. and Ruiz, C. (1967). "Pressure Vessel Design and Analysis." Macmillan, London.

Collacott, R. A. (1976). "Mechanical Failure—Diagnosis and Monitoring." CME.

Evans, U. R. (1975). "An Introduction to Metallic Corrosion." Edward Arnold, London.

Fisher, J. W. and Struick, J. H. A. (1974). "Guide to Design Criteria for Bolted and Rivetted Joints." John Wiley and Sons, Chichester.

"Fulmer Materials Optimiser." Fulmer Research Institute Ltd. (1974 and 1976).

Green, J. (1969–70). "Systematic Design Review and Fault Analysis." Paper No. 2/F/2c; *Proc. I. Mech. E.*, Vol. 184, Pt 3B.

Haugen, E. B. (1975). "Probabilistic Design." Machine Design.

Haugen, E. B. "Probabilistic Mechanical Design." John Wiley and Sons, Chichester (in press).
Kulak, C. L. (1972). "Safety Reliability of Metal Structures." A.S.C.E.
Lessels, J. M. (1954). "Strength and Resistance of Metals." John Wiley and Sons Ltd., Chichester.
Material Properties Data Centre, Traverse City, Michigan, U.S.A.
"Materials of Construction for the Chemical Industry." (December 1962.) Supplement; *Journal of Chemical Processing.*
Mechanical Properties Data Center (MPDC), Battelle, Columbus Laboratories, Columbus, Ohio, U.S.A.
Neale, M. J. (1973). "Tribology Handbook." Butterworths, London.
Niebel, B. H. and Draper, A. B. (1974). "Product Design and Process Engineering." McGraw-Hill Book Co., New York.
Osgood, C. C. (1969). Damage-tolerant design, *Journal Machine Design.* Oct. 30, 1969.
Peterson, R. E. (1965). "Stress Concentration Design Factors." John Wiley and Son Limited, Chichester.
Polar, J. P. (1961). "A Guide to Corrosion Resistance." The Climax Molybdenum Co. Ltd.
Pook, L. P. (1973). "Fracture Mechanics—How it can help Engineers." N.E.L., East Kilbride, Scotland.
Pook, L. P. and Frost, N. E. (1973). A fatigue crack growth theory, *Int. Journal of Fracture*, **9**, No. 1.
Pronikov, A. S. (1973). "Dependability and Durability of Engineering Products." (Translation by T. T. Furman.) Butterworths, London.
Richardson, R. C. D. (1967). "A Pilot Survey of Durability of Farm Machinery." Symposium, 14 September 1967, The Institute of Agricultural Engineers.
Stulen, F. B., Cummings, H. N. and Schulte, W. C. (1956). "Conference on Fatigue of Metals." I.Mech.E.
Whyte, R. R. (1975). "Engineering Progress Through Trouble." I.Mech.E.

Chapter 9

Modelling

In a lecture at the University of Manchester, on 13 May 1976, Professor Archer of the Royal College of Art spoke of the three R's—Reading, Reckoning and Wroughting—as the three essential components of our civilized life:
(1) The Humanities were expressed and communicated through *literacy* (reading and writing);
(2) The Sciences were expressed and communicated through *numeracy* (reckoning);
(3) The Arts and Crafts were expressed and communicated through *modelling* (wroughting).

The arts and crafts are now encompassed in the concept of "design", and modelling takes several forms depending on the area of activity. In engineering and architecture, modelling is essentially graphic in form. Because of the many interactions, constraints and performance requirements, the graphic form often has to be supplemented by alpha-numeric, as well as, iconic models.

The reasons for using models is usually self-evident and often an economic necessity, but it does imply, *per se*, the acceptance of a degree of approximation, i.e. a degree of unreliability. The more precisely the model can be made to represent the final artefact, the more accurate will be the forecast of the latter's true form and behaviour. On the other hand, the use of complex models or sophisticated techniques are unjustifed when the need is to estimate orders of magnitude only.

Modelling and approximations are the preoccupation of the engineering designer throughout all phases of his work:
(1) He must interpret customer's requirements into meaningful and realizable specifications.
(2) He must visualize in his mind models of possible solutions.
(3) He must examine the models using geometrical, mathematical and even iconic idealizations.
(4) He must convey all information about his model to ensure that the final product can be produced as accurately to the model as he requires.

9. MODELLING

(5) Finally he must *interpret* the feedback information from tests and field experience to enable him to modify and improve his models and the product.

The examples in this chapter illustrate mainly the type of problems that arise during the analysis stage of the design process. Not only is this one of the most important and demanding phases of engineering design, but the academic training of our young engineers often mesmerizes them with the idealized beauties of science and mathematics which seduces them away from the realities of engineering practice. Playing games with words, numbers, diagrams, graphs and computer programmes has become an end in itself. The fact that they are models (approximations) only of the real physical world has often been lost sight of. It is hoped that the examples which follow will help to mend this situation.

9.1 SHEAR STRESS DISTRIBUTION IN A THIN TUBULAR CYLINDER (Fig. 9.1)

We wish to establish a correlation which will enable the assessment of the shear stress induced by a radial load on a tubular cylinder. In order to apply conventional stress analysis which will yield a good approximation of the actual shear stress that may arise, the following simplifying assumptions are usually made:

Plane surfaces remain plane under strain and stress.

Wall thickness is small compared to tube diameter; stress across the wall thickness of the tube is assumed constant.

The tube remains circular; there is no departure from the perfect circular form in spite of localized loading.

Referring to Fig. 9.1a we may consider a longitudinal element ($dx \cdot t \cdot d\alpha$). The differential longitudinal (axial) load on this element is

$$d\sigma \cdot dA = \frac{dM \cdot y}{I} \cdot dA \tag{9.1}$$

Next consider an annular segment, AOB, symmetrical about the y-axis. The differential longitudinal (axial) load on this segment is

$$F_{AOB} = 2 \cdot \frac{dM}{I} \cdot \int_{\alpha_1}^{\pi/2} y \cdot dA \tag{9.2}$$

This force is balanced by an equal but opposite force acting as a shear force on the radial faces at A and B of the segment, i.e.

$$F_{AOB} = \tau_{yx} \cdot 2 \cdot t \cdot dx \tag{9.3}$$

APPROXIMATE METHODS IN ENGINEERING DESIGN

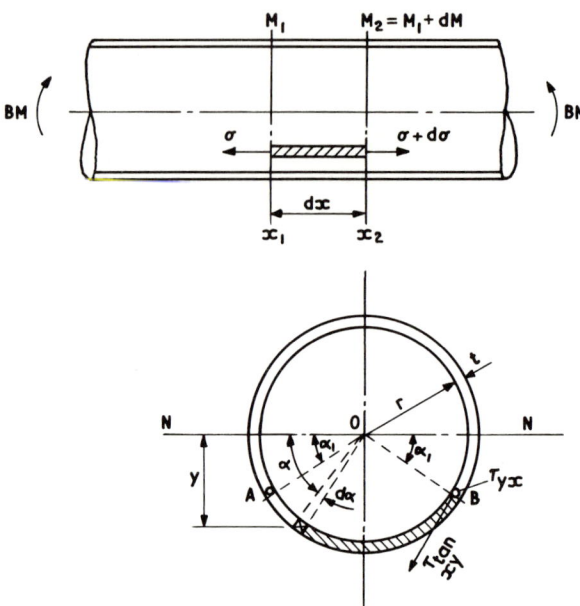

Fig. 9.1.

Equating (9.2) and (9.3) we get

$$\tau_{yx} = \left(\frac{dM}{dx}\right) \cdot \frac{1}{t \cdot I} \cdot \int_{\alpha_1}^{\pi/2} y \cdot dA \tag{9.4}$$

Now,

$$y = r_{mean} \cdot \sin\alpha$$

and

$$dA = r_{mean} \cdot d\alpha \cdot t$$

$$\therefore y \cdot dA = r_{mean}^2 \cdot t \cdot \sin\alpha \cdot d\alpha, \tag{9.5}$$

and since $dM/dx = $ SF (shear force), equation (9.4) may be written as

$$\tau_{yx} = (SF) \cdot \frac{r_{mean}^2}{I} \cdot \int_{\alpha_1}^{\pi/2} \sin\alpha \cdot d\alpha$$

$$= \frac{(SF) \cdot r_{mean}^2}{I} \cdot [(-\cos\alpha)]_{\alpha_1}^{\pi/2}$$

or

$$\tau_{yx\,axial} = \frac{(SF) \cdot r_{mean}^2}{I} \cdot \cos\alpha_1 \tag{9.6}$$

9. MODELLING

The complementary shear stress across the face of the segment must be perpendicular to axial faces of A and B, i.e. tangential to the arc at any point on the shell section. Thus,

$$\tau_{tan} = \frac{(SF) \cdot r_{mean}^2}{I} \cdot \cos \alpha_1 \qquad (9.7)$$

The maximum shear stress, therefore, occurs when $\alpha_1 = 0$ and $\tau_{tan} = 0$ when $\alpha_1 = \pi/2$, i.e.

$$\tau_{tan\,max} = \frac{(SF) \cdot r_{mean}^2}{I} \qquad (9.7a)$$

If $t \ll r$, we have, approximately,

and
$$\left.\begin{array}{c} A = 2 \cdot \pi \cdot r_{mean} \cdot t \\ \\ I = \pi \cdot r_{mean}^3 \cdot t \end{array}\right\} \qquad (9.8)$$

Inserting in equation (9.7a), we get

$$\tau_{tan\,max} = 2 \cdot \frac{(SF)}{A} = 2 \cdot \tau_{av} \qquad (9.9)$$

existing across the diameter perpendicular to the bending moment and/or the shear force.

9.2 SHEAR STRESS DISTRIBUTION ACROSS A DIAMOND-SHAPED SECTION (Fig. 9.2)

If the bending moment along a beam varies, the induced direct and shear stresses may be estimated by conventional analysis. The assumptions made, however, are

  Plane sections remain plane.
  The material is homogeneous and isotropic.
  The stress level at any point remains within the elastic limit.

The analysis proceeds then as follows: Referring to Fig. 9.2.

$$\sigma_x \cdot dA = \left(\frac{M \cdot y}{I}\right) \cdot dA, \quad \text{on face A}$$

and
$$(\sigma_x + d\sigma_x) \cdot dA = \left[\frac{M + dM}{I} \cdot y\right] \cdot dA, \quad \text{on face B}$$

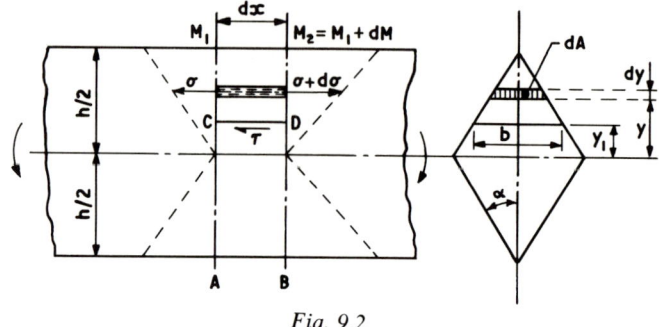

Fig. 9.2.

The net force (axial) on the element dA is

$$dF = \left(\frac{dM \cdot y}{I}\right) \cdot dA \tag{9.10}$$

The net longitudinal force acting on any area above a plane, such as C–D, is

$$F = \frac{dM}{I} \cdot \int_{y_1}^{h/2} y \cdot dA \tag{9.11}$$

For the part of the element dx above plane C–D to remain in equilibrium, the above force F is balanced by a shear force acting along face C–D. Thus,

$$\tau_{y_1} \cdot b \cdot dx = F = \frac{dM}{I} \cdot \int_{y_1}^{h/2} y \cdot dA$$

or

$$\tau_{y_1} = \frac{1}{b \cdot I} \frac{dM}{dx} \cdot \int_{y_1}^{h/2} y \cdot dA$$

Since dM/dx = shear force (SF), we have

$$\tau_{y_1} = \frac{(SF)}{b \cdot I} \cdot \int_{y_1}^{h/2} y \cdot dA \tag{9.12}$$

To evaluate equation (9.12) we need to find expressions for I and dA in terms of known parameters.

(i) $$dA = b \cdot dy$$

(NOTE: Again an approximation by treating the trapezium as a rectangle.) Now,

$$\tan \alpha = \frac{(\tfrac{1}{2}b)}{(\tfrac{1}{2}h - y)}$$

or

$$b = 2(\tfrac{1}{2}h - y) \cdot \tan \alpha \tag{9.13}$$

9. MODELLING

and

$$dA = (h-2y) \cdot \tan \alpha \cdot dy \tag{9.14}$$

(ii) $\quad I = 2 \cdot \int_0^{h/2} dA \cdot y^2 = 2 \cdot \int_0^{h/2} (h-2y) \cdot \tan \alpha \cdot dy \cdot y^2$

or

$$I = 2 \cdot \tan \alpha \cdot \int_0^{h/2} (hy^2 - 2y^3) \, dy$$

which yields

$$I = \frac{h^4}{48} \cdot \tan \alpha \tag{9.15}$$

N.B. For a square section the value of I across a diagonal is

$$I_{square} = \frac{h^4}{48} \tag{9.15a}$$

Returning to equation (9.12) and inserting (9.14) and (9.15), we obtain

$$\tau_{y_1} = \frac{(SF) \cdot \tan \alpha}{b \cdot I} \cdot \int_{y_1}^{h/2} (hy - 2y^2) \cdot dy$$

which on integration gives

$$\tau_{y_1} = \frac{(SF) \cdot \tan \alpha}{b \cdot I} \cdot \left[\frac{h^3}{24} - \frac{h \cdot y_1^2}{2} + \frac{2 \cdot y_1^3}{3} \right] \tag{9.16}$$

Inserting values for b and I from equations (9.13) and (9.15), we have

$$\tau_{y_1} = \frac{2(SF)}{h^4 \cdot \tan \alpha} \cdot [(h-2y_1)(h+4y_1)] \tag{9.17}$$

This represents the general expression for shear stress distribution for any symmetrical diamond-shaped section. It is of interest to obtain a graphical view of the distribution of τ_y across the section, and to establish the magnitude and position of maximum shear stress:

τ_{max} is found by putting $d\tau_y/dy = 0$; from equation (9.17), we obtain

$$\frac{d\tau_y}{dy} = \frac{(SF)}{h^4 \cdot \tan \alpha} [4h - 32y_1] = 0$$

which is true when

$$y = \frac{h}{8}$$

and then

$$\tau_{y_{max}} = \frac{(SF)}{h^2 \cdot \tan\alpha}\left[\frac{9}{4}\right] \qquad (9.18)$$

For a square section; $\alpha = 45°$ and $\tan\alpha = 1$. Then

$$\tau_{y_{max}} = \frac{9}{4} \cdot \frac{(SF)}{h^2} \qquad (9.18a)$$

Since the mean stress is

$$\frac{SF}{Area} = \frac{SF}{(h^2/2)} = \frac{(SF) \cdot 2}{h^2}$$

$$\tau_{y_{max}} = \tfrac{9}{8} \cdot \tau_{mean} \quad \text{for square section} \qquad (9.18b)$$

A computer program was written for equation (9.17) as applied to a square section, and numercial and graphical outputs were obtained as shown on Fig. 9.3.

9.3 SHEAR CENTRE OF CHANNEL SECTION BEAM

This is another example which is well known in strength of materials literature. However, as usual, the assumptions and approximations made are seldom clearly stated. The analysis below is valid only if the following approximations are accepted:

The section is sufficiently far away from the load to allow local effects to decay.
Plane sections remain plane, i.e. there is no warping.
The sections are thin enough to justify assumption of uniform stress distributions.
That shear is parallel to the shear force in the transverse plane.
Referring to Fig. 9.4 we see that

$$d\sigma_x = \frac{dM \cdot y}{I_{zz}} \quad \text{where } y = \tfrac{1}{2}h \qquad (9.19)$$

The net longitudinal force acting on the element in the x-direction is

$$(S_2 - S_1) = \Delta S_x = \frac{dM}{I_{zz}} \cdot \int_0^z y \cdot dA = \frac{dM}{I_{zz}} \cdot y \cdot \int_0^z t \cdot dz$$

or

$$\Delta S_x = \frac{dM \cdot y \cdot z \cdot t}{I_{zz}} \qquad (9.20)$$

```
10  V=1
20  D=20
30  T1=2*V/D↑2
40  PRINT "AVERAGE SHEAR STRESS, T1 = "; T1
50  PRINT "V", "T2", "R=T2/T1"
60  FOR Y=0 TO 10
70  T2=(2*V/D↑4)*(D↑2-(8*Y↑2)+(2*Y*D))
80  R=T2/T1
90  PRINT Y,T2,R;TAB(40)"."; TAB(40+20*R)"*"
100 NEXT Y
110 END
```

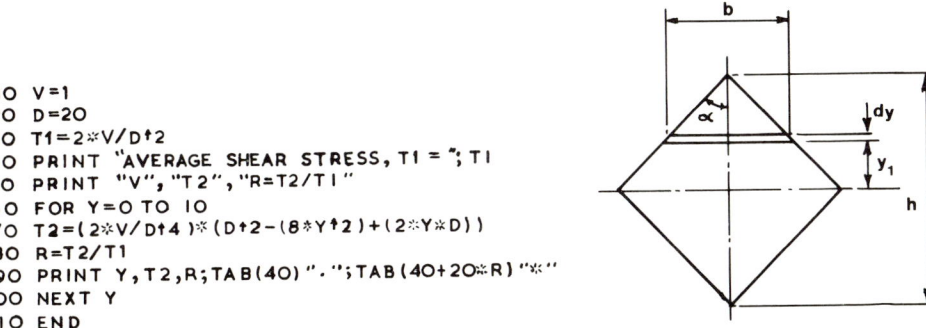

$$\tau = \frac{2V}{h^4}(h-2y_1)(h+4y_1)$$

$$= \frac{V}{h^4}(2h^2 + 4hy_1 - 16y_1^2)$$

AVERAGE SHEAR STRESS T1   .005

Y	T2	R T2/T1
0	.005	1
1	.0054	1.08
2	.0056	1.12
3	.0056	1.12
4	.0054	1.08
5	.005	1
6	.0044	.88
7	.0036	.72
8	.0026	.52
9	.0014	.28
10	0.000	0.00

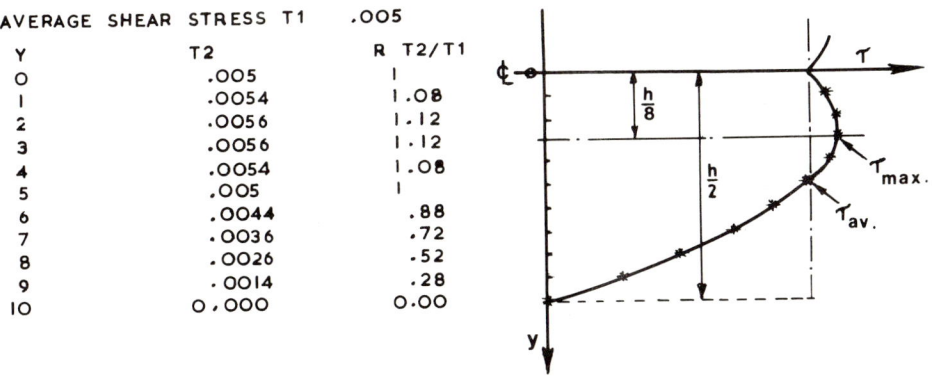

$$\tau_{av} = \frac{2V}{h^2} = 0.005$$

$$\tau_{max} = \frac{9}{4}\frac{V}{h^2} = 0.005625$$

$$\tau_{max} = 12.5\% > \tau_{av}$$

Fig. 9.3. Computer analysis of shear stress distribution across square diamond section.

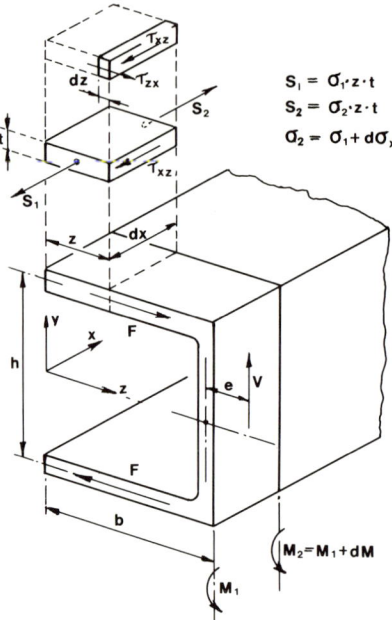

Fig. 9.4. Shear centre of channel section.

This longitudinal force on element z (wide) and dx (long) is held in equilibrium by the shear force along the interface at $z-x$, i.e.

$$\tau_{zx} \cdot dx \cdot t = \frac{dM \cdot y \cdot t \cdot z}{I_{zz}}$$

$$\therefore \tau_{zx} = \frac{dM \cdot h \cdot z}{dx \cdot 2 \cdot I_{zz}}; \quad y = \tfrac{1}{2}h$$

or

$$\tau_{zx} = (SF) \cdot \frac{h}{2} \cdot \frac{z}{I_{zz}} \tag{9.21}$$

The complementary shear stress acting along face z–y is also $= (SF) \cdot (h/2) \cdot (z/I_{zz})$ at the $z-x$ corner of the element. Thus, at any position z on the flange face, the complementary shear force is

$$\Delta F_{zx} = (SF) \cdot \frac{h}{2} \cdot \frac{Z}{I_{zz}} \cdot dz \cdot t$$

and the total shear force over the flange area is

$$F_{zx} = (SF) \cdot \frac{h}{2} \cdot \frac{t}{I_{zz}} \cdot \int_0^b z \cdot dz \tag{9.22}$$

Neglecting complications that arise at the junction between the web and the flange (note!), we obtain, total shear force along flange faces,

$$F_{zx} = (SF) \cdot \frac{h}{2} \cdot \frac{t}{I_{zz}} \cdot \frac{b^2}{2}$$

or

$$F_{zx} = \frac{(SF) \cdot h \cdot t \cdot b^2}{4 I_{zz}} \tag{9.23}$$

These shear forces may be considered as acting along the midline of the flanges in opposite directions and thus produce a moment

$$F_{zx} \cdot h = V \cdot e$$

or

$$e = \frac{(SF) \cdot h^2 \cdot b^2 \cdot t}{V \cdot 4 \cdot I_{zz}}$$

and since $V = (SF)$ the local shear force,

$$e = \frac{h^2 \cdot b^2 \cdot t}{4 \cdot I_{zz}} = \text{shear centre} \tag{9.24}$$

The approximation introduced by the assumptions initially stated make the e value thus calculated (equation (9.24)) larger than the actual value. This is because the stiffness of the web has been neglected, warping will tend to reduce the moment caused by the shear forces along the flanges (F) and the effect of the flange-web corners will tend to stiffen the section.

9.4 ESTIMATING THE INTERFACE PRESSURE BETWEEN A SPRING-CLIP AND SHAFT

A simple free-wheel clutch (Fig. 9.5) can be designed using a helical spring round a shaft or sleeve. When driving, friction will tighten the coil around the shaft, but when reversing, the spring uncoils. The problem is to dimension the first one or two coils so that their interference on the shaft will provide the essential initiation of tightening while offering a minimum of frictional drag when free-wheeling.

An appreciation of the modelling problem involved may be gained by recalling the analysis of a piston-ring. For such a ring to give a uniform radical pressure distribution, Resal (1874) showed that (see Fig. 9.6)

$$M_\phi = -2 \cdot p \cdot b \cdot r^2 \cdot \sin^2(\tfrac{1}{2}\phi)$$

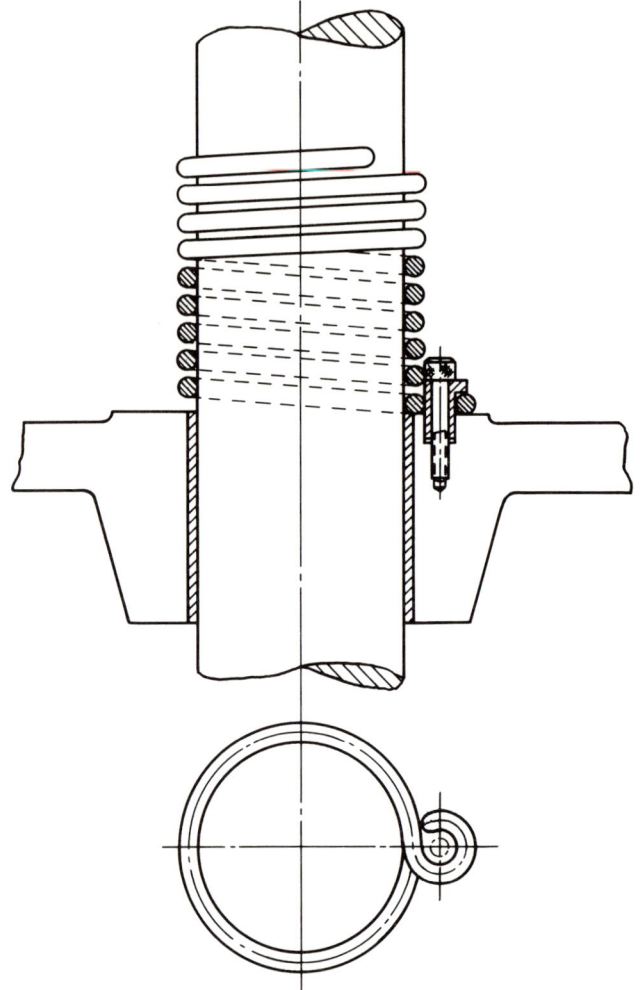

Fig. 9.5. Helical spring free-wheel coupling.

and

$$h^3 = \frac{p}{E} \cdot \frac{24 \cdot r^4}{\delta} \cdot \sin^2(\tfrac{1}{2}\phi) \\ h^3_{\max} = \frac{p}{E} \cdot \frac{24 \cdot r^4}{\delta} \Bigg\} \quad (9.25)$$

9. MODELLING

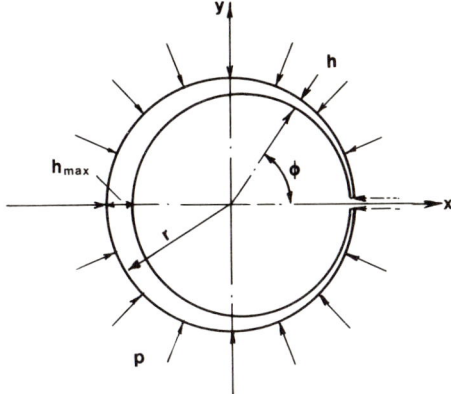

Fig. 9.6. C-clip for uniform radial pressure. $b = C$-ring thickness; δ = radial interference ($\delta \ll r$); p = circumferential pressure.

then

$$\sigma = \frac{M \cdot \tfrac{1}{2}h}{I} = \frac{12 \cdot p \cdot r^2}{h^2} \cdot \sin^2(\tfrac{1}{2}\phi)$$

$$\therefore \sigma_{max} = \frac{12 \cdot p \cdot r^2}{h_{max}^2}$$

(9.26)

We note from equations (9.25) and (9.26) that the section height h has to vary from zero to h_{max} in proportion to $(\sin \tfrac{1}{2}\phi)^{2/3}$ in order to produce a constant radial pressure between the piston-ring and the cylinder wall.

Conversely, if a piston-ring is of uniform section the pressure on its circumferences would vary. An approximate analysis to determine such variation in order to allow an assessment of the total interference pressure, and hence the frictional torque between such a C-clip and a shaft is shown below.

Simplified Model of Spring Coil (Fig. 9.7)

In place of the non-uniform (unknown) pressure distribution, we will use the resultant (point) loads in the X- and Y-directions.

Consider half of coil only (symmetry about X):

for $\alpha = 0$ to $\frac{\pi}{2}$, $\quad BM_{(1)} = \tfrac{1}{2} \cdot P_x \cdot R \cdot \sin \alpha$ \hfill (9.27)

for $\alpha = \frac{\pi}{2}$ to π, $\quad BM_{(2)} = \tfrac{1}{2} \cdot P_x \cdot R \cdot \sin \alpha + P_y \cdot R \cdot \cos \alpha$ \hfill (9.28)

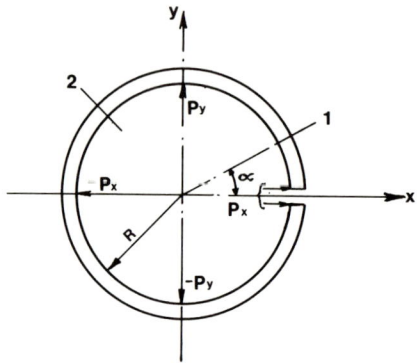

Fig. 9.7. C-clip of uniform cross-section.

The strain energy induced by bending in a short length of coil is given by

$$U_{0\to\alpha} = \int_0^\alpha \frac{M^2 \cdot R \cdot d\alpha}{2E \cdot I} \qquad (9.29)$$

Therefore, total strain energy induced in half coil is

$$U_{0\to\pi} = \int_0^{\pi/2} \frac{BM_1^2 \cdot R \cdot d\alpha}{2EI} + \int_{\pi/2}^{\pi} \frac{BM_2^2 \cdot R \cdot d\alpha}{2EI} \qquad (9.30)$$

By Castigliano's theorem, the deflection of any point relative to a fixed datum $(-P_x)$ is given by

$$\delta_p = \frac{\partial U}{\partial P} = \frac{1}{2EI}\int_0^\alpha .2M \frac{\partial M}{\partial P}. R.d\alpha \qquad (9.31)$$

Applying equation (9.31) in our case for P_x and P_y in both the X- and Y-directions (deflections) we get

X-direction

$$\delta_{P_x} = \frac{\partial U}{\partial P_x} = \frac{1}{EI}\Biggl\{\int_0^{\pi/2} (\tfrac{1}{2}P_x.R.\sin\alpha)(\tfrac{1}{2}R.\sin\alpha).R.d\alpha$$
$$+\int_{\pi/2}^{\pi}[(\tfrac{1}{2}P_x.R.\sin\alpha)+(P_y.R.\cos\alpha)].\tfrac{1}{2}R.\sin\alpha.R.d\alpha\Biggr\} \qquad (9.32)$$

which simplifies to

$$\delta_{P_x} = \frac{R^3}{8EI}[\pi.P_x - 2P_y] \qquad (9.32a)$$

N.B. δ_{P_x} = *dimetral* deflection

Y-direction

$$\delta_{P_y} = \frac{\partial U}{\partial P_y} = \left\{ \int_0^{\pi/2} (\tfrac{1}{2} P_x . R . \sin\alpha).(0). R . d\alpha \right.$$
$$\left. + \int_{\pi/2}^{\pi} (\tfrac{1}{2} P_x . R . \sin\alpha + P_y . R . \cos\alpha). R . \cos\alpha . R . d\alpha \right\} \quad (9.33)$$

which simplifies to

$$\delta_{P_y} = \frac{R^3}{4EI}[\pi . P_y - P_x] \quad (9.33a)$$

N.B. δ_{P_y} = *radial* deflection.

Since the coil is fitted on to a shaft with a *small* interference and assuming contact is maintained all round the circumference, then

$$\delta_{P_x} = 2 . \delta_{P_y}$$

i.e.

$$\tfrac{1}{8} . [\pi . P_x - 2 . P_y] = \tfrac{1}{2}[\pi . P_y - P_x]$$

or

$$P_x = 2 \cdot 04 . P_y \quad (9.34)$$

The initiating or free-wheeling (slip) torque of the first one or two coils is now obtained from

$$T_{init} \simeq 6 . P_y . \mu . R$$

or $\quad (9.35)$

$$T_{init} \simeq 3 . P_x . \mu . R$$

For *design* one can either select an initial torque and use equation (9.35) to calculate P_x or P_y and thence the dimetral interference ($= \delta_{P_x}$) from equation (9.32a)
or
one selects a practical minimum interference and thence calculates P_y and T_{init}.

Q: On what side have we erred?
Cantilever analogy. Deflection due to point load > deflection due to same distributed load. *Conversely*, for a given deflection the total distributed load > point load. We shall, therefore, be under-estimating the initiating torque for any given interference.

9.5 REVERSIBILITY OR NON-REVERSIBILITY OF MACHINES

(1) Consider first a simple wedge or slope.

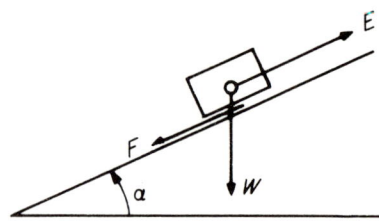

(a) For the case of moving the body up the slope we have:

W = load (or weight)

E = effort

$\therefore E = W.\sin\alpha + F$

$ = W.\sin\alpha + \mu.N$

$\therefore E = W.\sin\alpha + \mu.W.\cos\alpha$

or

$$E = W.\cos\alpha.(\tan\alpha + \tan\phi) \qquad (9.36)$$

where ϕ = friction angle.

(b) For the case of pulling the body down the slope, we have

$$E + W.\sin\alpha = F$$

i.e.

$$E = W.\cos\alpha(\tan\phi - \tan\alpha) \qquad (9.37)$$

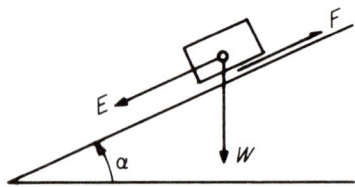

Equation (9.37) shows that if $\phi > \alpha$, an effort is required to pull the load down against friction; gravity alone cannot move the body down.

If $\phi = \alpha$, no effort is required; the load is a state of equilibrium, i.e. gravity just balances frictional resistance.

9. MODELLING

If $\phi < \alpha$, a negative effort is required, i.e. if such an effort is not applied the body will slide down by its own weight; friction is too low to retain the load in its position and the mechanism is self-reversible.

As an example, we may require to determine the maximum wedge angle to avoid self-ejection when the coefficient of friction is $\mu = 0\cdot 10$

Now, $\tan \phi = 0\cdot 10$; i.e. $\phi = 6°$. Thus from equation (9.37), we see that the wedge angle α must be $\leq 6°$ to stay in place, i.e. be non-reversible.

(ii) Consider next the relationship between the efficiency and reversibility of a machine system.

In the normal mode, the efficiency

$$\eta = \frac{(\text{load}).(\text{velocity of load})}{(\text{effort}).(\text{velocity of effort})} = \frac{\text{output work}}{\text{input work}}$$

Now, if we denote

$$A = \frac{\text{load}}{\text{effort}} = (\text{mechanical advantage})$$

and

$$R = \frac{\text{velocity of effort}}{\text{velocity of load}} = (\text{velocity ratio})$$

then

$$\eta = \frac{A}{R}$$

A machine is said to be self-reversing if the load can drive the effort, however small. If it cannot do so, the load will cause the machine to self-lock, or self-brake. This self-braking characteristic is often a desirable and "safe" feature required for such equipment as lifts, cranes, conveyors, elevators, etc. This must not be confused with specially non-reversible mechanisms such as ratchets, free-wheels, harpoons, lobsterpots, mousetraps, etc. The former can be driven by reverse efforts, backwards, whereas the latter are designed not to permit reverse driving.

Self-reversing criterion

In normal (forward) driving

$$\eta_N = \frac{\text{Output}_{(N)}}{\text{Input}_{(N)}} = \frac{\text{Input}_{(N)} - \text{Friction}_{(N)}}{\text{Input}_{(N)}}$$

or

$$\eta_N = \frac{O_N}{I_N} = \frac{I_N - F_N}{I_N} = \left(1 - \frac{F_N}{I_N}\right)$$

$$\therefore \frac{F_N}{I_N} = (1 - \eta_N) \tag{9.38}$$

310 APPROXIMATE METHODS IN ENGINEERING DESIGN

If the machine is self-reversing, the output could drive the input. Assuming the friction loss is the same as for normal forward driving (an approximation often justified), then

$$I_{Rev} = \eta_N \cdot I_N \tag{9.39}$$

From equation (9.38), for normal forward driving, we may write

$$F_{Rev} = (1-\eta_N) \cdot I_N \tag{9.38a}$$

Therefore, combining (9.39) and (9.38a) we obtain

$$\frac{F_{Rev}}{I_{Rev}} = \frac{(1-\eta_N)}{\eta_N} \tag{9.40}$$

Hence,

$$\eta_{Rev} = \frac{O_{Rev}}{I_{Rev}} = 1 - \left(\frac{1-\eta_N}{\eta_N}\right)$$

i.e.

$$\eta_{Rev} = \left(2 - \frac{1}{\eta_N}\right) \tag{9.41}$$

Thus, if $\eta_N \leqslant 50\%$, self-reversing is *not* possible.

Effort for reverse driving

$$E_{Rev} \times v_{effort} = (W_{load} \times v_{load}) - (\text{friction work})$$
$$= (W_{load} \times v_{load}) - (1-\eta_N) \times E_N \times v_{effort}$$

or

$$\frac{E_{Rev}}{E_N} = \frac{W_{load} \times v_{load}}{E_N \times v_{load}} - (1-\eta_N) \times \frac{v_{effort}}{v_{effort}}$$

which reduces to

$$\frac{E_{Rev}}{E_N} = (2\eta_N - 1) \tag{9.42}$$

Thus the effort required to reverse drive any given load equals *zero*, when $\eta_N = 0.50$.

For $\eta_N < 50\%$, a reverse drive effort is required and if $\eta_N > 50\%$, a negative effort has to be supplied to prevent load itself driving—i.e. the machine is self-reversing.

An interesting example where the above criteria is of significance is on test rigs for large gear-boxes. In order to avoid having to use full capacity driving motors and brakes, a closed loop set-up, as shown on Fig. 9.8, is often used.

9. MODELLING

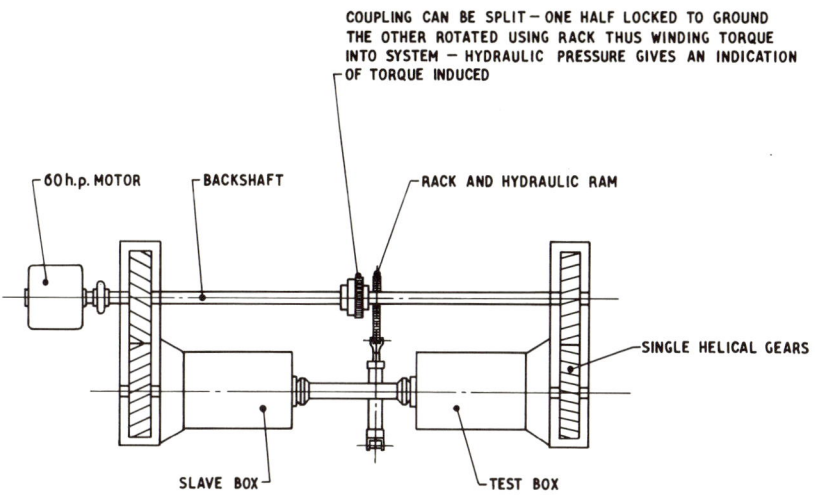

Fig. 9.8. Schematic arrangement of gearbox test rig.

Two "identical" gear-boxes are connected through a pre-torqued shaft so that the input needs to supply only the friction power required. There are, however, limitations to the advantages gained with this arrangement, as shown in Table 9.1. In the table η_{Rev} has been calculated using equation (9.41).

Table 9.1. Tandem drive system (Fig. 9.8)

η_N%	η_{Rev}%	η_{Tand}%
90	89	80
80	75	60
75	66·5	50[a]
70	57	40
60	33	19·8
50	0	

[a] No gain in power below this value of η_{Tand}.

9.6 FATIGUE FAILURES OF BURNISHING ROLLERS

Three roller units (Fig. 9.9) are mounted on a rotating head (Fig. 9.10). During operation a slightly oversized bar is drawn axially through the centre of the head, whilst the rollers "burnish" the bar to the finished size required. Early failures were experienced with these rollers breaking at the root of the 38·1 mm diameter shaft (Fig. 9.11).

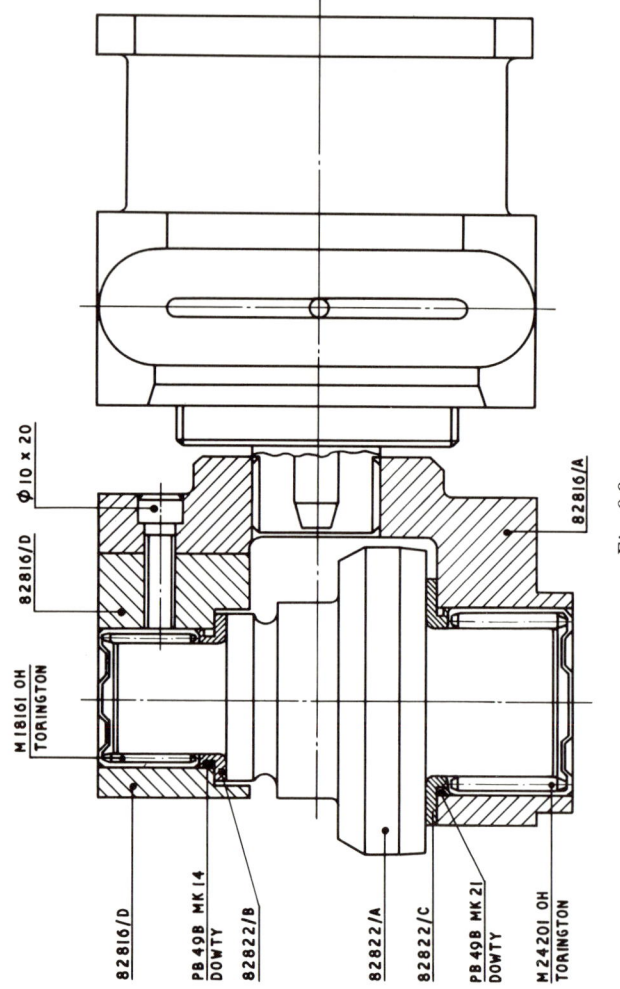

Fig. 9.9.

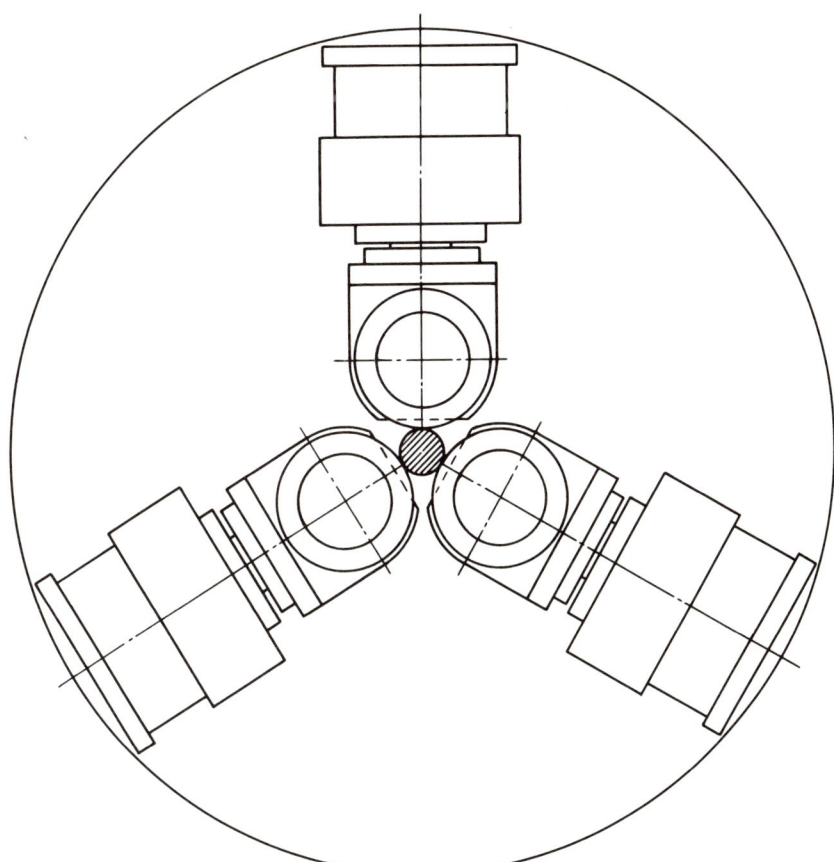

Fig. 9.10. Burnishing roller head.

For the purpose of fatigue failure analysis it is necessary to establish the stresses caused by the loads $P_{RAD} = 16\,800$ lbf and $P_{AX} = 2500$ lbf.

Figure 9.12 shows a possible load distribution. The system is clearly statically indeterminate, and a rigourous elastic analysis is also difficult because numerical stiffness values cannot be easily determined.

An approximate method using "extreme value" analysis may be employed here. *Model 1* represents the case of a very stiff thrust race (Fig. 9.13). *Model 2* represents the case of a very flexible thrust race (Fig. 9.14).

9.7 BASE PLATE BOLT LOADS

An extrusion machine for elastomeric materials comprised three independently floor-anchored units; an extruder, a reduction gear and a motor drive.

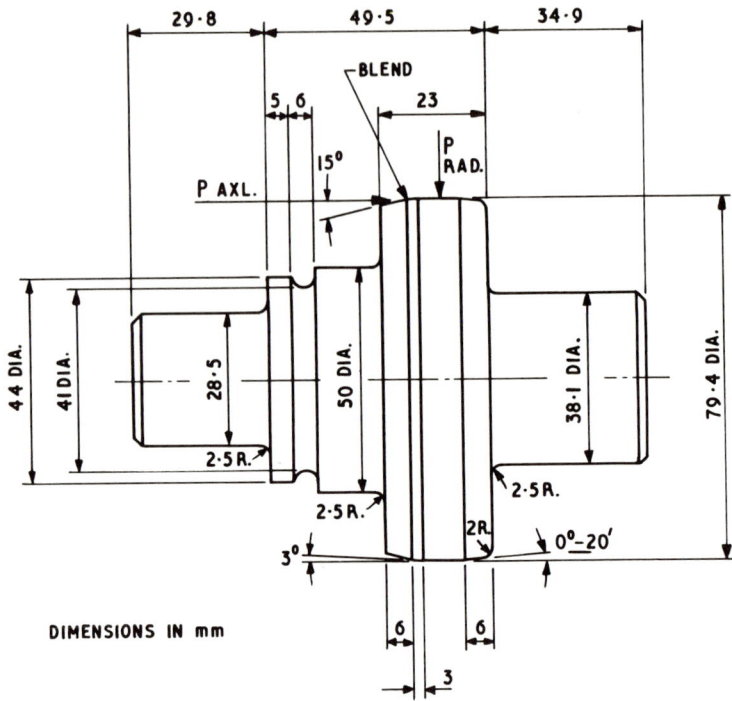

Fig. 9.11. Roller. Analysis: 2·0% C, 0·3% Si, 0·3% Mn, 13% Cr, 0·25% V. Harden to 60–62 Rockwell C; harden to 950 °C; soak 30–45 min; oil quench; temper at 160 °C; grind to ra. 0·2–0·4 (c.l.a. 8-16). Material: Darwin's Neor.

The latter posed some problems in assessing safe-bolt loadings, especially as the system was to be installed on an intermediate factory floor. It can be seen from Fig. 9.15 that the motor and its cooler are bolted to a sole-plate which has three pairs of floor-bolts. A simplified representation of this assembly is shown in Fig. 9.16.

9.7.1 Modelling

For further steps in the modelling procedure we have to make a few simplifying assumptions:

(i) The flange connections between the motor and cooler are relatively flexible.
(ii) The motor and cooler are both near enough centrally positioned on the sole-plate to justify treating the loading on the three bolts on each side as symmetrical.

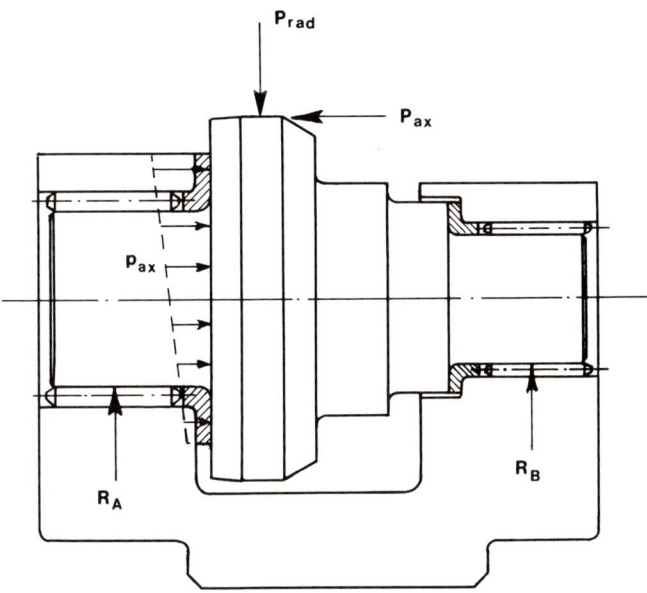

Fig. 9.12. Load system. Statically indeterminate assembly. Support reactions depend on: shaft stiffness, needle bearing stiffness, thrust bearing stiffness, all of which are difficult to quantify.

(iii) Motor and cooler produce point loads only on the sole-plate.
(iv) As it is usual to place packing plates under each foundation bolt to obviate distortion of base-plates, it may be assumed that the floor supports of the sole-plate are at the bolt positions only.

We thus derive at a line diagram as shown on Fig. 9.17a. Information needed for a rigorous fully elastic analysis of this system would be

(a) sole-plate stiffness,
(b) motor and cooler stiffness,
(c) bolting stiffness between (a) and (b),
(d) floor stiffness.

Practical engineers would immediately appreciate that such information is not usually available. However, the designer can still make useful approximate estimates by considering possible "upper" and "lower" limits for the foundation loadings. To do so we may consider two models:

Model (a) = very stiff motor/sole-plate assembly (Fig. 9.17b)

Model (b) = relatively flexible motor/sole-plate assembly (Fig. 9.17c)

316   APPROXIMATE METHODS IN ENGINEERING DESIGN

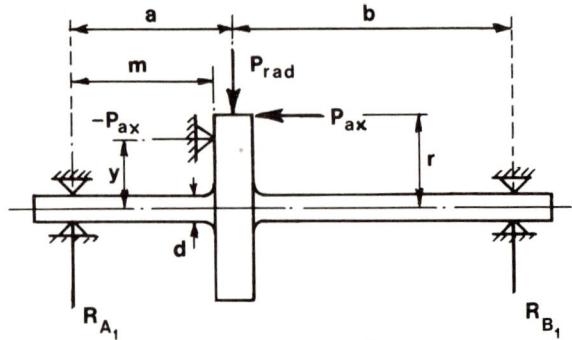

Fig. 9.13. Model 1. Very stiff thrust race.

$$R_{A_1} = \frac{P_{RAD}\,b + P_{AX}(r-y)}{(a+b)}$$

$$BM_1 = R_{A_1} \cdot m$$

$$\sigma_{b_1} = \frac{R_{A_1} \cdot m}{\dfrac{\pi}{32} d^3}$$

$$\tau = \frac{4}{3} \frac{R_{A_1}}{\dfrac{\pi}{4} d^2}$$

Data given

$P_{RAD} = 16\,800\,\text{lbf}$ $d = 1\cdot 50\,\text{in}$
$P_{AX} = 2500\,\text{lbf}$ $a = 1\cdot 34\,\text{in}$
$r = 1\cdot 5625\,\text{in}$ $b = 2\cdot 44\,\text{in}$
$y = 1\cdot 25\,\text{in}$ $m = 0\cdot 867\,\text{in}$

Calculations yield

$R_{A_1} = 11\,010\,\text{lbf}$
$BM_1 = 9560\,\text{lbf in}$
$\sigma_{b_1} = 28\,850\,\text{lbf in}^{-2}\ (\simeq 12\cdot 9\,\text{ton in}^{-2})$
$\tau_1 = 8330\,\text{lbf in}^{-2}$

9.7.2 Loads on sole-plate (Fig. 9.17)

(i) *Motor*

Stalling torque $= T = P(l_{p1} - l_{p2})$ \hfill (i)

Weight of motor $= M$

∴ Loads on feet (pair)

$P_1 = P + \dfrac{M}{2}$ (down)

$P_2 = P - \dfrac{M}{2}$ (up) \hfill (ii)

9. MODELLING

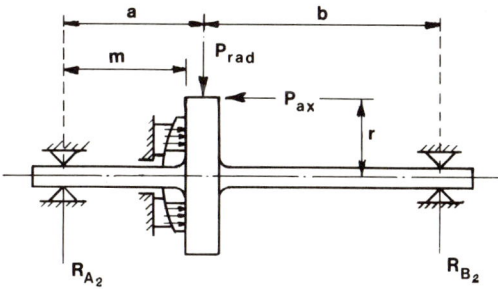

Fig. 9.14. Model 2. Very flexible thrust race

$$R_{A_2} = \frac{P_{RAD} \cdot b + P_{AX} \cdot r}{(a+b)}$$

$$BM_2 = R_{A_2} \cdot m$$

$$\sigma_{b_2} = \frac{R_{A_2} \cdot m}{\frac{\pi}{32}d^3}$$

$$\tau_2 = \frac{4}{3} \cdot \frac{R_{A_2}}{\pi/4 \cdot d^2}.$$

Calculations yield

$$R_{A_2} = 11\,730\,\text{lbf}$$

$$BM_2 = 10\,170\,\text{lbf in}$$

$$\sigma_{b_2} = 30\,650\,\text{lbf in}^{-2}\,(\simeq 13\cdot7\,\text{ton in}^{-2})$$

$$\tau_2 = 8850\,\text{lbf in}^{-2}$$

$$\text{Ratio}\frac{(2)}{(1)} = 1\cdot061 = 6\cdot1\%$$

(ii) *Cooler*
Weight $= W$
Loads on feet (pair) $= W_1$ and W_2, respectively (iii)

(iii) *Initial bolt tightening*
Assume equal for all six bolts
Initial load $= I_T$ (iv)

9.7.3 Foundation and bolt loads

(a) Stiff motor/sole-plate assembly
Bolt loading (without pre-tension). Moments about R_0:

$$R_2 \cdot l_2 + W_2\,l_{w2} + W_1\,l_{w1} + R_1\,l_1 + P_1\,l_{p1} - P_2\,l_{p2} = 0 \qquad \text{(v)}$$

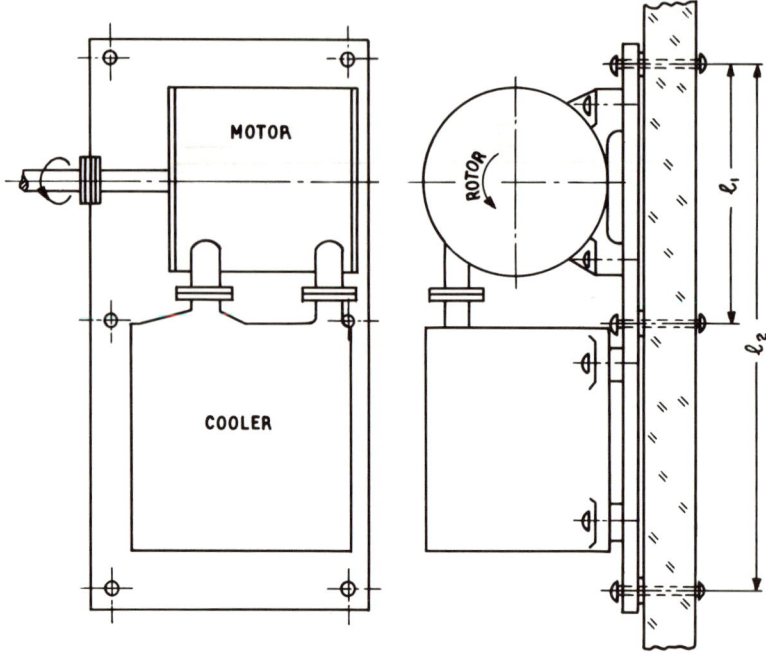

Fig. 9.16. Motor drive assembly.

Equilibrium in *Y*-direction:
$$R_2 + R_1 + W_2 + W_1 + P_1 - P_2 + R_0 = 0 \qquad \text{(vi)}$$

Due to lever ratio:
$$\frac{R_2}{R_1} = \left(\frac{l_2}{l_1}\right); \quad R_2 = R_1 \cdot \left(\frac{l_2}{l_1}\right) \qquad \text{(vii)}$$

∴ from (v) and (vii):
$$R_1 = \left(\frac{l_1}{l_2 + l_1^2}\right) \cdot (P_2 l_{p2} - P_1 l_{p1} - W_2 l_{w_2} - W_1 l_{w_1}) \qquad \text{(viii)}$$

Thence, R_1 and R_0 may be calculated from equations (iii) and (iv).

Notes

If a reaction $R_i = -\text{ve}$ (= upward on sole-plate) the local floor pressure is increased, but bolt tension remains virtually = initial tension.

If any reaction $R_i = +\text{ve}$ (= downward on sole-plate) bolt tension is increased and local floor pressure slightly reduced.

Floor "lifting" or "depressing" is a function of R_i reactions only (i.e. not of the initial bolt tension).

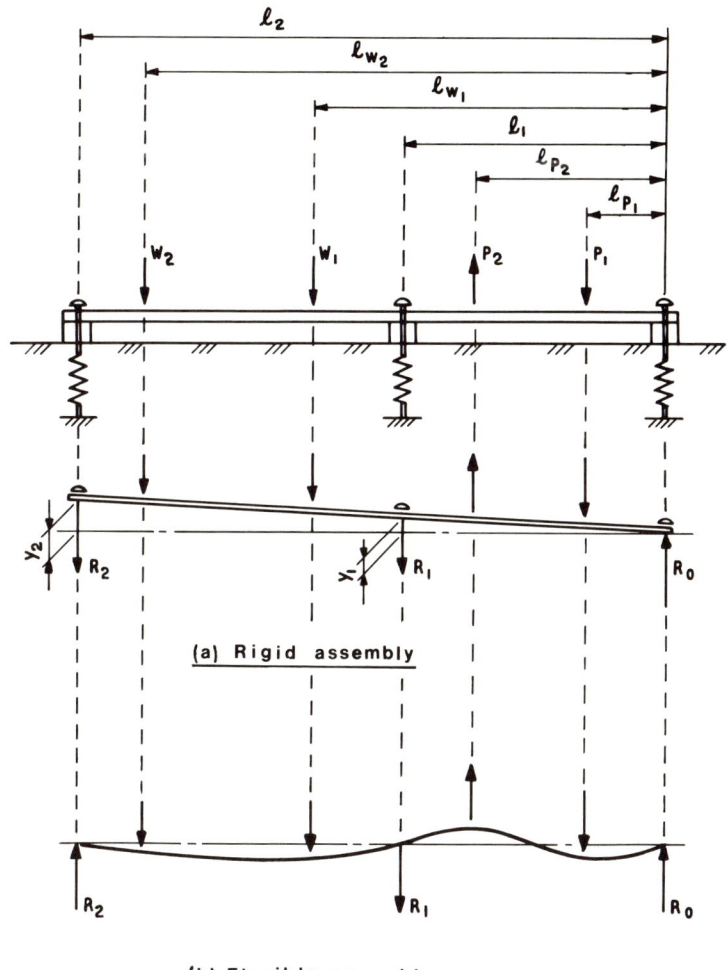

Fig. 9.17. Beam model of motor drive assembly.

(b) Flexible motor/sole-plate assembly

In this model we shall assume a relatively flexible sole-plate of uniform stiffness.

For bolt-loading (without pre-tension) we have: Taking moments about R_2,

$$EI\frac{d^2y}{dx^2} = R_0 \cdot x - P_1(x-l_{p_1}) + P_2(x-l_{p_2}) - R_1(x-l_1) - W_1(x-l_{w_1}) - W_2(x-l_{w_2})$$

or

$$EI\frac{d^2y}{dx^2} = [R_0 - P_1 + P_2 - R_1 - W_1 - W_2] \cdot x$$
$$- [P_1 \cdot l_{p_1} - P_2 \cdot l_{p_2} + R_1 \cdot l_1 + W_1 \cdot l_{w_1} + W_2 \cdot l_{w_2}] \quad \text{(i)}$$

which may be written in simplified form

$$EI\left(\frac{d^2y}{dx^2}\right) = A \cdot x - B \quad \text{(ia)}$$

Integrating, we get

$$EI\left(\frac{dy}{dx}\right) = A\frac{x^2}{2} - Bx + C \quad \text{(ii)}$$

and further

$$EI(y) = A\frac{x^3}{6} - B\frac{x^2}{2} + C \cdot x + D \quad \text{(iii)}$$

Our end conditions are

$$y = 0, \quad \text{at } x = 0 \quad \text{and} \quad x = l_2$$

Thus

$$D = 0 \quad \text{and} \quad C = -\left(A\frac{l_2^2}{6} - B\frac{l_2}{2}\right) \quad \text{(iv)}$$

Inserting (iv) into (iii) we get

$$EI(y) = A \cdot \frac{x^3}{6} - B \cdot \frac{x^2}{2} - \left(A\frac{l_2^2}{6} - B\frac{l_2}{2}\right) \cdot x \quad \text{(v)}$$

The three unknown reactions R_0, R_1 and R_2 are linked together by the flexibility or otherwise of the sole-plate assembly. Since in this model we are considering a relatively flexible sole-plate, we may take the deflection at R_1 also as zero. Considering the R_1 position, we must leave out all terms in equation (i) which are beyond l_1. Thus,

and
$$\left.\begin{array}{l} A' = (R_0 - P_1 + P_2) \\[6pt] B' = (P_1 \cdot l_{p_1} - P_2 \cdot l_{p_2}) \end{array}\right\} \quad \text{(vi)}$$

From (v) we therefore have

$$0 = A' \cdot \frac{l_1^3}{6} - B' \cdot \frac{l_1^2}{2} - \left(A' \cdot \frac{l_2^2}{6} - B'\frac{l_2}{2}\right) \cdot l_1$$

which becomes

$$0 = \frac{R_0}{6}(l_1^2 - l_2^2) - \left(\frac{P_1 + P_2}{2}\right)(l_1^2 - l_2^2) - \left(\frac{P_1 l_{p_1} - P_2 l_{p_2}}{2}\right) \cdot (l_1 - l_2)$$

or

$$R_0 = 3\left[P_1 + P_2 + (P_1 \cdot l_{p_1} - P_2 \cdot l_{p_2})\left(\frac{1}{l_1 + l_2}\right)\right] \quad \text{(vii)}$$

To determine R_1 and R_2, we have to set up two further equations for

(a) linear equilibrium
(b) rotational equilibrium

For (a) we have

$$W_2 + W_1 + P_1 + R_1 - R_0 - R_2 + P_2 = 0$$

or (viii)

$$(R_2 - R_1) = (W_2 + W_1 + P_1 - P_2 - R_0)$$

and for (b) (moments about R_0)

$$R_2 \cdot l_2 - W_2 \cdot l_{w_2} - W_1 \cdot l_{w_1} - R_1 \cdot l_1 + P_2 \cdot l_{p_2} - P_1 \cdot l_{p_1} = 0$$

or

$$R_2 = R_1 \frac{l_1}{l_2} + \frac{1}{l_2}[W_2 \cdot l_{w_2} + W_1 \cdot l_{w_1} - P_2 \cdot l_{p_2} + P_1 \cdot l_{p_1}] \quad \text{(ix)}$$

Thus with the assumptions made of a flexible motor/sole-plate assembly all bolt loads or floor reactions may be estimated.

9.8 ASSESSMENT OF SPRING FORCE FOR POPPET VALUE ASSEMBLY

A poppet valve arrangement for a gas expander engine is shown on Fig. 9.18. High pressure gas enters underneath the mushroom head and passes into the cylinder when the valve is opened about 5° before top-dead-centre of the piston. Some 50° of crank movement after TDC the valve is closed, the pressure having practically equalized on both sides of the valve head. An engine had to be designed to produce about 1 horse power at 600 r.p.m. with a gas inlet of 100 bars ($\simeq$ 1500 p.s.i.). How would the designer proceed with the design of such a valve unit?

First, it is necessary to select possible basic operational features. These would be essentially based on past experience and familiarity with these types of engines. Figure 9.18 shows one such arrangement, which comprises

(i) a mush-head poppet valve.
(ii) a nest of chevron lip seals with facility for initial nipping of the seals,

(iii) a helical spring,
(iv) spring locations in housing and at the spindle end.

Next the designer has to carry out a series of calculations in order to estimate geometries and dimensions that will give satisfactory performance.

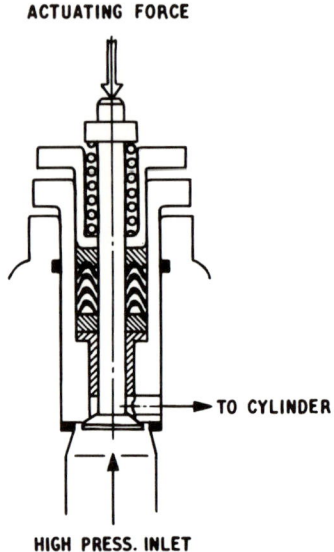

Fig. 9.18. Poppet valve assembly.

9.8.1. Valve head and lift dimensions

Given: Supply pressure = 100 bars ($\simeq 1500$ p.s.i)
Supply temperature = 0 °C ($\simeq 273$ K)
Engine speed = 600 r.p.m.
Power output required $\simeq 1$ h.p.

The expander engine was designed for an adiabatic expansion/compression cycle and assuming a 75% ideal indicator diagram efficiency, it was estimated that a gas consumption of some 4.6×10^{-6} m^{-3} of gas was required per engine stroke.

(i) The mean flow area of the valve would, therefore, have to be

$$a_{\text{mean}} = \frac{4.60 \times 10^{-6}}{C_D \cdot v_m \cdot t} \tag{i}$$

where
C_D = orifice coefficient (vena contracta)
v_m = mean gas velocity
t = valve opening time

9. MODELLING

From gas-engine experience, one obtains

$C_D = 0.5$
$v_m = 50 \text{ m s}^{-1}$ to avoid excessive pressure drop through the valve

For $N = 600$ r.p.m.; 1 cycle = 0·10 seconds. Therefore,

$$t = \frac{55°}{360°} \times 0.10 = 0.0152 \text{ seconds}$$

Thus equation (i) above yields

$$a_{mean} = \frac{4.6 \times 10^{-6}}{0.5 \times 50 \times 0.0152} = 12 \times 10^{-6} \text{ m}^2$$

or

$$a_{mean} = 12 \text{ mm}^2 \qquad \text{(ii)}$$

(ii) C. F. Taylor (1966) reviewed the performance characteristics of a large number of IC engines and concluded that the volumetric efficiency falls off rapidly when the factor

$$Z = \left(\frac{b}{d}\right)^2 \cdot \frac{s}{C_i \cdot a} > 0.5 \qquad \text{(iii)}$$

where
s = piston speed (mean).
b = bore diameter of the cylinder
d = diameter of the valve head
C_i = inlet valve flow coefficient

$$\left(\simeq 1.5 \cdot \frac{\text{valve lift}}{\text{area of head}}\right)$$

a = acoustic velocity

($a = 337 \text{ m s}^{-1}$ at 273 K)

Using Taylor's (1966) correlation, a number of valve dimensions were tried: A piston diameter of $b = 23$ mm had been established. Try valve head diameter $d = 8$ mm: Taking a mean piston speed of $s = 1 \text{ m s}^{-1}$ and trying with $Z = 0.5$, we obtained $C_{i_{min}} = 0.0493$, which means that the valve lift (i) should be 0·262 mm.

This is clearly too small and would demand extremely accurate machining, assembly and setting. Very little wear could be tolerated. For a valve diameter of $a = 6$ mm, the lift required was $l = 0.48$ mm. The annular area of the fully open valve would then be 9 mm².

Making allowances for the small size of engine and valve-gear required in this case and also moving further away from the critical value of $Z = 0.50$, the final choice for the flow area was made = 14 mm², which for the 6 mm diameter valve head represented a valve lift of 0·70 mm.

9.8.2 Valve stem diameter

We note from Fig. 9.18 that the valve has to open against the full supply pressure. Ignoring the frictional resistance of the chevron seals at this stage, the valve stem has to withstand the compressive load acting on the valve head. Thus,

$$\text{axial load to open valve} = 100 \cdot \frac{\pi}{4} \cdot 6^2 = 28 \cdot 3 \text{ kgf}$$

$$= 277 \cdot 54 \text{ N}$$

In this case buckling is the critical factor. Using Euler's equation,

$$P = \frac{c \cdot \pi^2 \cdot E \cdot I}{f \cdot l^2} \qquad \text{(iv)}$$

where

P = maxim safe working load
E = Young's modulus ($= 20 \times 10^4$ N mm^{-2})
$I = \frac{\pi}{64} d^4$

f = factor of safety relative to crippling load
l = effective length
c = constant depending on end-fixings

In this case:

$l \simeq 35$ mm
$c = 1$
$P = 277.54$ N (from above)
$f = 6$ (for long fatigue life)

These yield a safe stem diameter of $d = 2$ mm. However, taking account of the undercut for the spring collar and the smallest standard chevron seals available, a stem diameter of 3 mm was chosen.

9.8.3 Valve seal resistance

Seal manufacturers were consulted about suitable seal materials and arrangements. Their recommendation was a nest of four PTFE chevron double lip seals as shown on Fig. 9.18.

9. MODELLING

Assuming a minimal leakage past the seal, a reasonable proposition is to assume that the gas pressure will fall linearly through the seal assembly. Thus, the pressure drop per seal ring would be $100/4 = 25$ bars $\simeq 245 \text{ N mm}^{-2}$.

The frictional force opposing axial gliding will depend on the radial forces applied by the seal lips on the stem and the coefficient of sliding friction:

$$\text{radial force} = \text{pressure} \times \text{area of contact}$$

The seal lip contact area has to be estimated. One guide line is that the contact pressure could not exceed the compressive yield strength of PTFE. A safe contact width was taken as 2 mm. Thus the contact area for each lip on the valve stem would be

$$\text{contact area per lip} = \pi \times 3 \times 2 = 18.85 \text{ mm}^2$$

Now what contact pressure would exist between lip and stem? Clearly for sealing the contact pressure must be at least equal to the pressure difference across the lip, but for effective sealing a contact pressure of $1.25 . \Delta p$ is usually considered, thus allowing for the initial nip in the seal. The frictional force opposing motion of the valve spindle in the seal is thus

$$F = \mu \frac{18.85}{100} \times 1.25 \times 100 \times 9.807$$

and if $\mu = 0.10$;

$$F = 0.10 \times 0.1885 \times 100 \times 9.807$$

$$F = 20.5 \text{ N}$$

9.8.4 Spring force

The spring force to close the valve has to overcome the frictional resistance to sliding in the seals plus the inertia force due to acceleration.

From results already established, we have

$$\text{valve lift} = 0.7 \text{ mm}$$
$$\text{opening time} = 0.0152 \text{ s}$$
$$\therefore \omega = \frac{\pi}{0.0152} = 206.68 \text{ rad s}^{-1}$$

Assuming SHM the maximum acceleration will occur at the start of closing the valve,

$$\alpha_{max} = \tfrac{1}{2} . \omega^2 . l$$
$$\therefore \alpha_{max} = \tfrac{1}{2}(206.68)^2 \times 0.7 \times 10^{-3}$$
$$= 14.951 \text{ m s}^{-2}$$

Valve mass

The volume of the valve (plus collars and about one-third of spring) was estimated to be $284{\cdot}74 \text{ mm}^3$. Taking density of steel $= 7{\cdot}8 \text{ g cm}^{-3}$ the mass of the moving parts of the valve $= 2{\cdot}21 \times 10^{-3} \text{ kg}$:

$$\therefore \text{ inertia force needed} = 2{\cdot}21 \times 10^{-3} \times 14{\cdot}951$$
$$= 33{\cdot}04 \times 10^{-3} \text{ N}$$

During the closing period the gas pressure is practically equal on both sides of the valve head, but the head area is less on the stem side. Therefore, the gas pressure effectively assists in closing the valve by

$$F_{\text{gas press.}} = 100 \cdot \frac{\pi (3)^2}{4\ 100} = 7{\cdot}875 \text{ N}$$

The net force to initiate valve closing, therefore, is

$$F_{\text{spring}} = 33 \times 10^{-3} + 20{\cdot}50 - 7{\cdot}875$$
$$= 12{\cdot}6 \text{ N}$$

This force is applied when the spring is at its maximum compression. It will be seen from the above that seal friction offers by far the greatest resistance, and this remains practically constant during the full lift movement. In the fully closed position the spring load has its lowest value. The spring should, therefore, be designed to have a minimum built-in load of $20{\cdot}50 \text{ N}$.

The above calculations cater only for the valve unit. Any additional inertia and frictional resistance in the rocker mechanism must be added to the spring or a separate spring-member designed for these parts.

9.9 INEQUALITY OF FORCED AND NATURAL FREQUENCIES

A shaft assembly for testing rolling element bearing performance in an ultra-high-vacuum chamber had been designed for operation at about 500 r.p.m. It was necessary to ensure that the natural frequency (NF) of the assembly did not coincide with the running speed. Figure 9.19 represents the shaft, the components resting on it and the support bearings. Figure 9.20 is a pictorial view of the shaft assembly showing also some additional, but independently supported parts. The capital letters in bubbles represent the following items:

- A = support bearings
- B = test bearings with radial loadings
- C = axial thrust rings
- D = axial thrust adjustments
- E = disc-springs
- F = spacer between springs

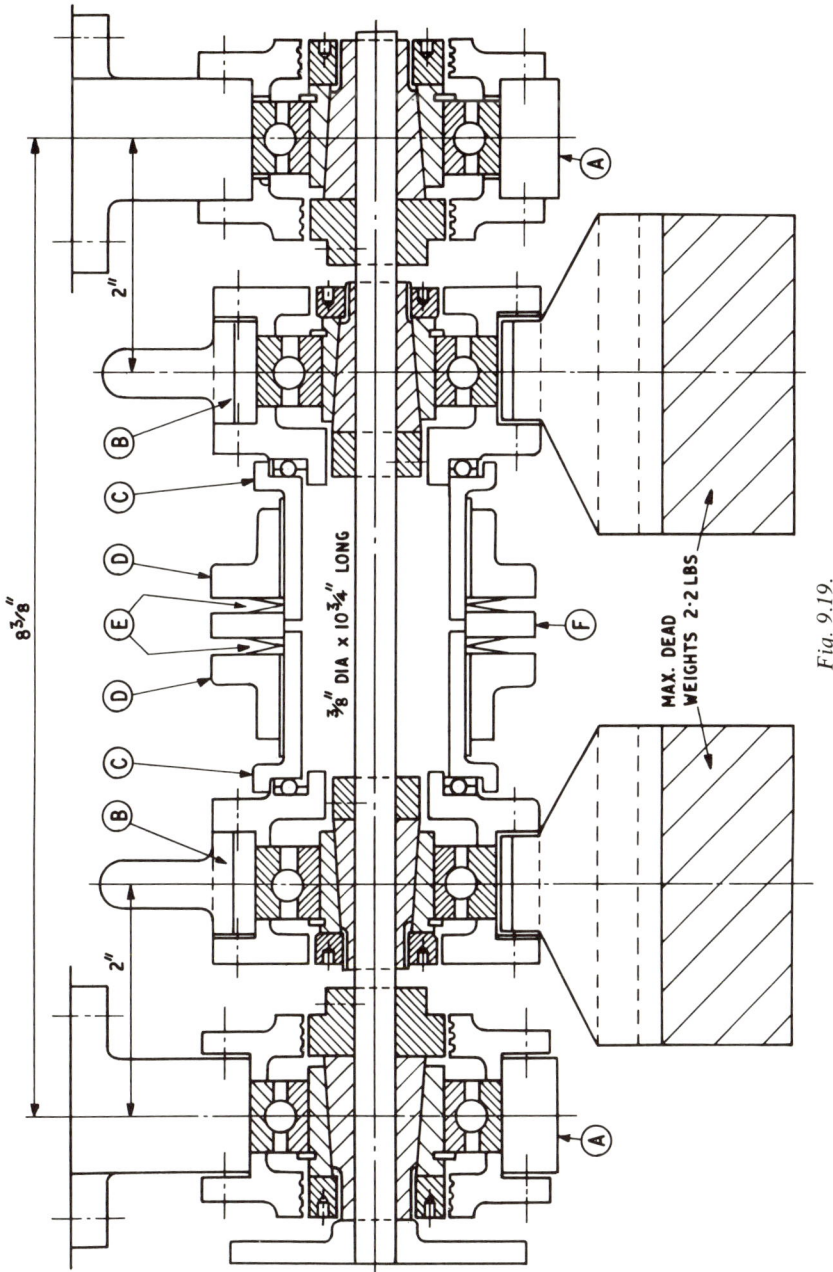

Fig. 9.19.

328 APPROXIMATE METHODS IN ENGINEERING DESIGN

Note. The axial thrust components do not add to the radial load as they are independently supported on the side bars shown in Fig. 9.20.

How should the designer deal with this problem?

(a) Should he attempt to carry out a rigorous analysis of the first, second, etc. modes of natural frequencies?

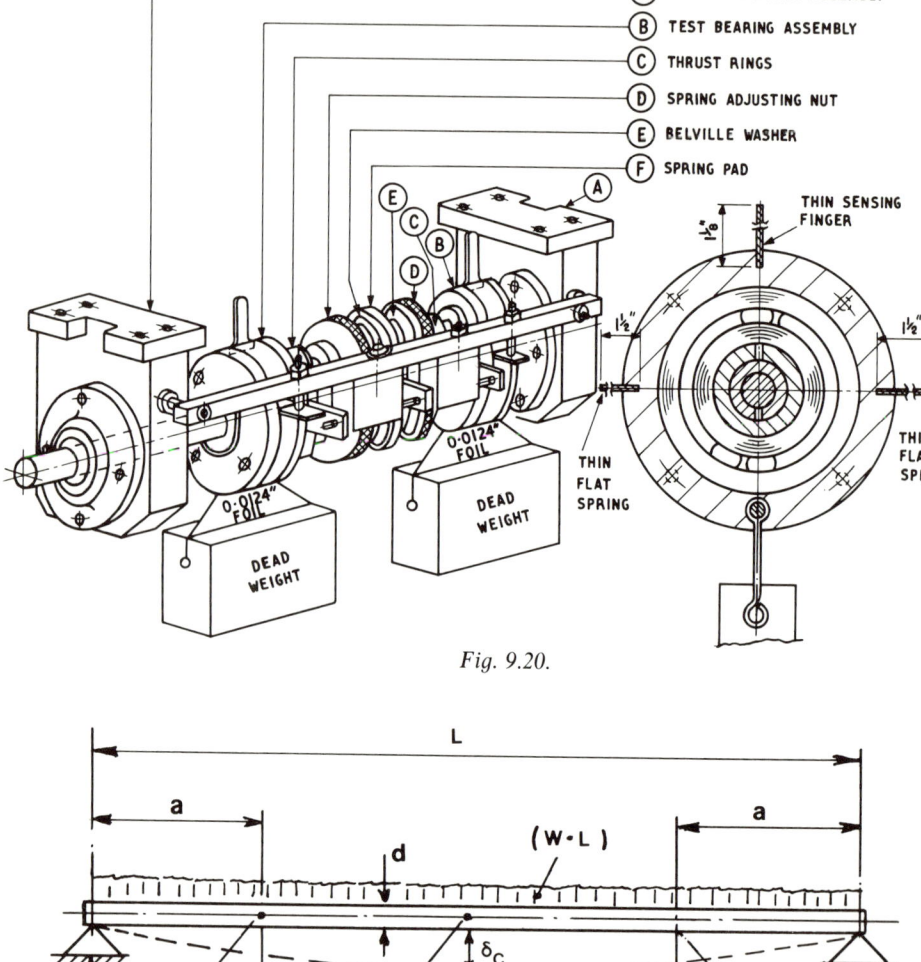

Fig. 9.20.

Fig. 9.21. Simplified model. Data available: $L = 8.375$ in; $a = 2$ in; $d = 0.375$ in; $W_B = 5.084$ lbf; $(WL) = 0.261$ lbf; $I = 97 \times 10^{-5}$ in^2; $E = 24.8 \times 10^6$ lbf in^{-2} at 300 °C.

9. MODELLING

(b) Should he confine himself to an *approximate* assessment of the lowest natural frequency of the system?

If the second approach is used, it will be important to know on which side the approximation is in error—i.e. whether higher or lower than the true value. Since in this case one is not interested in the natural frequencies as such, but only in their non-coincidence with the forced frequency, i.e. the operational speed, it is sensible to use the approximate approach, at least to start with.

Simplified model (Fig. 9.21)	Effect on NF estimate
1. Shaft is simply supported in bearings (A)	Reduced NF
2. All loads considered as point loads	Reduced NF
3. Local stiffening of shaft in the bearings ignored	Reduced NF
4. Effects of axial tension ignored	Reduced NF
5. Coupling and shaft overhangs at support bearings neglected	Small increase in NF

The natural frequency (primary mode) estimates based on the above simplifications would yield an NF value *lower* than the true natural frequency of the shaft assembly.

Referring to Fig. 9.22 we note that the natural frequency of a mass supported by an elastic member (weightless) is given by

$$\text{NF} = \frac{187 \cdot 8}{\sqrt{\delta_{\text{stat}}}} \text{ r.p.m.} \tag{i}$$

and for a distributed mass the NF is given by

$$\text{NF} = \frac{211 \cdot 4}{\sqrt{\delta_{\text{stat}}}} \text{ r.p.m.} \tag{ii}$$

where δ_{stat} = the static deflection (inches) at the point of load application. Thus, if we establish the deflections δ_C and δ_B (Fig. 9.21) and ignore the "distributed" mass effect we will again obtain a lower than true NF value.

Deflections at C and B

At C

(1) Due to (wl)

$$\delta_{C_1} = \frac{5}{384} \cdot \frac{(wl) \cdot l^3}{EI}$$

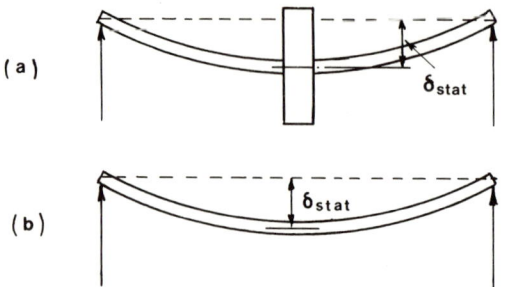

Fig. 9.22. Natural frequencies of transverse vibrations of a shaft.
(a) Concentrated mass. Critical speed $= \dfrac{187\cdot 7}{\sqrt{\delta_{\text{stat}}}}$ r.p.m., for δ_{stat} in inches.
(b) Distributed mass. Critical speed $= \dfrac{211\cdot 4}{\sqrt{\delta_{\text{stat}}}}$ r.p.m., for δ_{stat} in inches.

yielding $\delta_{C_1} = 0\cdot000\,082\,6$ in.
(2) Due to W_B

$$\delta_{C_2} = \frac{W_B \cdot a}{24 \cdot EI}(3l^2 - a^2)$$

yielding $\delta_{C_2} = 0\cdot003\,36$ in.
∴ Total deflection at centre of shaft is

$$\delta_C = 0\cdot00345 \text{ in.}$$

At B
(1) Due to (wl),

$$\delta_{B_1} = \frac{(wl)\cdot a \cdot (l-a)}{24EI \cdot l}[l^2 + a(l-a)]$$

yielding $\delta_{B_1} = 0\cdot000\,005\,68$ in.
(2) Due to W_B,

$$\delta_{B_2} = \frac{W_B \cdot a^2}{6 \cdot EI}[3l - 4a]$$

yielding $\delta_{B_2} = 0\cdot002\,412$ in.
∴ Total deflection at test bearings is

$$\delta_B = 0\cdot002\,42 \text{ in.}$$

Inserting these deflection values into equation (1) above, we obtain

$$\text{NF}_{(C)} = \frac{187\cdot 7}{\sqrt{0\cdot003\,45}} \simeq 3195\cdot 6 \text{ r.p.m.}$$

and

$$NF_{(B)} = \frac{187{\cdot}7}{\sqrt{0{\cdot}00242}} \simeq 3815{\cdot}5 \text{ r.p.m.}$$

Thus the ratios of forced over natural frequencies are

$$\text{Ratio}_{(C)} = \frac{500}{3195{\cdot}6} = 0{\cdot}156$$

$$\text{Ratio}_{(B)} = \frac{500}{3815{\cdot}5} = 0{\cdot}131$$

Referring to Fig. 9.23, the amplification ratio for undamped vibrations is given by Baker (1975)

$$T_D = \frac{1}{1-(\omega/\Omega)^2}$$

where ω = forced frequency and Ω = natural frequency.

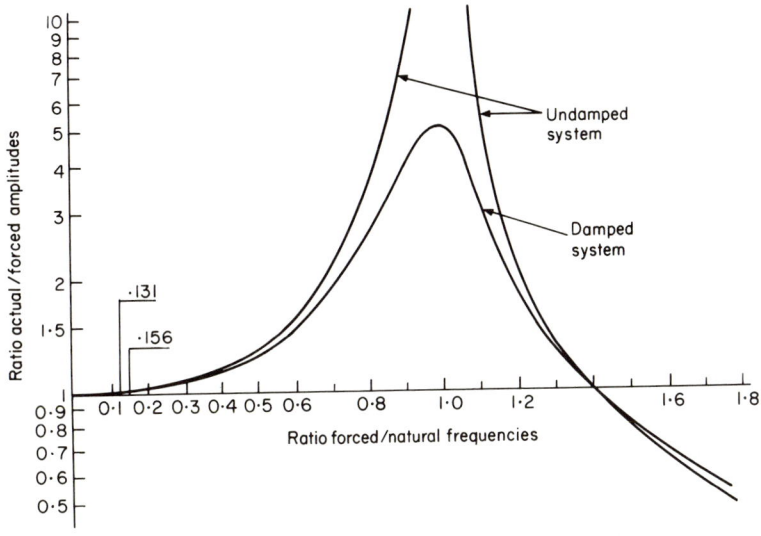

Fig. 9.23. T_D = amplitude ratio = $\dfrac{1}{1-\left(\dfrac{\omega}{\Omega}\right)^2}$ undamped

ω = forced frequency
Ω = natural frequency

$$T_D \leqslant \frac{1}{1-(0{\cdot}156)^2} \leqslant 1{\cdot}025, \text{ i.e., } \leqslant 2{\cdot}5\%$$

For the two cases considered above we obtain

$$T_{D_{(C)}} = 1{\cdot}025 \leqslant 2{\cdot}5\%$$
$$T_{D_{(B)}} = 1{\cdot}0175 \leqslant 1{\cdot}75\%$$

Since the model assumptions were shown to yield lower values than the actual natural frequency of the system, we see that the design is well on the safe side and no coincidence with the forced frequency needs to be expected.

9.10 MANUFACTURING TOLERANCES AND PROBABILITY OF PERFORMANCE ACHIEVEMENT
(Furman and Ellis, 1977)

9.10.1 Introduction

In the design of engineering products, performance analysis is generally carried out on a deterministic basis. It is at the detailing stage that tolerances are introduced, and their choice is governed by manufacturing processes, interchangeability, and whether adjustability features can be incorporated. The case study described in this paper demonstrates the significance of tolerance analysis with respect to performance and this leads inevitably to the assessment of probability of achieving specified performance criteria.

9.10.2 Problem brief

The principal features of an ink-roller assembly of a printing machine are shown on Fig. 9.24. The roller is free to rotate while resting on three wheels mounted on a support shaft. The ends of the ink-roller can slide radially relative to the support shaft. Ideally the ink-roller should exert a uniform pressure against a printing screen. Satisfactory performance has been obtained if the difference between the highest and lowest contact points does not exceed 0·035 mm.

Data available:

$L_R = 1500 \pm 0{\cdot}50$ mm
$L_S = 1600 \pm 2{\cdot}0$ mm (between bearing centres)
$D_R = 38 \pm 0{\cdot}05$ mm
$D_{S_o} = 54 \pm 0{\cdot}05$ mm
$D_{S_i} = 32 \pm 0{\cdot}30$ mm
$\phi_1 = 91 \pm 0{\cdot}021$ mm (same tolerance on ϕ_2, but nominal value to be established)
$Q = 0{\cdot}90$ N mm^{-1} (assumed constant)
$E = 20 \times 10^4$ N mm^{-2} (Young's modulus, assumed constant)

9. MODELLING   333

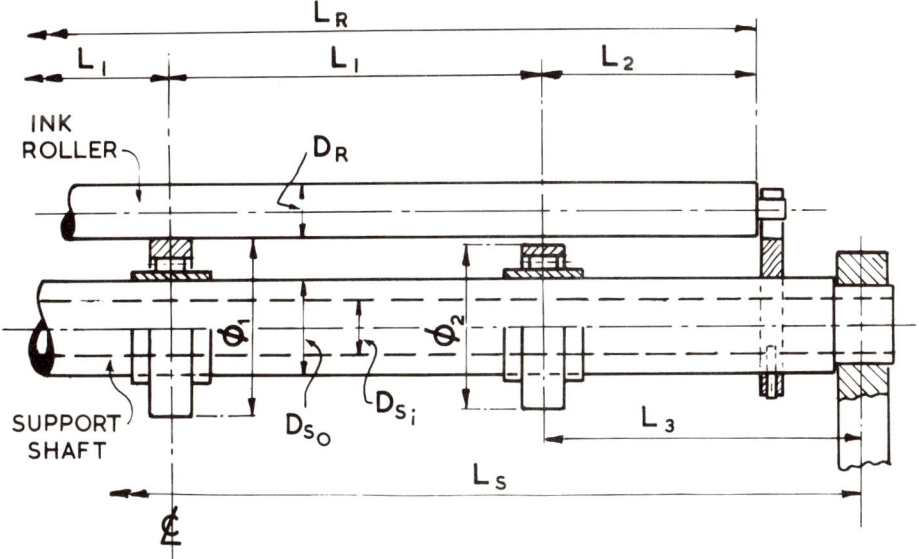

Fig. 9.24. Ink-roller assembly.

The assembly is assumed to be symmetrical about the centre wheel.
Determine:
(a) The best possible location of outer wheels (L_3).
(b) The ideal diameter of the outer wheels (ϕ_2).
(c) The probability of not exceeding the 0·035 mm maximum contact point difference due to deflection, and manufacturing and assembly tolerances.

9.10.3 Deterministic analysis

Figure 9.25 is a conventional idealized representation of the real roller assembly in diagrammatic form.

9.10.3.1 Initial approximation

Ink-roller

Using Macaulay's theorem for the forces and bending moments acting on the ink-roller, and assuming the supports at P_1 and P_2 remain at the same level, we derive the following relationships:
End deflection,

$$y_5 = \frac{Q(L)^4}{48EI_R} \cdot a[6a^2 - (1-a)^3] \tag{1}$$

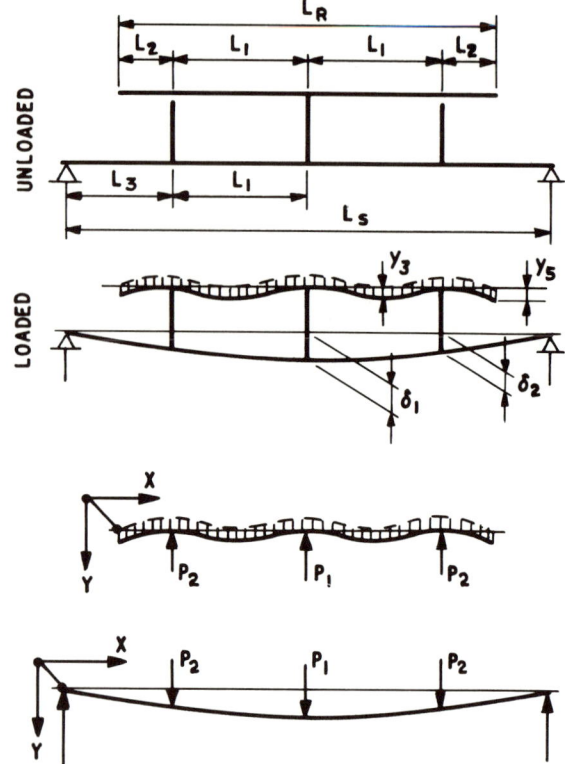

Fig. 9.25. Diagram of ink-roller assembly.

and the max. deflection between P_1 and P_2,

$$y_3 = \frac{Q(L)^4}{48EI_R} \cdot \{2 \cdot Z_3^4 - 8R(Z_3-a)^3 + Z_3[1-3a(1+a+a^2)] + a[6a^2-(1-a)^3]\} \qquad (2)$$

where

$$L = \tfrac{1}{2}L_R; \quad I_R = \frac{\pi}{64}(D_R)^4 \qquad (3a)$$

$$\left.\begin{array}{l} a = \dfrac{L_2}{L} \\[2mm] R = \left[\dfrac{3}{4}\left(\dfrac{1}{1-a}\right)-\left(\dfrac{a+3}{8}\right)\right] \\[2mm] Z_3 = \dfrac{x_3}{L} = \left(\dfrac{3R-1}{2}\right)+\left[\left(\dfrac{3R-1}{2}\right)^2+(3R-1)-6R \cdot a\right]^{1/2} \end{array}\right\} \qquad (3b)$$

and

$$P_1 = Q.L.(1-2R) \tag{4}$$

$$P_2 = Q.L.R \tag{5}$$

Intuitively, the ideal position for P_2 should be such as to make the deflections y_3 and y_5 equal. An interactive computer program was written based on equations (1), (2) and (3a), (3b) allowing y_3 and y_5 to be determined for an entered a value. The value of a which equalized the two deflections was found to be $a = 0.285$. Equations (3a), (3b), (4) and (5) then enabled the corresponding L_2, L_3, P_1 and P_2 values to be calculated.

Support shaft
Again using Macaulay's theorem we established the deflections at P_1 and P_2, their difference being

$$(\delta_1 - \delta_2) = \frac{QL_R L_1^3}{12EI_s}.(2+3C-2R) \tag{6}$$

where

and

$$\left.\begin{array}{c} C = \left(\dfrac{L_3}{L_1}\right); \quad R = \dfrac{P_2}{QL} \\[2mm] I_s = \dfrac{\pi}{64}(D_{S_o}^4 - D_{S_i}^4) \end{array}\right\} \tag{7}$$

$$\therefore \phi_2 = \phi_1 - 2(\delta_1 - \delta_2) \tag{8}$$

For the data given and the values established under (3a) the following nominal results were obtained:

$$L = \frac{L_R}{2} = 750 \text{ mm}$$

$$L_2 = 0.285 \times 750 = 213.75 \text{ mm}$$

$$L_1 = \tfrac{1}{2}L_R - L_2 = 536.25 \text{ mm}$$

$$L_3 = \tfrac{1}{2}L_S - L_1 = 263.75 \text{ mm}$$

$$\phi_2 = \phi_1 - 2(\delta_1 - \delta_2) = 89.957 \text{ mm diameter}$$

$$y_3 = y_5 = 0.01 \text{ mm}$$

9.10.3.2 More rigorous analysis

The initial approximate analysis encouraged the belief that the ink-roller assembly as designed could be expected to perform satisfactorily. However, the assumptions of non-deflection of P_2 relative to P_1 may not be justified. Macaulay's analysis was therefore applied again, but without the initial assumptions. It was convenient in this case to start with the support shaft.

Support shaft
We obtained

$$(\delta_1 - \delta_2) = \frac{1}{EI_s}\left[Q \cdot L \cdot \left(\frac{L_3^3}{6} + \frac{L_s^3}{24} - \frac{L_s^2 L_3}{8}\right) - \frac{P_2 L_1^3}{3}\right] \quad (9)$$

Ink-roller
The general expression for deflection at any point is

$$y = \frac{1}{EI_R} \cdot \left[\frac{Q \cdot x^4}{24} - \frac{P_2(x - L_2)^3}{6} + \left(\frac{P_2 L_1^2}{2} - \frac{QL^3}{6}\right)x + P_2 L_1^2 \left(\frac{L_1}{6} - \frac{L}{2}\right) + \frac{QL^4}{8}\right] \quad (10)$$

When $x = L_2$, equation (10) yields

$$y_2 = \frac{1}{EI_R}\left[\frac{Q \cdot L_2^4}{24} + L_2 \cdot \left(\frac{P_2 L_1^2}{2} - \frac{QL^3}{6}\right) + P_2 L_1^2 \left(\frac{L^1}{6} - \frac{L}{2}\right) + \frac{Q \cdot L^4}{8}\right] \quad (10a)$$

From Fig. 9.26, neglecting infinitesimal tilting of ϕ_2, we note that

$$y_2 = \left(\frac{\phi_1 - \phi_2}{2}\right) - (\delta_1 - \delta_2) \quad (11)$$

which when combined with equations (9) and (10a) yields

$$P_2 = \frac{1}{8L_1^3(I_s + I_R)}\{Q[I_R \cdot L(4L_3^3 + L_s^3 - 3L_s^3 L_s^2 L_3) - I_s(4L^3 L_2 - 3L^4 - L_2^4)] - 12EI_R I_s(\phi_1 - \phi_2)\} \quad (12)$$

The deflections at points 1, 2, 3 and 4 (see Fig. 9.27) are now obtained as follows:

Point 1. $y_1 = 0 =$ datum.
Point 2. For y_2, use equations (12) and (10a).
Point 3. This point lies between P_2 and P_1, where $dy/dx = 0$. Differentiating equation (10) for $x > L_2$, we get

$$\frac{dy}{dx} = 0 = \left(\frac{Qx^3}{6} - \frac{P_2(x - L_2)^2}{2} + \frac{P_2 L_1^2}{2} - \frac{QL^3}{6}\right) \quad (13)$$

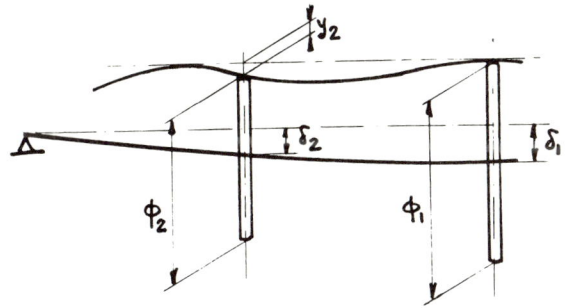

Fig. 9.26.

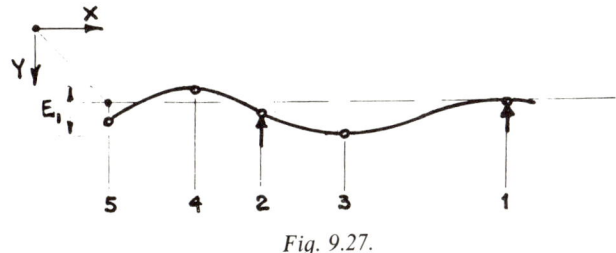

Fig. 9.27.

Since one root of this equation must be $x = L$, the other roots for $x > L_2$ are

$$x_3 = \frac{(3P_2 - Q.L) \pm [(3P_2 - Q.L)^2 - 4Q(Q.L^2 - 3P_2 L + 6P_2 L_2)]^{1/2}}{2Q} \quad (14)$$

and y_3 is obtained by inserting x_3 into equation (10). The larger x_3 value will correspond to the position of maximum deflection, and the smaller x_3 (if $> L_2$) will represent the reverse curvature stationary point on the roller.

Point 4. When this point is real it will lie between the end and P_2, where

$$\frac{dy}{dx} = 0 \quad \text{for } x < L_2$$

Differentiating equation (10) and omitting the second term, we get

$$\frac{dy}{dx} = 0 = \frac{1}{EI_R}\left(\frac{Qx^3}{6} + \frac{P_2 L_1^2}{2} - \frac{QL^3}{6}\right)$$

or

$$x_4 = \left(L^3 - \frac{3P_2 L_1^2}{Q}\right)^{1/3} \quad (15)$$

y_4 is obtained by putting $x = x_4$ in equation (10) and omitting the second term.

Point 5. At $x = 0$, we get y_5 from equation (10), omitting the second term.

338   APPROXIMATE METHODS IN ENGINEERING DESIGN

A short interactive computer program was prepared for equations (9) to (15). Using data given and the values for (L_2 and L_3) and (ϕ_1) calculated under 9.10.3.1, the deflections at points 1 to 5 were computed for (a) nominal parameter values, and (b) extreme tolerance limit values.

The results are given in Table 9.2. There is good agreement with the initial conventional approximation for nominal parameter values, but we can see now that the sure-fit tolerances may completely overwhelm the deflection

Table 9.2. Phase I deflection calculations

Input data: $E = 20E4$; $Q = 0.90$; $L_R = 1500 \pm 0.50$; $L_s = 1600 \pm 2.0$; $L_3 = 263.75 \pm 1.0$
$D_R = 38 \pm 0.05$; $D_{s_o} = 54 \pm 0.05$; $D_{s_i} = 32 \pm 0.3$; $\phi_1 = 91 \pm 0.021$;
$\phi_2 = 89.9575215 \pm 0.021$

$a = \dfrac{L_2}{\frac{1}{2}L_R} = 0.285$		(5) ← y↓ x→	(4)	(2) ↑ P_2	(3)	(1) ↑ P_1
$\phi_2 = 89.9575215$	x	0	206.458	213.75		478.817
	y	$0.100620E-1$	$-0.222203E-4$	$-0.95867E-7$		$0.991239E-2$
$\phi_2 = 89.9785215$	x	0	178.973	213.75		508.857
	y	$-0.304872E-2$	$-0.868846E-2$	$-0.820504E-2$		$0.755486E-2$
$\phi_2 = 89.9365215$	x	0	231.868	213.75		443.474
	y	$0.231727E-1$	$0.80622E-2$	$0.820485E-2$		$0.128335E-1$
$D_R = 38.05$	x	0	206.958	213.75		478.817
	y	$0.100092E-1$	$-0.221042E-4$	$-0.957566E-7$		$0.986039E-2$
$D_R = 37.95$	x	0	206.958	213.75		478.817
	y	$0.101151E-1$	$-0.223373E-4$	$-0.959772E-7$		$0.996473E-2$
$D_{s_o} = 54.04$	x	0	211.939	213.75		471.970
	y	$0.128163E-1$	$0.171576E-2$	$0.171737E-2$		$0.104726E-1$
$D_{s_o} = 53.95$	x	0	201.705	213.75		485.468
	y	$0.730713E-2$	$-0.179152E-2$	$-0.172414E-2$		$0.937464E-2$
$D_{s_i} = 32.30$	x	0	200.243	213.75		487.223
	y	$0.656568E-1$	$-0.227211E-2$	$-0.218816E-2$		$0.923396E-2$
$D_{s_i} = 31.70$	x	0	213.045	213.75		470.372
	y	$0.134334E-1$	$0.210957E-2$	$0.210981E-2$		$0.106041E-1$
$L_R = 1500.5$	x	0	207.507	214.0		479.278
	y	$0.10154E-1$	$-0.376271E-4$	$-0.17324E-4$		$0.988554E-2$
$L_R = 1499.5$	x	0	206.408	213.50		478.357
	y	$0.997051E-1$	$-0.680762E-5$	$0.172187E-4$		$0.993926E-2$
$L_s = 1602$	x	0	196.213	212.75		486.772
	y	$0.558856E-1$	$-0.255891E-2$	$-0.24367E-2$		$0.934904E-2$
$L_s = 1598$	x	0	216.749	214.75		470.279
	y	$0.145456E-1$	$0.24251E-2$	$0.242709E-2$		$0.105149E-1$
$L_3 = 264.75$	x	0	213.857	214.75		474.529
	y	$0.128919E-1$	$0.139428E-2$	$0.139467E-2$		$0.101704E-1$
$L_3 = 262.75$	x	0	199.586	212.75		482.872
	y	$0.724655E-1$	$-0.147569E-2$	$-0.139655E-2$		$0.966661E-2$
$\phi_1 = 91.021$	x	0	231.868	213.75		443.474
	y	$0.231727E-1$	$0.80622E-2$	$0.820485E-2$		$0.128335E-1$
$\phi_1 = 90.979$	x	0	178.973	213.75		508.857
	y	$-0.304872E-1$	$-0.868846E-2$	$-0.820504E-2$		$0.755486E-2$

limits required for satisfactory operation of the ink-roller assembly. Figure 9.28 shows clearly the influence of the wheel diameter tolerances on the ink-roller deflections at points 3 and 5. It is also noted that the individual y-values do not satisfy the criterion in the original brief, i.e. that "the difference between the highest and lowest contact points does not exceed 0·035 mm". To verify this we must establish the envelope within which the deflected centre-line of the ink-roller lies.

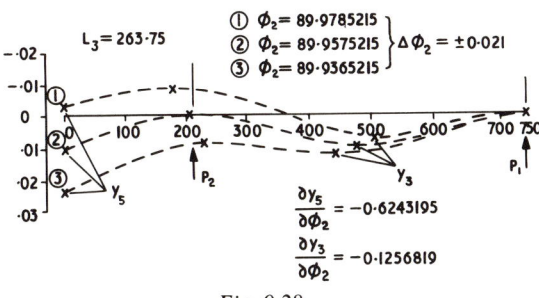

Fig. 9.28.

9.10.4 Sure-fit tolerance analysis

The computer program written for equations (9) to (15) was extended to compute the deflection envelope (Fig. 9.29, Phase II). The results obtained by using the extreme values of any one parameter at a time while the rest were maintained at their nominal values is shown in Table 9.3. Also tabulated are the values of the approximate partial derivatives in order to illustrate the envelope sensitivity relative to each parameter. For example,

$$\frac{\partial E_1}{\partial \phi_2} = \frac{0{\cdot}023\,172\,7 - 0{\cdot}010\,084\,2}{-0{\cdot}021} = -0{\cdot}6233$$

i.e. as the diameter of the outer support wheel ϕ_2 is reduced the envelope value relative to the nominal (minimum) is increased. Increasing ϕ_2 also increases E_1, although less so. (N.B. Clearly E_1 is most sensitive to ϕ_1 and ϕ_2.) These relationships are illustrated graphically in the last column of Table 9.3.

By noting the changes in the envelope magnitude as each variable was adjusted within its tolerance band, the following conditions were anticipated (using Table 9.3).

(a) Lowest E_1 values may be expected to occur when

$\phi_2 = 89{\cdot}957\,521\,5$ (nominal value)
$D_R = 38{\cdot}05$ (maximum value)
$D_{s_o} = 54{\cdot}00$ (nominal value)
$D_{s_i} = 32{\cdot}00$ (nominal value)

340 APPROXIMATE METHODS IN ENGINEERING DESIGN

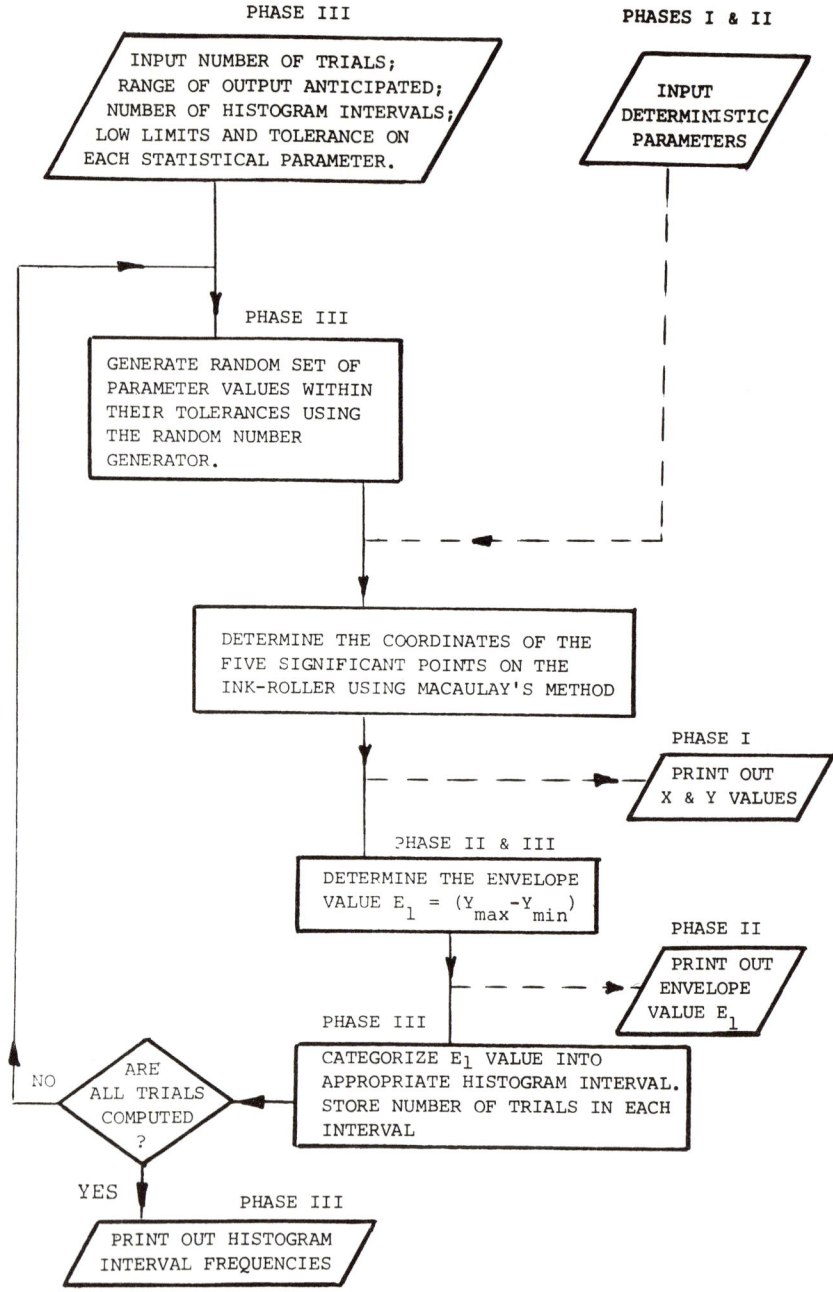

Fig. 9.29. Computer algorithm.

Table 9.3. Phase II envelope calculations

Basic conditions: $E = 20E4$; $Q = 0.90$; $L_R = 1500$; $L_s = 1600$; $L_3 = 263.75$
$D_R = 38$; $D_{s_o} = 54$; $D_{s_i} = 32$; $\phi_1 = 91$; $\phi_2 = 89.9575215$
Envelope $= (y_{max} - y_{min}) = 0.0100842$ Nominal

Tolerance band	Parameter	Envelope	Stationary point	Approximate slopes	
0.042	$\phi_2 = \dfrac{89.9785215}{89.9365215}$	$E_1 = \dfrac{0.0162433}{0.0231727}$	Not between end and P_2	$\dfrac{\partial E_1}{\partial \phi_2} = \dfrac{0.2933}{-0.6233}$	
0.10	$D_R = \dfrac{38.05}{37.95}$	$E_1 = \dfrac{0.0100313}{0.0101375}$	Ref. point 4, Table 9.2.	$\dfrac{\partial E_1}{\partial D_R} = -0.00106$	
0.10	$D_{s_o} = \dfrac{54.05}{53.95}$	$E_1 = \dfrac{0.0128063}{0.0111662}$	Ref. point 4, Table 9.2	$\dfrac{\partial E_1}{\partial D} = \dfrac{0.0544}{-0.02164}$	
0.60	$D_{s_i} = \dfrac{32.30}{31.70}$	$E_1 = \dfrac{0.0115061}{0.0134334}$	Ref. point 4 Table 9.2	$\dfrac{\partial E_1^{s_o}}{\partial D_{s_i}} = \dfrac{0.00474}{-0.01116}$	
1.0	$L_R = \dfrac{1500.5}{1499.5}$	$E_1 = \dfrac{0.0101916}{0.0099773}$	Ref. point 4 Table 9.2	$\dfrac{\partial E_1}{\partial L_R} = 0.00021$	
4.0	$L_s = \dfrac{1602}{1598}$	$E_1 = \dfrac{0.011908}{0.0145456}$	Not between end and P_2	$\dfrac{\partial E_1}{\partial L_s} = \dfrac{0.00091}{-0.00223}$	
2.0	$L_3 = \dfrac{264.75}{262.75}$	$E_1 = \dfrac{0.0128919}{0.0111423}$		$\dfrac{\partial E_1}{\partial L_3} = \dfrac{0.0028}{-0.00106}$	
0.042	$\phi_1 = \dfrac{91.021}{90.979}$	$E_1 = \dfrac{0.0231727}{0.0162433}$	Not between end and P_2	$\dfrac{\partial E_1}{\partial \phi_1} = \dfrac{0.6233}{-0.2933}$	

$L_R = 1499.5$ (minimum value)
$L_{s_i} = 1600$ (nominal value)
$L_3 = 263.75$ (nominal value)
$\phi_2 = 91.0$ (nominal value)

Computing with these values, one obtains; $E_{1_{min}} = 0.00992501$ mm. It will be noted that the minimum E_1 corresponds principally to the parameters being at their nominal values.

(b) Highest E_1 values can be expected to occur when

$\phi_2 = 89.9365215$ (minimum value)
$D_R = 37.95$ (minimum value)
$D_{s_0} = 54.05$ (maximum value)
$D_{s_i} = 31.70$ (minimum value)
$L_R = 1500.5$ (maximum value)
$L_s = 1598$ (minimum value)
$L_3 = 264.75$ (maximum value)
$\phi_1 = 91.021$ (maximum value)

Computing with these values one obtains $E_{1_{max}} = 0.0500882$ mm. If this result is correct, one would have to expect that a number of ink-roller assemblies may well be outside the limits for satisfactory operation. The probability of this happening in normal production and assembly needs to be ascertained.

9.10.5 Monte Carlo simulation

The probability that all worst parameter values would occur simultaneously must be low. The method chosen to establish a statistical forecast of such an occurrence is the Monte Carlo technique. This is a direct simulation of building a large number of assemblies by using random values of the various parameters involved (within their tolerance limits).

Table 9.4. Phase III Monte Carlo simulation

Number of trials = 1000; Number of intervals = 24
Lowest $E_1 = 0$; Highest $E_1 = 0.06$; anticipated

(1) 0	(5) 237	(9) 79	(13) 29	(17) 0
(2) 0	(6) 204	(10) 58	(14) 14	(18) 0
(3) 0	(7) 157	(11) 44	(15) 8	(19) 0
(4) 0	(8) 130	(12) 35	(16) 5	(20) 0

A standard software package generates random numbers between 0 and 1 of uniform distribution. Using suitable multipliers a random set of values for the parameters is obtained. It should be appreciated that machined dimensions do not usually have uniform (i.e. rectangular) distributions, but this would, if anything, tend to give pessimistic results.

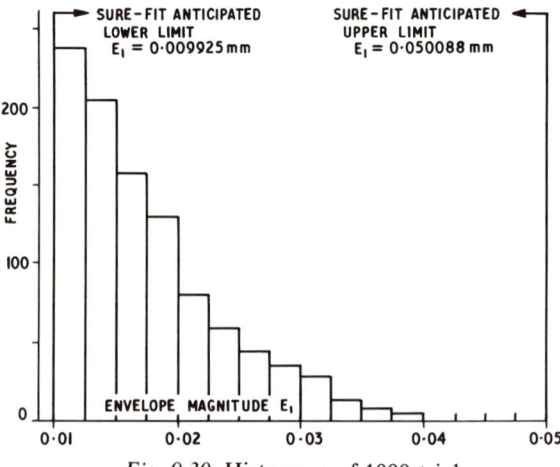

Fig. 9.30. Histogram of 1000 trials.

9. MODELLING 343

The random number generator program was looped into the Phase II program, previously referred to, which was also extended to collect and count the E_1 values within 24 preselected intervals (Fig. 9.29, Phase III); 1000 trials within the tolerance band of each parameter were computed and tabulated (Table 9.4); the results are plotted as a histogram in Fig. 9.30.

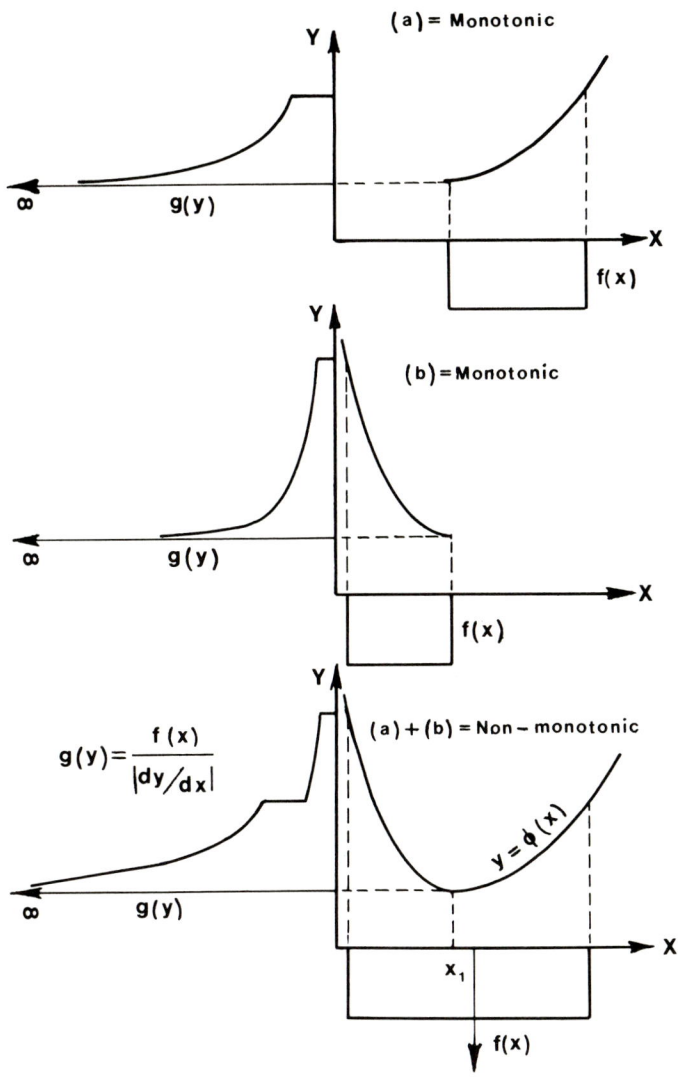

Fig. 9.31. Probability of a function.

It will be seen that of 1000 simulated random assemblies only 13 have E_1 values greater than 0·035 mm. Thus the roller assembly design, if manufactured and assembled to the specified tolerances, will have a 98·7% probability of giving satisfactory performance.

The totally skewed distribution of E_1 (Fig. 9.30) begs an explanation. Consider the case of a function of a single random variable, $y = \phi(x)$. The probability density at any value of $x = f(x)$, and the corresponding probability density of $y = g(y)$. It can be shown that,

$$g(y) = \frac{f(x)}{|dy/dx|}$$

where $|dy/dx|$ is the absolute value of slope at (y, x) on the curve (Breipohl, 1970).

If one now considers a distribution which spans a non-monotonic function, with a stationary point (minimum or maximum) at x_1 (Fig. 9.31), we note that

(i) where $\dfrac{dy}{dx} = 0, \quad g(y) = \infty$

and the density distribution of y is totally skewed such that

$$g(y) = \infty \quad \text{at } y = y_{\min} \text{ or } y_{\max}$$

We observed in Table 9.3 that E_1 is related non-monotonically to most of the influential variables, especially ϕ_1 and ϕ_2. Thus although E_1 is a function of several variables its overall relationship to them clearly can be expected to be non-monotonic. In addition it should be noted that the minimum envelope value is obtained when $(\phi_1 - \phi_2)$ is at its nominal value, and this difference can occur over the whole range of ϕ_1 and ϕ_2. Thus the skewed probability distribution of E_1 is as would be expected.

The case described above illustrates only too well the limitations of the well-known Central Limit Theorem accepted by statistical theory.

REFERENCES

Baker, J. K. (1975). "Vibration Isolation". Engineering Design Guides, Design Council.
Breipohl, A. M. (1970). "Probabilistic System Analysis". J. Wiley and Sons, Chichester.
Furman, T. T. and Ellis, J. (1977). "Manufacturing Tolerances and Probability of Performance Achievement (Design Case Study)". II International Conference on "Increase of Automobile Reliability". Prague.
Taylor, C. F. (1966). "Internal Combustion Engines", Vol. I. p. 174.
(Resal. H.) (1874). "Strength of Materials".
Whyte, R. R. (1975). "Engineering Progress Through Trouble". I.Mech.E.

Solutions

Chapter One

Example 1.7.2

$$x^3 + x = b \qquad \text{(i)}$$

Take $x_0 = x_1 + \varepsilon_1$ then $(x_1+\varepsilon_1)^3 + (x_1+\varepsilon_1) = b$. First approximation

$$x_1^3 + \varepsilon_1(3x_1^2+1) + x_1 \simeq b$$

$$\therefore \varepsilon_1 \simeq \left(\frac{b - x_1^3 - x_1}{3x_1^2 + 1}\right) \qquad \text{(ii)}$$

Thus generally,

$$\varepsilon_n \simeq \left(\frac{b - x_n^3 - x_n}{3x_n^2 + 1}\right) \qquad \text{(iii)}$$

Second approximation

$$x_2 \simeq x_1 + \varepsilon_1 \simeq x_1 + \left(\frac{b - x_1^3 - x_1}{3x_1^2 + 1}\right).$$

Thus,

$$x_{n+1} \simeq \left(\frac{2x_n^3 + b}{3x_n^2 + 1}\right) \qquad \text{(iv)}$$

Example 1.8.1

Point	x_i	x_i^2		(y_i)	$(x_i)(y_i)$	(From equation (iii) below) $\phi(x_i)$	$[y_i - \phi(x_i)]^2$
0	0	0	...	0·82	0	0·85	0·0009
1	1	1	...	3·46	3·46	3·41	0·0025
2	2	4	...	5·92	11·84	5·97	0·0025
3	3	9	...	8·57	25·71	8·53	0·0016
4	4	16	...	11·07	44·28	11·09	0·0004
5	5	25	...	13·64	68·20	13·65	0·0001
6	15	55	...	43·48	153·49	43·5	0·0080
n	$\sum x_i$	$\sum(x_i)^2$	...	$\sum(y_i)$	$\sum(x_i)(y_i)$	$\sum \phi(x_i)$	$\sum[y_i - \phi(x_i)]^2$

348   APPROXIMATE METHODS IN ENGINEERING DESIGN

Inserting into equation (1.18), we get

$$6a_0 + 15a_1 \simeq 43.48 \rightarrow (\times 2) \rightarrow 30a_0 + 75a_1 = 217.40 \qquad \text{(i)}$$
$$15a_0 + 55a_1 \simeq 153.49 \rightarrow (\times 5) \rightarrow 30a_0 + 110a_1 = 306.98 \qquad \text{(ii)}$$

which yield $a_1 = 2.56$ and $a_0 = 0.85$, ∴ best line is

$$y = 0.85 + 2.56 \cdot x \qquad \text{(iii)}$$

$$\text{RMS of error} = \left[\frac{0.0080}{6}\right]^{\frac{1}{2}} = 0.0365 \simeq 0.037$$

$$y_{\text{mean}} = \frac{43.48}{6} = 7.247$$

$$\therefore \% \text{ RMS of error} = \frac{0.037 \times 100}{7.247} = 0.504\%$$

Example 1.8.2
Given Table 1.

Table 1.

x	0	1	2	3	4	5	6
y	−0.01	0.37	0.71	0.98	1.28	1.52	1.69

It is convenient to change x into $z = (x-3)$.

Table 2.

Point	z_i	z_i^2	z_i^3	z_i^4	...	y_i	$y_i(z_i)$	$y_i(z_i)^2$
0	−3	9	−27	81	...	−0.01	+0.03	−0.09
1	−2	4	−8	16	...	0.37	−0.74	+1.48
2	−1	1	−1	1	...	0.71	−0.71	+0.71
3	0	0	0	0	...	0.98	0	0
4	+1	1	+1	1	...	1.28	1.28	1.28
5	+2	4	+8	16	...	1.52	3.04	6.08
6	+3	9	+27	81	...	1.69	5.07	15.21
7	0	28	0	196	...	6.54	7.97	24.67
n	$\Sigma(z_i)$	$\Sigma(x_i^2)$	$\Sigma(z_i)^3$	$\Sigma(z_i)^4$	...	Σy_i	$\Sigma(y_i \cdot z_i)$	$\Sigma(y_i \cdot z_i^2)$

(a) For best line we have
   (i) $7a_0 + 0a_1 = 6.54$
   (ii) $0a_0 + 28a_1 = 7.97$

Solving these two simultaneous equations yields
$a_0 = 0.934$ and $a_1 = 0.285$
Thus for best line

$$y = 0.934 + 0.285z$$

and changing back to x $\left. \begin{array}{c} \\ z = (x-3) \end{array} \right\} \to y = 0.08 + 0.285 \cdot x$

To determine the RMS value we set up Table 3.

Table 3.

Point	x_i	y_i	$\phi(x_i)$	$\overline{y_i - \phi(x_i)}$	$[y_i - \phi(x_i)]^2$
0	0	−0.01	0.080	0.090	0.0081
1	1	0.37	0.365	0.005	0.000 025
2	2	0.71	0.660	0.060	0.0036
3	3	0.98	0.935	0.075	0.002 025
4	4	1.28	1.22	0.006	0.000 036
5	5	1.52	1.505	0.015	0.000 225
6	6	1.69	1.790	0.10	0.01
7	7	$\sum y_i = 6.54$			0.024
n		$\bar{y}_i = 0.934$ mean			$\sum [y_i - \phi(x_i)]^2$

$$\therefore \text{RMS}_{(a)} = \left(\frac{0.024}{7} \right)^{\frac{1}{2}} = 0.059$$

$$\therefore \% \text{ RMS}_{(a)} = \frac{0.059 \times 100}{0.934} = 6.27\%$$

(b) Best quadratic curve
From Table 2, we have
  (i) $7a_0 + 0a_1 + 28a_2 = 6.54$
  (ii) $0a_0 + 28a_1 + 0a_2 = 7.97$
  (iii) $28a_0 + 0a_1 + 196a_2 = 24.67$
Solving these three simultaneous equations yields

$$\phi(z) = 1.0052 + 0.2846 \cdot z - 0.017\,74 \cdot z^2$$

converting z back into x, $z = (x - 3)$, we get

$$\phi(x) = -0.0083 + 0.391 \cdot x - 0.018 \cdot x^2$$

as the "best quadratic polynomial". To find RMS for this polynomial, we set up Table 4.

Table 4.

Point	x_i	y_i	$\phi(x_i)$	$\overline{y_i - \phi(x_i)}$	$[y_i - \phi(x_i)]^2$
0	0	−0·01	−0·0083	0·0017	0·000 003
1	1	0·37	0·365	0·005	0·000 025
2	2	0·71	0·702	0·008	0·000 064
3	3	0·98	1·003	0·0023	0·000 053
4	4	1·28	1·268	0·020	0·000 40
5	5	1·52	1·497	0·023	0·000 53
6	6	1·69	1·690	0·00	0·000

$\sum = 6{\cdot}54$
y mean $= 0{\cdot}934$ $\qquad\qquad\qquad\qquad\qquad\qquad\qquad \sum = 0{\cdot}001\,075$

$$\therefore \text{RMS}_{(b)} = \left[\frac{0{\cdot}001075}{7}\right]^{\frac{1}{2}} = 0{\cdot}0124$$

and $\%\ \text{RMS}_{(b)} = \dfrac{0{\cdot}0124 \times 100}{0{\cdot}934} = 1{\cdot}33\% = $ much closer fit than straight line.

(See Fig. 1.5 p. 19.)

Chapter Two

Example 2.1
$$V = r^2 \times h$$
Using rules (4) and (6), we get
$$\Delta V = V.\,\delta V = V(2\delta r + \delta h)$$
$$\therefore \delta v = 2\delta r + \delta h$$

Example 2.2
$$z = \sqrt{\{x/(1+y)\}}; \quad x \text{ and } y \text{ are independent variables.}$$
Using rules (5) and (6), we get
$$\delta z = \tfrac{1}{2}[\delta x \pm \delta(1+y)]$$
or
$$\frac{\Delta z}{z} = \tfrac{1}{2}\left[\frac{\Delta x}{x} \pm \frac{\Delta y}{(1+y)}\right]$$
For independent variables the worst combination must be expected, i.e.
$$\Delta z = \tfrac{1}{2}.z\left[\frac{\Delta x}{x} + \frac{\Delta y}{(1+y)}\right]$$

Example 2.3
$$y = x^3 - 2x^2 + x + 2$$
Using rules (1), (2) and (6), we get
$$\Delta y = (3.\,\delta x.\,x^3 - 2.\,\delta x.\,x^2 + \delta x.\,x)$$
or
$$\Delta y = (3x^3 - 2x^2 + x).\,\delta x$$
Since $\delta x = \Delta x/x$, we get
$$\Delta y = (3x^2 - 2x + 1)\Delta x$$

Note. The minus sign is retained between the first and second terms because x^3 is directly correlated with x^2 and with x, of course.

Example 2.4
The relationship between r, x and y is
$$r^2 = x^2 + y^2$$
Using rules (6) and (1), we get
$$\delta(x^2) = 2\delta x; \quad \therefore \Delta(x^2) = 2\delta x \cdot x^2$$
$$\delta(y^2) = 2\delta y; \quad \therefore \Delta(y^2) = 2\delta y \cdot y^2$$
$$\delta(r^2) = 2\delta r; \quad \therefore \Delta(r^2) = 2\delta r \cdot r^2$$
Thus
$$2\delta r \cdot r^2 = 2\delta x \cdot x^2 + 2\delta y \cdot y^2$$
or
$$\delta r = \frac{x^2 \cdot \delta x + y^2 \cdot \delta y}{r^2}$$
or
$$\delta_r = \frac{x \cdot \Delta x + y \cdot \Delta y}{r^2}$$
or
$$\Delta_r = \frac{x \cdot \Delta x + y \cdot \Delta y}{r}$$

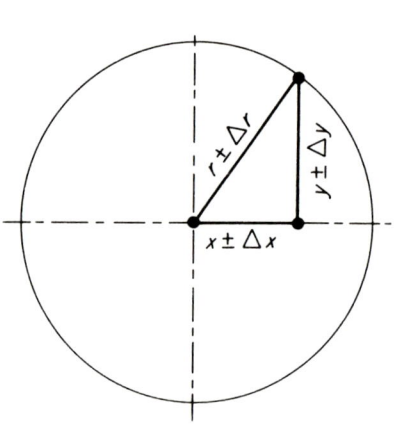

Example 2.5
Circle dia. = $5 \cdot 92 \pm 0 \cdot 005$.
  (a) Circumference, $C = \pi \cdot D$. By rule (4)
$$\delta C = (0) + \delta D$$
$$\therefore \delta C = \delta D$$
or
$$\frac{\Delta C}{C} = \frac{\Delta D}{D}$$
$$\therefore \Delta C = \frac{\pi \cdot D}{D} \cdot \Delta D = \pi \cdot \Delta D$$
$$\therefore \Delta C = \pm \pi(0 \cdot 005) = \pm 0 \cdot 0157$$

(b) Area, $A = (\pi/4)D^2$.

$$\delta A = 0 + \delta(D^2) = 2\delta D$$

$$\frac{\Delta A}{A} = \frac{2\Delta D}{D}$$

$$\therefore \Delta A = \frac{A}{D} \cdot 2 \cdot \Delta D = \frac{\pi}{4} D \cdot 2 \cdot \Delta D$$

or

$$\Delta A = \frac{\pi}{2} \cdot D \cdot \Delta D$$

Therefore,

$$\Delta A = \frac{\pi}{2}(5 \cdot 92) \cdot (0 \cdot 005)$$

i.e.

$$\Delta A = \pm 0 \cdot 047$$

Example 2.6
Area, $A = \frac{1}{2}b \cdot h$, b and h being independent variables. By rule (4):

$$\delta A = 0 + \delta b + \delta h$$

or

$$\frac{\Delta A}{A} = \frac{\Delta b}{b} + \frac{\Delta h}{h}$$

or

$$\Delta A = A\left(\frac{\Delta b}{b} + \frac{\Delta h}{h}\right)$$

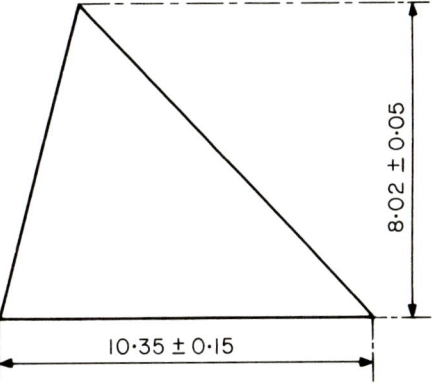

Therefore,

$$\Delta A = \frac{1}{2}(10 \cdot 35 \times 8 \cdot 02)\left(\frac{0 \cdot 15}{10 \cdot 35} + \frac{0 \cdot 05}{8 \cdot 02}\right)$$

$$\therefore \Delta A = \pm 0 \cdot 86 \text{ square units,}$$

i.e. the relative error = $2 \cdot 07\%$.

Example 2.7

$$\tan \alpha = 0.818 \pm 0.002$$

As a first step we must take help from differentiation: Let $z = \tan \alpha$, then

$$dz = (1 + \tan^2 \alpha) \cdot d\alpha$$

$$\therefore d\alpha = \frac{dz}{1 + \tan^2 \alpha} \text{ radians}$$

The equivalent difference equation is

$$\Delta \alpha = \frac{\Delta z}{1 + \tan^2 \alpha} = \frac{\Delta z}{1 + z^2}$$

$$\therefore \Delta \alpha = \frac{(\pm 0.002)}{1 + (0.818)^2}$$

which yields $\Delta \alpha = \pm 0.11983 \times 10^{-2}$ radians or $\Delta \alpha = \pm (0°, 4.1')$ and $\alpha = (39°, 17' \pm 4.1')$.

Example 2.8
Leaf spring

$$l = 100 \pm 2.0 \text{ mm}$$
$$b = 20 \pm 0.5 \text{ mm}$$
$$t = 3 \pm 0.10 \text{ mm}$$
$$E = 21 \times 10^4 \, N\,mm^{-2} \pm 3.33\%$$

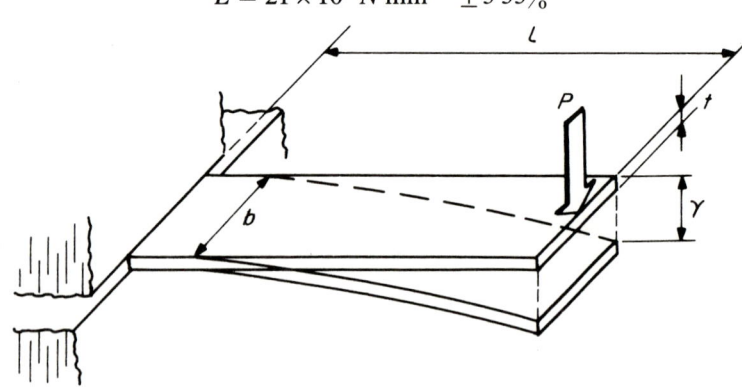

Deflection for a cantilever is

$$\gamma = \frac{Pl^3}{3IE}$$

SOLUTIONS TO CHAPTER TWO

where
$$I = \frac{b \cdot t^3}{12}$$

$$\therefore \text{stiffness} = \frac{P}{\gamma} = \frac{bt^2 E}{4l^3} = k \tag{1}$$

By rules (4) and (6), we have

$$\delta k = \left(\frac{\Delta b}{b} + 3\frac{\Delta t}{t} + \frac{3\Delta l}{l} + \frac{\Delta E}{E}\right) \tag{2}$$

Inserting the given numerical values in equation (2) above we have

$$\delta k = \left(\frac{0.5}{20} + 3\frac{0.10}{3} + 3\frac{2.0}{100} + \frac{1}{30}\right)$$

i.e.
$$\delta k = 0.1583 \equiv 15.83\%$$

$$k = 28.35 \text{ N mm}^{-1} \quad \text{(from equation (1))}.$$

$$\therefore \Delta k = \pm 4.48 \text{ N mm}^{-1}$$

Example 2.9
*Technical information, tables and formulae**

The equations developed originally by Almen and Laszlo lead to very tedious calculations in the design of these springs, but tables of constants for the ratios of height to thickness and outside diameter to inside diameter have been developed, which shorten them considerably. The characteristics required of the spring have first to be selected from consideration of the load/deflection curves. The shape of the curve depends on the ratio h/t but it can be shown that the load required to flatten the spring is independent of this ratio and is given by

$$P_{(\text{flat})} = K_1 \times 10^6 \left(\frac{ht^3}{D^2}\right) \tag{i}$$

where K_1 is a constant depending on the material used and the ratio D/d. The values of K_1 for various ratios D/d are given for steel springs given in Table 1. The load required to deflect the spring to some other state than flat $= C \times P_{\text{flat}}$, where C is a multiplier depending on the deflection f and the ratio h/t. The values of C for various ratios of h/t are given in Table 2, and other values if required can be calculated from the equation

$$C = a\left[\left(1 - \frac{3a}{2} + \frac{a^2}{2}\right)\frac{h^2}{t^2} + 1\right] \tag{ii}$$

where $a = f/h$.

* This information was obtained from data supplied by Wallingford (Spring Pressings) Ltd.

Table 1. Values of K for various ratios of D/d.

D/d	K_1	K_2	K_3	D/d	K_1	K_2	K_3
1·4	1·934	1·072	1·135	2·75	1·157	1·378	1·652
1·45	1·810	1·085	1·157	2·8	1·153	1·388	1·670
1·5	1·716	1·088	1·178	2·85	1·150	1·398	1·688
1·55	1·638	1·111	1·199	2·9	1·146	1·407	1·704
1·6	1·567	1·124	1·219	2·95	1·144	1·417	1·721
1·65	1·514	1·136	1·239	3·0	1·141	1·420	1·739
1·7	1·467	1·149	1·260	3·05	1·139	1·435	1·756
1·75	1·430	1·161	1·280	3·1	1·137	1·445	1·773
1·8	1·394	1·173	1·300	3·15	1·136	1·454	1·790
1·85	1·367	1·185	1·320	3·2	1·134	1·464	1·806
1·9	1·338	1·197	1·339	3·25	1·132	1·473	1·823
1·95	1·317	1·208	1·350	3·3	1·131	1·482	1·840
2·0	1·296	1·220	1·378	3·35	1·130	1·491	1·856
2·05	1·279	1·231	1·397	3·4	1·128	1·500	1·873
2·1	1·261	1·242	1·416	3·45	1·128	1·509	1·869
2·15	1·249	1·253	1·435	3·5	1·127	1·518	1·906
2·2	1·235	1·264	1·454	3·55	1·126	1·527	1·922
2·25	1·225	1·275	1·472	3·6	1·126	1·535	1·933
2·3	1·214	1·286	1·491	3·65	1·126	1·544	1·955
2·35	1·206	1·296	1·509	3·7	1·126	1·553	1·971
2·4	1·196	1·307	1·527	3·75	1·126	1·561	1·987
2·45	1·190	1·317	1·545	3·8	1·126	1·570	2·003
2·5	1·182	1·328	1·563	3·85	1·126	1·578	2·019
2·55	1·176	1·339	1·581	3·9	1·126	1·587	2·035
2·6	1·171	1·348	1·599	3·95	1·125	1·595	2·051
2·65	1·166	1·358	1·617	4·0	1·125	1·604	2·067
2·7	1·161	1·368	1·635				

Table 2.

Multiplier C h/l	Values of f/h							
	0·2	0·4	0·6	0·8	1·0	1·2	1·4	1·6	1·8
0·2	0·21	0·41	0·61	0·80	1·0				
0·4	0·22	0·43	0·63	0·82	1·0				
0·6	0·25	0·47	0·66	0·81	1·0				
0·8	0·29	0·52	0·71	0·86	1·0				
1·0	0·34	0·59	0·77	0·90	1·0	1·10	1·23	1·41	1·66
1·2	0·41	0·63	0·84	0·94	1·0	1·06	1·16	1·32	1·59
1·4	0·48	0·78	0·93	0·99	1·0	1·01	1·07	1·22	1·51
1·6	0·57	0·89	1·03	1·05	1·0	0·95	0·97	1·11	1·43
1·8	0·67	1·02	1·14	1·11	1·0	0·89	0·86	0·98	1·34
2·0	0·78	1·17	1·27	1·18	1·0	0·82	0·73	0·83	1·22
2·2	0·90	1·33	1·41	1·26	1·0	0·74	0·59	0·67	1·10

Finally, it is necessary to check the stress between the maximum and minimum deflections to be used.

The formula can be written:

Since
$$S = K_1 \frac{f}{D^2} 10^6 \left[(2K_3 - K_2)\left(h - \frac{f}{2}\right) + K_3 t \right]$$

$$\frac{K_1}{D^2} \times 10^6 = \frac{P_{\text{flat}}}{ht^3}$$

(iii)

and f is a function of h, the calculations may be shortened by the use of the equations (i) to (iv). Values for K_1, K_2 and K_3 may be found in Table 1.

When $f = 0.4h$

$$S = 0.4 \frac{P\,\text{flat}}{t^2}\left[0.8 \frac{h}{t}(2K_3 - K_2) + K_3\right]$$ (iv)

When $f = 0.6h$

$$S = 0.6 \frac{P\,\text{flat}}{t^2}\left[0.7 \frac{h}{t}(2K_3 - K_2) + K_3\right]$$ (v)

When $f = 0.8h$

$$S = 0.8 \frac{P\,\text{flat}}{t^2}\left[0.6 \frac{h}{t}(2K_3 - K_2) + K_3\right]$$ (vi)

When $f = h$

$$S = \frac{P\,\text{flat}}{t^2}\left[0.5 \frac{h}{t}(2K_3 - K_2) + K_3\right]$$ (vii)

With oscillating loads the life of the spring will depend upon the stress range as well as the maximum stress. In general, if no preload is used, the stress should not exceed 950 N mm^{-2} for a life of 100 000 cycles. For longer lives it will be necessary to limit the maximum stress further or to reduce the stress range by using a preload.

From equation (i) we get

$$\delta P_{\text{flat}} = \pm(\delta h + 3.\delta t + 2\delta D)$$ (a)

and from (iii) using $P_{\text{flat}} = (P/C)$, we have

$$\delta s_{(x)} = (\delta P_{\text{flat}} + 3.\delta t + \delta h)$$ (b)

assuming,

$$(f/h) = \text{constant}$$

∴ Combining (a) and (b), we get
$$\delta s_{(x)} = \pm(2.\delta.D + 6.\delta t + 2.\delta h) \tag{c}$$

The stiffness of these disc springs varies with deflection and has to be estimated for selected values of P or f or f/h; thence use equations (iv) to (vii) as appropriate. Select a particular disc spring, numerical values of which are given in the manufacturer's catalogue.

Example 2.10
Shrink fit:
$$\text{Radial pressure} = \frac{E.\gamma_{rad}(D^2-d^2)}{d.D^2} = P_r \tag{i}$$

Torque $= p_r.\mu.\pi.d.d/2$ per unit length of hub
or
$$T_1 = p_r.\mu.\pi.d^2/2 \text{ per unit length of hub} \tag{ii}$$

Taking E and μ as constants and combining (i) and (ii) we get
$$T_1 = (\tfrac{1}{2}.\mu.\pi.E).\frac{D^2-d^2}{D^2}.d.\gamma_{rad} \tag{iii}$$

In this expression d and γ_{rad} are dependent variables. We may write (iii) in the form
$$T_1 = (\tfrac{1}{2}.\mu.\pi.E).d.\gamma_{rad}.\left(1-\frac{d^2}{D^2}\right)$$

We wish to find $\Delta T_1 = (\delta T_1).T_1$
Now
$$\delta T_1 = \delta d + \delta \gamma_{rad} + \delta\left(1-\frac{d^2}{D^2}\right) \tag{iv}$$

Considering the last term, we note that
$$\delta\left(1-\frac{d^2}{D^2}\right) = \frac{\Delta\left(1-\frac{d^2}{D^2}\right)}{\left(1-\frac{d^2}{D^2}\right)}$$

or
$$\delta\left(1-\frac{d^2}{D^2}\right) = \frac{\left(0+\Delta\frac{d^2}{D^2}\right)}{(D^2-d^2)}.D^2$$
$$= \frac{D^2\left(\frac{d^2}{D^2}.\delta\frac{d^2}{D^2}\right)}{(D^2-d^2)}$$

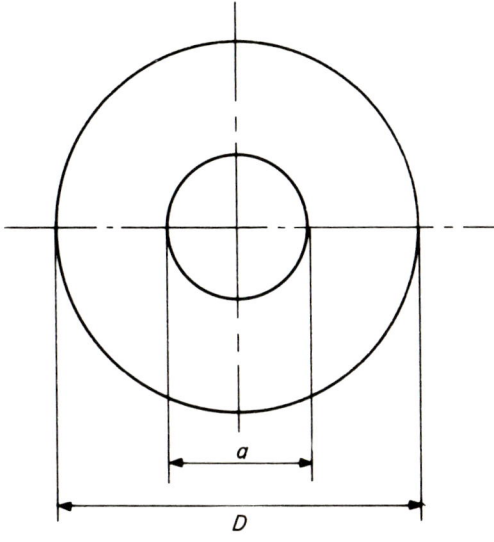

or

$$\delta\left(1 - \frac{d^2}{D^2}\right) = \frac{2 \cdot d^2(\delta d + \delta D)}{D^2 - d^2}$$

or

$$= \frac{2\left(\dfrac{d}{D}\right)[D \cdot \Delta d + d \cdot \Delta D]}{D^2 - d^2} \quad \text{(v)}$$

Thus substituting this into equation (iv) we have

$$\delta T_1 = \delta d + \delta \gamma_{\text{rad}} + \frac{2 \cdot d^2 \cdot (\delta d + \delta D)}{D^2 - d^2} \quad \text{(vi)}$$

The numerical case
From ISO data sheet 4500 A (1970) we obtain for interference fit (H7) and (S6) the following dimensions:

$$\text{Shaft diameter} = \left(50 \, {}^{+0.072}_{+0.053}\right) \text{mm}$$

$$\text{Hole diameter} = \left(50 \, {}^{+0.030}_{+0.000}\right) \text{mm}$$

$$\therefore \gamma_{\text{rad}} = \tfrac{1}{2} \cdot \left[\left({}^{+0.072}_{+0.053}\right) - \left({}^{+0.030}_{+0.000}\right)\right]$$

i.e.
$$\gamma_{\text{rad}} = + \begin{bmatrix} 0.036 \\ 0.0115 \end{bmatrix}$$

which can be presented as
$$\gamma_{\text{rad}} = 0.0238 \pm 0.0123$$

and
$$\delta\gamma_{\text{rad}} = \pm 0.516 \qquad \text{(vii)}$$

The mean bore and shaft diameters will differ since they always remain an interference fit. The shaft diameter appears to have the larger tolerance range, so we will use these for d in equations (iii) and (vi), thus

$$\left. \begin{array}{l} d = (50.0625 \pm 0.0095)\,\text{mm} \\ \therefore \delta d = \dfrac{0.0095}{50.0625} = 1.898 \times 10^{-4} \end{array} \right\} \qquad \text{(viii)}$$

Given:
Hub diameter $D = (98 \pm 0.120)\,\text{mm}$, i.e.

$$\delta D = 0.001\,225; \quad E = 21 \times 10^{4}\,\text{N\,mm}^{-2}; \quad \mu = 0.25 \qquad \text{(ix)}$$

Thus, from equation (iii) we get

$$T_1 = (\tfrac{1}{2} \times 0.25 \times \pi \times 21 \times 10^4 \times 50.0625 \times 0.0238) \cdot \left(\frac{(98)^2 - (50.0625)^2}{(98)^2} \right)$$

or
$$T_1 = 72.63 \times 10^3\,\text{N\,mm}$$

or
$$T_1 = 72.63\,\text{N\,m}$$

And from equation (vi) we get

$$\delta T_1 = 1.898 \times 10^{-4} + 0.516 + \frac{2 \times (50.0625)^2 \times (1.898 \times 10^{-4} + 0.001\,225)}{(98)^2 - (50.0625)^2}$$

or
$$\delta T_1 = 0.518 = \pm 51.8\%$$

It is noted that the main influence on the variation in the transmissible torque is the tolerance of the interference fit between the shaft and the bore in the hub.

It is useful to confirm the above results by direct calculations of the maximum and minimum torque one might expect. The reason for this check is the large relative error of the radial interference. We have

$$T_1 = \underbrace{(\tfrac{1}{2} . \mu . \pi . E)}_{\text{const}} . d . \gamma_{\text{rad}} . \left(1 - \frac{d^2}{D^2}\right)$$

Now,
$T_1 = $ maximum when

$$d = \text{max}; \quad \gamma = \text{max}; \quad D = \text{min}$$

and
$T_1 = $ minimum when

$$d = \text{min}; \quad \gamma = \text{min}; \quad D = \text{max}$$

Now,
$$d_{\text{max}} = 50.072; \quad d_{\text{min}} = 50.053$$
$$D_{\text{max}} = 98.120; \quad D_{\text{min}} = 97.880$$
$$\gamma_{\text{max}} = 0.072; \quad \gamma_{\text{min}} = 0.023$$

These yield

$$T_{1\,\text{min}} = \text{constant} \times 50.053 \times 0.023 \left[1 - \left(\frac{50.053}{98.120}\right)^2\right]$$
$$= C \times [1 - 0.2603] \times 1.151$$
$$= 0.8517 . C$$

$$T_{1\,\text{max}} = \text{constant} \times 50.072 \times 0.072 \left[1 - \left(\frac{50.072}{97.88}\right)^2\right]$$
$$\therefore T_{1\,\text{max}} = 2.6640 . C$$

Calculations for $T_{1\,\text{mean}}$ yield

$$T_{1\,\text{mean}} = \frac{0.8517 + 2.6640}{2} . C$$

or

$$T_{1\,\text{mean}} = 1.758 . C$$
$$\therefore \delta T_1 = \frac{0.9062}{1.758} \simeq \pm 51.5\%$$

which compares well with the value predicted with the theory of error method.

Example 2.11
Given

$$Q = 0.027 \cdot \left(\frac{\mu_{bf}}{\mu_s}\right)^{0.14} \cdot \text{Re}^{0.80} \cdot \text{Pr}^{0.33} \cdot \frac{k}{d} \cdot (t_w - t_{bf})$$

$$\mu_{bf}, \mu_s, k, C_p = \text{constant}$$

Using rules (2), (4), (5) and (6), we get

$$\delta Q = \delta d + 0.8 \cdot \delta \text{Re} + 0.33 \cdot \delta \text{Pr} + \frac{t_w \cdot \delta t_w + t_{bf} \cdot \delta t_{bf}}{(t_w - t_{bf})}$$

Now, $\text{Pr} = C_p \cdot \mu/k = $ constant by assumptions made

$$\therefore 0.33 \cdot \delta \text{Pr} = 0$$

$$\text{Re} = \frac{v \rho d}{\mu} \quad \text{(all independent variables)}$$

$$\therefore \delta \text{Re} = \delta v + \delta \rho + \delta d + \delta \mu$$

and since μ is assumed constant

$$\delta \text{Re} = \delta v + \delta \rho + \delta d$$

Therefore, variation in heat transmission is

$$\delta Q = 1.8 \delta d + 0.8 \, (\delta v + \delta \rho) + \frac{t_w \cdot \delta t_w + t_{bf} \cdot \delta t_{bf}}{(t_w - t_{bf})},$$

and the absolute variation is then obtained from

$$\Delta Q = Q \cdot \delta Q$$

Chapter Four

Example 4.5

Average occurrence of errors = 1 per 1000 dimensions on a drawing.

∴ Average occurrence of errors = $\dfrac{n.1}{1000}$ for n dimensions.

For at least *one* error,

$$p_{f \atop n \geqslant 1} = (1 - p_{f \atop n = 0}); \quad \therefore \ p_{f \atop n = 0} = 1 - p_{f \atop n \geqslant 1} = 1 - 0{\cdot}05$$

i.e.

$$p_{f_0} = 0{\cdot}95$$

Now,

$$p_{f_0} = e^{-\tfrac{n}{1000}} = 0{\cdot}95$$

$$\therefore \left(-\dfrac{n}{1000}\right) = \log 0{\cdot}95$$

or

$$\left(-\dfrac{n}{1000}\right) = (-0{\cdot}0513)$$

i.e. $n \leqslant 51$ dimensions.

Example 4.8

$$\tfrac{1}{2} I_x = \int_0^t (y \cdot dx) \cdot x_0^2$$

Now,

$$(h - y) = ax_0^2$$
$$y = (h - ax_0^2)$$

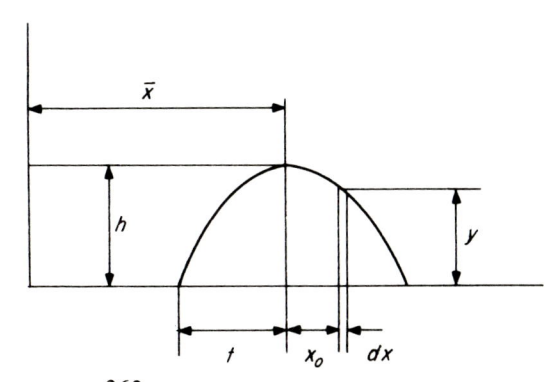

363

and
$$h = a.t^2$$
$$\therefore \tfrac{1}{2}I_x = \int_0^t (h - ax_0^2).x_0^2.dx = \tfrac{1}{3}h.t^3 - \tfrac{1}{5}at^5 = \tfrac{1}{3}at^5 - \tfrac{1}{5}at^5$$

or
$$\tfrac{1}{2}I_x = \tfrac{2}{15}.at^5$$

Half area of parabola is
$$(\tfrac{1}{2}A) = \int_0^t y\,dx = \tfrac{2}{3}a.t^3$$

$$\therefore \sigma^2 = \frac{2 \times 3}{15 \times 2} \cdot \frac{a.t^5}{a.t^3} = \tfrac{3}{15}.t^2 = \tfrac{1}{5}.t^2$$

$$\therefore \sigma = \sqrt{\tfrac{1}{5}}.t = 0.447.t$$

Example 4.9

(i) To find $\bar{x}$. We know that for a triangle the centre of gravity is $\tfrac{1}{3}$ from the base,

$$\therefore \bar{x} = 1 + \tfrac{2}{3} = 1\tfrac{2}{3} = 1.666$$

(ii) The standard deviation σ is equal on both sides of $\bar{x}$. It is simpler to transfer the origin to $x = \bar{x}$. The ordinate $f(x)$ at any point of x_0 from $\bar{x}$ is

the area of the triangle = 1

Transferring the origin to $\bar{x}$, we have

$$\frac{f(x)}{(\tfrac{4}{3} - x_0)} = \frac{1}{2}; \quad f(x) = \tfrac{1}{2}(\tfrac{4}{3} - x_0) = \tfrac{1}{6}(4 - 3x_0)$$

The moment of inertia of the triangle about the $\bar{x}$ ordinate is

$$I_x = \int_{-2/3}^{+4/3} f(x).x_0^2.dx_0 = \tfrac{1}{6}.\int_{-2/3}^{+4/3} (4 - 3x_0).x_0^2.dx_0$$

which yields
$$I_{\bar{x}} = \frac{16}{72}\left[\frac{72-45}{27}\right] = \frac{16}{72} = \frac{2}{9}$$

Alternatively, using known relationships for second moments of areas we have
$$I_{\bar{x}} = \frac{b \cdot h^3}{36} = \frac{1 \times (2)^3}{36} = \frac{8}{36} = \frac{2}{9}$$

Now, since $A = 1$, we get
$$\sigma^2 = \tfrac{2}{9}$$

and
$$\sigma = \frac{\sqrt{2}}{3} = 0.4714$$

Example 4.10
   (i) From symmetry $\bar{x} = 1.5$ units
   (ii) Area of trapezium $= 3 \times \tfrac{1}{5} + 2\tfrac{1}{5} = 1$
   (iii) Second moments of areas are

$$\text{rectangle } (\tfrac{1}{5} \times 3) = I_1 = \frac{\tfrac{1}{5} \times (3)^3}{12} = \frac{9}{20}$$

$$\text{rectangle } (\tfrac{1}{5} \times 1) = I_2 = \frac{\tfrac{1}{5} \times (1)^3}{12} = \frac{1}{60}$$

$$2 \times \text{triangles: } I_3 = 2\left[\frac{\tfrac{1}{5} \times 1^3}{36} + \frac{1}{2} \times \left(\frac{1}{5} \times 1\right) \times \left(\frac{1}{3} + \frac{1}{2}\right)^2\right]$$

$$= 2\left[\frac{1}{180} + \frac{1}{10} \times \left(\frac{5}{6}\right)^2\right]$$

$$I_3 = \frac{27}{180}$$

$$\therefore \text{Total } I_{\bar{x}} = I_1 + I_2 + I_3 = \tfrac{9}{20} + \tfrac{1}{60} + \tfrac{27}{180}$$

$$I_x = \frac{111}{180}$$

Thus since $A = 1$,
$$\sigma^2 = \frac{111}{180} = 0.618$$

and
$$\sigma = 0.785$$

Example 4.11
(a) Equation (ii) reads

$$\bar{x} = x_0 + c \cdot \frac{\sum f \cdot t}{\sum f} = x_0 + c \cdot \bar{t}$$

and equation (4.30) gives

$$\sigma = c \cdot \left[\frac{\sum (f \cdot t^2)}{\sum f} - \frac{(\sum f \cdot t)^2}{\sum f} \right]^{\frac{1}{2}}$$

where

x_0 = approximately guessed mean value
c = cell, or block interval ($= 0.05$ mm)
f = frequency of each cell mean value
$t = \dfrac{x - x_0}{c} = \dfrac{\Delta x}{c}$, say

i.e. $t = 20 \cdot (\Delta x)$ in this case.

Table 1.

	Length x	t	f	$f \cdot t$	$f \cdot t^2$
	42·23	−4	1	−4	16
	42·28	−3	7	−21	63
	42·33	−2	7	−14	28
	42·38	−1	12	−12	12
$x_0 =$	42·43	0	17	0	0
	42·48	+1	15	15	15
	42·53	+2	8	16	32
	42·58	+3	7	21	63
	42·63	+4	5	20	80
	42·68	+5	1	5	25
	$n = 10$	—	80	26	334

From Table 1 we get

$$\bar{x} = 42 \cdot 43 + 0 \cdot 05 \cdot \tfrac{26}{80}$$

or

$$\bar{x} = 42 \cdot 446$$

And

$$\sigma = 0 \cdot 05 \left[\frac{334}{80} - \left(\frac{26}{80}\right)^2 \right]^{1/2}$$

$$= 0 \cdot 05 \times 2 \cdot 0173$$

i.e.

$$\sigma = 0 \cdot 1008$$

SOLUTIONS TO CHAPTER FOUR

(b) Given: Lower limit = 42·20
    Upper limit = 42·60
Probability of exceeding upper limit (for a normal distribution) is

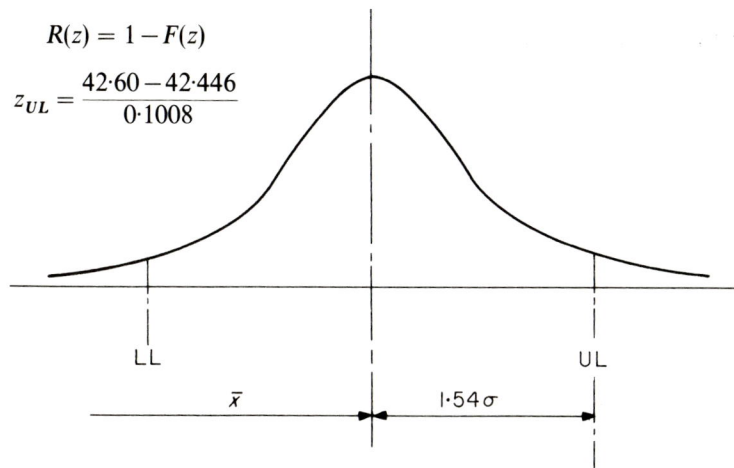

or
$$z_{UL} = 1·54 \text{ standard deviations}$$

Thus from standard probability tables (or interpolating between 1·5 and 1·6 in table, Fig. 4.5) we obtain

$$F(1·54) = 0·9382$$

i.e.
$$R(1·54) = 0·0618 = 6·18\%$$

(c) Construction of "normal" distribution curve: Referring to p. 94, we note that

$$y = \frac{N \cdot C}{\sigma} \cdot f(z); \quad y = \frac{80 \cdot 0·05}{0·1008} \cdot f(z)$$

or
$$y = 39·68 \cdot f(z)$$

y = scaled ordinate of the density function
N = total number of frequencies in histogram
c = cell interval
σ = standard deviation

z	0	0·5	1·0	1·5	2·0	2·5	3·0	3·5
$f(z)$	0·399	0·352	0·242	0·130	0·054	0·018	0·0044	0·0008
y	15·83	13·96	9·60	5·16	2·14	0·714	0·159	0·0317

See Fig. 4.7 (p. 95)

Chapter Five

Example 5.1
(i) $\phi = A \cdot y \cdot \log x$
Using partial derivative method we obtain:
$$\frac{\partial \phi}{\partial y} = A \cdot \log \bar{x}$$
$$\frac{\partial \phi}{\partial x} = A \frac{\bar{y}}{\bar{x}}$$
$$\therefore \sigma_\phi \simeq \left[(A \cdot \log \bar{x})^2 \cdot \sigma_y^2 + \left(A \cdot \frac{\bar{y}}{\bar{x}}\right)^2 \cdot \sigma_x^2\right]^{1/2}.$$

(ii) $\phi = A \cdot \log x$;
$$\frac{\partial \phi}{\partial x} \simeq \frac{A}{\bar{x}}; \quad \therefore \sigma_\phi \simeq \left[\left(\frac{A}{\bar{x}}\right)^2 \cdot \sigma_x^2\right]^{1/2}$$
or
$$\sigma_\phi \simeq \frac{A}{\bar{x}} \cdot \sigma_x$$

(iii) $\phi = c \cdot y \cdot \sin x$
$$\frac{\partial \phi}{\partial x} = c \cdot \bar{y} \cdot \cos \bar{x}$$
$$\frac{\partial \phi}{\partial y} = c \cdot \sin \bar{x}$$
$$\therefore \sigma_\phi \simeq c[\bar{y}^2 \cdot \cos^2 \bar{x} \cdot \sigma_x^2 + \sin^2 \bar{x} \cdot \sigma_y^2]^{1/2}$$
or
$$\sigma_\phi \simeq c \cdot \bar{y} \left[\cos^2 \bar{x} \cdot \left(\sigma_x^2 - \frac{\sigma_y^2}{\bar{y}^2}\right) + \frac{\sigma_y^2}{\bar{y}^2}\right]^{1/2}$$
If $\sigma_y \ll \bar{y}$, then $\sigma_\phi \simeq c \cdot \bar{y} \cdot \cos \bar{x} \cdot \sigma_x$.

(iv) $\phi = C \cdot \sin x$
$$\frac{\partial \phi}{\partial x} = C \cdot \cos x$$
$$\therefore \sigma_\phi = C \cdot \cos \bar{x} \cdot \sigma x$$

Example 5.2
Generally $P = f \cdot A$, i.e.

$$(\bar{P}, \sigma_p) = (\bar{f}; \sigma_f) \cdot (\bar{A}; \sigma_A).$$

Given

$$\bar{A} = 10 \text{ in}^2 \text{ and } \sigma_A = 0.4 \text{ in}^2$$
$$\bar{f} = 5 \times 10^4 \text{ p.s.i.}; \quad \sigma_f = 3 \times 10^3 \text{ p.s.i.}$$
$$\therefore \bar{P} = 5 \times 10^5 \text{ lbf}$$

Using Table 5.1, p. 104–105, we have

$$\sigma_p = [(\bar{f}^2 \cdot \sigma_A^2 + (\bar{A})^2 \cdot \sigma_f^2 + \sigma_f^2 \cdot \sigma_A^2]^{1/2}$$
$$= [(5 \times 10^4)^2 (0.4)^2 + (10)^2 \times (3 \times 10^3)^2 + (0.4)^2 (3 \times 10^3)^2]^{1/2}$$
$$= 10^3 [16 \cdot 25 + 900 + 1 \cdot 44]^{1/2} = 10^3 \times [1301 \cdot 44]^{1/2}$$
$$\therefore \sigma_p = 3 \cdot 61 \times 10^4$$

i.e.

$$(\bar{P}; \sigma_p) = (5 \times 10^5; 3 \cdot 6 \times 10^4) \text{ lbf}$$

Example 5.3
Thermal expansion per unit length = $\alpha(t_1 - t_0)$:

$$\alpha = \text{coefficient of thermal expansion}$$

$$\text{strain} = \frac{f_c}{E} \text{ compression per unit length}$$

$$f_c = \text{compressive stress}$$
$$E = \text{Young's modulus}$$

$$\therefore \frac{f_c}{E} = \alpha(t_1 - t_0) \tag{i}$$

Now,
$$\Delta t = (t_1 - t_0)$$
$$\therefore \bar{\Delta t} = 50\,°F$$

and

$$\sigma_{\Delta t} = [\sigma_{t_1}^2 + \sigma_{t_0}^2]^{1/2} = [(2 \cdot 5)^2 + (3)^2]^{1/2}$$

i.e.

$$\sigma_{\Delta t} = 3 \cdot 91.$$

From (i) we have

$$\alpha = \frac{f_c}{\Delta t \cdot E} \quad \text{(ii)}$$

$$\therefore \bar{\alpha} = \frac{\bar{f_c}}{\overline{\Delta t} \cdot \bar{E}} = \frac{10\,500}{50 \times 30 \times 10^6}$$

or

$$\bar{\alpha} = 7 \times 10^{-6} \text{ per } °F$$

To estimate σ_α. The denominator of equation (ii) is

$$D = \Delta t \cdot E \quad \text{(iii)}$$

$$\therefore \bar{D} = \overline{\Delta t} \cdot \bar{E} = 15 \times 10^8 \text{ p.s.i. } °F$$

and

$$\sigma_D \simeq [\overline{\Delta t}^2 \cdot \sigma_E^2 + \bar{E}^2 \cdot \sigma_{\Delta t}^2]^{1/2}$$
$$= [50^2 \times (1 \cdot 05 \times 10^6)^2 + (30 \times 10^6)^2) \times (3 \cdot 91)^2]^{1/2}$$

or

$$\sigma_D = 12 \cdot 75 \times 10^7 \text{ p.s.i. } °F.$$

Combining (ii) and (iii), we can write

$$\alpha = \left(\frac{\bar{f_c}}{\bar{D}}\right) \quad \text{(iv)}$$

and, therefore,

$$\sigma_\alpha = \frac{1}{(\bar{D})^2}[f_c^2 \cdot \sigma_D^2 + \bar{D}^2 \cdot \sigma_{f_c}^2]^{1/2} \quad \text{(v)}$$

Thus,

$$\sigma_\alpha = \left(\frac{10^4}{15 \times 10^8}\right)^2 [(1 \cdot 05 \times 10^4)^2 (1 \cdot 275)^2 + (1 \cdot 5)^2 (6 \cdot 50 \times 10^2)^2]^{1/2}$$

or

$$\sigma_\alpha = 7 \cdot 35 \times 10^{-7} \text{ per } °F.$$

Thus, from before we have

$$(\bar{\alpha}; \sigma_\alpha) = (7 \times 10^{-6}; 7 \cdot 35 \times 10^{-7}) \text{ per } °F$$

Example 5.4

The end-deflection of a cantilever (the load acting at the free end) is given by

$$y = \frac{P \cdot l^3}{3 \cdot EI}$$

$$\therefore \text{ stiffness } k = \frac{P}{y} = \frac{3 \cdot EI}{l^3} \quad \text{(i)}$$

I for a rectangular section is

$$I = \frac{b \cdot t^3}{12}$$

where b = width and t = thickness in direction of bending. Thus equation (i) may be written as

$$k = \frac{b \cdot t^3 \cdot E}{4 \cdot l^3} \quad \text{(ii)}$$

From (ii) we obtain

$$\bar{k} = \frac{\bar{b} \cdot (\bar{t})^3 \cdot \bar{E}}{4 \cdot (\bar{l})^3}$$

which yields

$$\bar{k} = 28 \cdot 35 \text{ N mm}^{-1}$$

By partial derivatives,

$$\left(\frac{\partial k}{\partial b}\right)^2 = \left[\frac{(\bar{t})^3 \cdot \bar{E}}{4 \cdot (\bar{l})^3} 3\right]^2$$

$$\left(\frac{\partial k}{\partial t}\right)^2 = \left[\frac{\bar{b} \cdot \bar{E}}{4 \cdot (\bar{l})^3} \cdot 3(\bar{t})^2\right]^2$$

$$\left(\frac{\partial k}{\partial E}\right)^2 = \left[\frac{\bar{b} \cdot (\bar{t})^3}{4 \cdot (\bar{l})^3}\right]^2$$

$$\left(\frac{\partial k}{\partial l}\right)^2 = \left[\frac{\bar{b} \cdot (\bar{t})^3 \cdot \bar{E}}{4} \cdot \left\{-3\left(\frac{1}{\bar{l}}\right)^4\right\}\right]^2$$

Thus we have,

$$\sigma_k^2 = (\tfrac{1}{4})^2 \left[\left(\frac{(\bar{t})^3 \cdot \bar{E}}{(\bar{l})^3}\right)^2 \cdot \sigma_b^2 + \left(\frac{\bar{b} \cdot \bar{E} \cdot 3(\bar{t})^2}{(\bar{l})^3}\right)^2 \cdot \sigma_t^2 \right.$$

$$\left. + \left(\frac{\bar{b} \cdot (\bar{t})^3}{(\bar{l})^3}\right)^2 \cdot \sigma_E^2 + \left(\frac{\bar{b} \cdot (\bar{t})^3 \cdot \bar{E} \cdot 3}{(\bar{l})^4}\right)^2 \cdot \sigma_l^2\right] \quad \text{(iii)}$$

SOLUTIONS TO CHAPTER FIVE 373

Data given:

$$b = 20 \text{ mm} \pm 0.05 \text{ mm} \quad \therefore \sigma_b = 0.17 \text{ mm}$$

$$l = 100 \text{ mm} \pm 2 \text{ mm} \quad \therefore \sigma_l = 0.66 \text{ mm}$$

$$t = 3 \text{ mm} \pm 0.1 \text{ mm} \quad \therefore \sigma_t = 0.033 \text{ mm}$$

$$E = 21 \times 10^4 \text{ N mm}^{-2}$$

$$\frac{\Delta E}{E} = \pm \frac{1}{30}; \quad \therefore \sigma_E = 2.33 \times 10^3 \text{ N mm}^{-2}$$

N.B. Standard deviations are taken to be one-third of sure-fit tolerance limits.

Inserting the numerical values into equation (iii), we obtain

$$\sigma_k = \tfrac{1}{4} \times 4.643$$

or

$$\sigma_k = 1.161 \text{ N mm}^{-1}$$

We thus have for the stiffness of the cantilever,

$$(\bar{k}; \sigma_k) = (28.35; 1.161) \text{ N mm}^{-1}$$

The results obtained for Example 2.8, using error analysis, were

$$k = (28.35 \pm 4.48) \text{ N mm}^{-1}$$

We note that for $\Delta k = \pm 3\sigma_k$, we get $\Delta k = \pm 3.482$ (99·73% probability) and $\Delta k = 4 \cdot \sigma_k$ we obtain $\Delta k = \pm 4.643$ (99·9937% probability).

Example 5.5
Given:

$$\text{Shaft diameter} = (1.048; 0.0013) \text{ in}$$

$$\text{Bearing diameter} = (1.054; 0.0017) \text{ in}$$

Normal distribution of tolerances assumed. First we must express clearance between shaft and bearing:

$$\Delta = d_b - d_{sh}$$

$$\therefore \bar{\Delta} = \bar{d}_b - \bar{d}_{sh} = 1.059 - 1.048 = 0.011 \text{ in}$$

From Table 5.1

$$\sigma_\Delta = [\sigma_b^2 + \sigma_{sh}^2]^{1/2} = [(0.0017)^2 + (0.0013)^2]^{1/2}$$

$$\therefore \sigma_\Delta = 0.00214 \text{ in.}$$

$\Delta = 0$ (zero clearance), when $d_b = d_{sh}$. Expressing Δ in terms of standard deviations, we have

$$\mathop{z}_{0 \to \bar\Delta} = \frac{0 - 0\cdot011}{0\cdot00214}$$

or

$$\mathop{z}_{0 \to \bar\Delta} = 5\cdot14$$

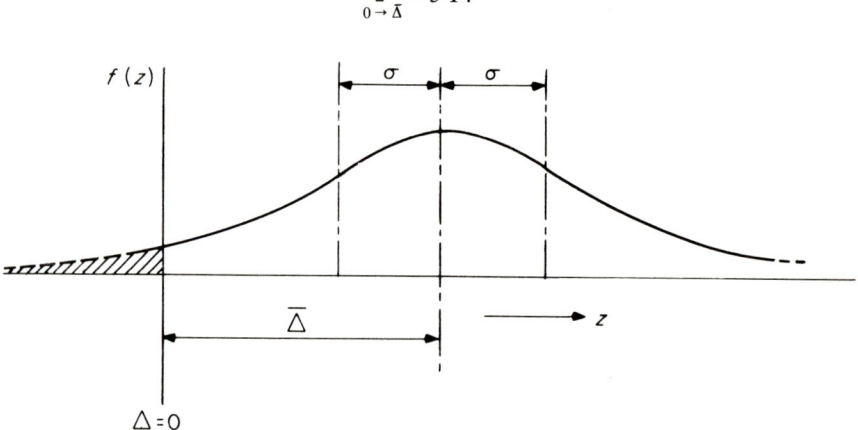

The probability that the clearance between shaft and bearing would be equal or less than zero (= interference) is represented by the area under the distribution curve to the left of $\Delta = 0$. From Table 4.1 (p. 89), for $z = -5\cdot1$, $F(z) = 5 \times 10^{-7}$ which is very small indeed. The probability that $\Delta \leqslant 0\cdot0010$ is obtained similarly:

$$z = \frac{0\cdot001 - 0\cdot011}{0\cdot00214} = -4\cdot673,$$

which from probability tables gives us

$$F(z) = 1\cdot5 \times 10^{-6}$$

representing the probability that the minimum running clearance will not be satisfied.

Example 5.6

$$(\bar H_1; \sigma_{H_1}) = (5500; 500)\,\text{BTU/hour}$$
$$(\bar H_2; \sigma_{H_2}) = (5500; 700)\,\text{BTU/hour}$$

Question: What is the probability that the heat absorbed by the air stream is $\geqslant 12\,750$ BTU/hour?

$$\bar{H}_1 + \bar{H}_2 = 11\,000 \text{ BTU/hour} = \bar{H}_{\text{total}}$$

$$\sigma_{H_{\text{total}}} = [\sigma_{H_1}^2 + \sigma_{H_2}^2]^{1/2}$$

$$= [25 \times 10^4 + 49 \cdot 10^4]^{1/2}$$

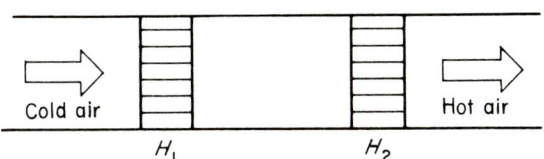

i.e.

$$\sigma_{H_{\text{total}}} = 860 \text{ BTU/hour}$$

The number of standard deviations between $H = 12\,750$ and $\bar{H} - 11\,000$ is

$$Z = \frac{12\,750 - 11\,000}{860}$$

or

$$z = 2 \cdot 035$$

From standard probability tables (Table 4.1, p. 89) for $z = 2 \cdot 035$ $R(z) = 0 \cdot 021 = 2 \cdot 1\%$. There is only a $2 \cdot 1\%$ chance that the heat absorbed by the air-stream would $\geqslant 12\,750$ BTU/hour.

Example 5.7

Data given in Example 2.10 are:

$$\text{Radial pressure} = p_r = \frac{E \cdot \gamma_{\text{rad}} \cdot (D^2 - d^2)}{d \cdot D^2} \qquad \text{(i)}$$

Torque = $T = p_r \cdot \mu \cdot \pi \cdot d^2 \cdot \frac{1}{2}$ per unit length of hub (ii)

$E = 21 \times 10^4$ N mm^{-2} = constant

$\mu = 0.25$ = constant

Shaft diameter = $\left(50 {+0.072 \atop +0.053}\right)$ mm

Hole diameter = $\left(50 {+0.030 \atop +0.000}\right)$ mm

$d = (50.0625 \pm 0.0095)$

Hub diameter = (98 ± 0.120) mm

$\gamma_{rad} = \frac{1}{2} \cdot \left[\left({+0.072 \atop +0.053}\right) - \left({+0.030 \atop +0.000}\right)\right]$

i.e.

$$\gamma_{rad} = (0.0238 \pm 0.0123)$$

Combining (i) and (ii) we get

$$T = \tfrac{1}{2} \cdot \pi \cdot \mu \cdot E \cdot \gamma_{rad} \cdot d \left[1 - \left(\frac{d}{D}\right)^2\right] \qquad \text{(iii)}$$

Inserting numerical values, we obtain

$$T = \tfrac{1}{2} \cdot \pi \cdot 0.25 \times 21 \times 10^4 \times 0.0238 \times 50.0625 \cdot \left[1 - \left(\frac{50.0625}{98}\right)^2\right]$$

i.e.

$$T = 72.63 \times 10^3 \text{ N mm}$$

or

$$T = 72.63 \text{ N.m}$$

To determine the standard deviation for T we use the partial derivative method (equation (5.8)):

$$\frac{\partial T}{\partial \gamma_{rad}} = (\tfrac{1}{2}\pi \cdot \mu \cdot E) \cdot d \cdot \left[1 - \left(\frac{d}{D}\right)^2\right] = 30.52 \times 10^5$$

$$\frac{\partial T}{\partial d} = (\tfrac{1}{2}\pi \cdot \mu \cdot E) \cdot \gamma_{rad} \cdot \left[1 - \frac{3d^2}{D^2}\right] = 4.26 \times 10^2$$

$$\frac{\partial T}{\partial D} = (\tfrac{1}{2}\pi \cdot \mu \cdot E) \cdot \gamma_{rad} \cdot \left[0 + \frac{2d^3}{D^3}\right] = 5.23 \times 10^2$$

By convention, we take $\pm 3\sigma = \pm$ limit tolerances for γ_{rad}, d and D respectively:

$$\sigma_{\gamma_{rad}} = 0.0041$$
$$\sigma_d = 0.0033$$
$$\sigma_D = 0.040$$

Inserting into equation (5.8) we obtain

$$\sigma_T^2 = [(30.52 \times 10^5)^2 \times (0.0041)^2 + (4.26 \times 10^2)^2 \times (0.0033)^2 \\ + (5.23 \times 10^2)^2 \times (0.040)2]$$

i.e.
$$\sigma_T^2 = 156.6 \times 10^6$$

or
$$\sigma_T = 12.512 \times 10^3 \text{ N mm}$$

or
$$\sigma_T = 12.512 \text{ Nm}$$

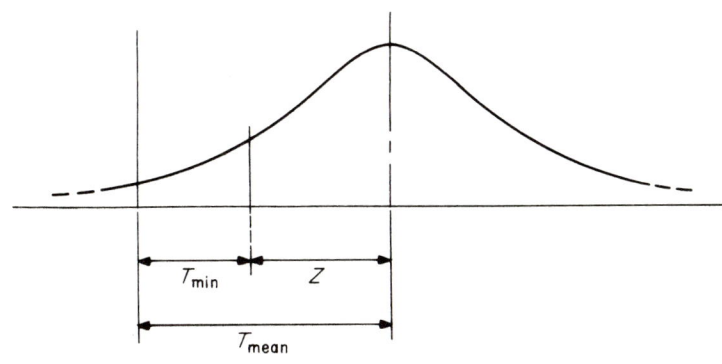

Now,
$$T_{mean} = 72.63 \text{ Nm}$$
for (a): $T_{min} = 36.37 \text{ Nm}$
for (b): $T_{min} = 0.7 \times 72.63 = 50.84 \text{ N m}$

So,
$$\text{for (a): } z = \frac{72.63 - 36.37}{12.512} = 2.9276$$

$\therefore R(z)_a = 0.0017 = 0.17\%$, from standard probability tables and for (b)

$$z = \frac{72 \cdot 63 - 50 \cdot 84}{12 \cdot 512} = 1 \cdot 7252$$

$\therefore R(z)_b = 0 \cdot 042 = 4 \cdot 2\%$, from standard probability tables.

Thus the probability of the transmissible torque due to the shrink fit falling below,

(a) 50% of mean value = $0 \cdot 17\%$
(b) 70% of mean value = $4 \cdot 2\%$

This is clearly very unlikely, assuming normal distribution.

It is of interest to compare the statistical estimations with those we obtained using sure-fit error analysis. The error analysis gave us

$$T_{mean} = 72 \cdot 63 \text{ Nm}$$

and

$$\Delta T = \pm 37 \cdot 8 \text{ Nm}$$

i.e.

$$T_{min} = 33 \cdot 83 \text{ Nm}$$

and

$$T_{max} = 110 \cdot 43 \text{ Nm}$$

The statistical results were

$$T_{mean} = 72 \cdot 63 \text{ Nm}$$

$$\sigma_T = 12 \cdot 512 \text{ Nm}$$

i.e. $3\sigma_T = 37 \cdot 536$, which results from the assumptions made for $\sigma_{\gamma_{rad}}$, σ_d and σ_D.

From the sure-fit analysis it would appear that the transmissible torque by the shrink fit has a large range of values, but the statistical analysis indicates that the likelihood of such large variations is extremely small.

It may be further noted that engineering components with dimensions above or below the specified limits are usually discarded. This implies that the actual population relative to the normal distribution is already reduced from 100 to 99·73 (for sure-fit tolerance = $\pm 3\sigma$).

Now, for

$$z = 3 \cdot 00, \quad R(z) = 0 \cdot 001\ 350$$

$\therefore [R(2 \cdot 9276) - R(3 \cdot 00)] = 0 \cdot 0017 - 0 \cdot 001\ 35 = 0 \cdot 000\ 35$

and
$$[R(1·7252) - R(3·00)] = 0·042 - 0·001\,25 = 0·040\,65$$

Taking these values as a ratio of all components passed inspection, the probability of the transmissible torque falling below

(a) 50% of mean value $= \dfrac{0·000\,35}{0·9973} = 0·000\,350\,9 = 0·0351\%$

(b) 70% of mean value $= \dfrac{0·040\,65}{0·9963} = 0·040\,76 = 4·076\%$

Example 5.8

$(\bar{R}_1;\ \sigma_{R_1}) = (10\,000;\ 300)\ \text{ohm}$

$(\bar{R}_2;\ \sigma_{R_2}) = (20\,000;\ 600)\ \text{ohm}$

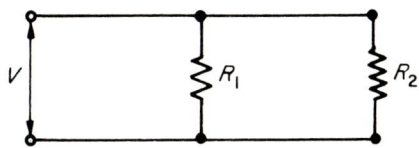

The overall resistance of the circuit is given by
$$\frac{1}{R_T} = \frac{1}{R_1} + \frac{1}{R_2}$$
$$\therefore R_T = \frac{R_1 \cdot R_2}{R_1 + R_2} \qquad\qquad (i)$$

Thus,
$$\bar{R}_T = \frac{10\,000 \times 20\,000}{30\,000} = 6666·66\ \text{ohm}$$

To determine σ_{R_T} we use the partial derivative method:

$$\frac{\partial R_T}{\partial R_1} = \frac{R_2^2}{(R_1+R_2)^2} = 0·444; \quad \sigma_{R_1} = 300$$

$$\frac{\partial R_T}{\partial R_2} = \frac{R_1^2}{(R_1+R_2)^2} = 0·111; \quad \sigma_{R_2} = 600$$

$$\therefore \sigma_{R_T} = [(0·444)^2 \times (300)^2 + (0·111)^2 \times (600)^2]^{1/2}$$

or
$$\sigma_{R_T} = 148·6\ \text{ohm}$$

Thus the resultant circuit resistance will be
$$(\bar{R}_T;\ \sigma_{R_T}) = (6·667 \times 10^3;\ 148·6)\ \text{ohm}.$$

Note: The coefficient of variation of the individual resistances was equal to 0·030, but the resultant coefficient of variation has decreased to 0·0223; i.e. for resistors in parallel the overall resistance is reduced and so also is the dispersion from the mean value.

Since equation (i) poses a problem of correlation between the numerator and the denominator, this can be circumvented by using the relationship

$$C_{tot} = C_1 + C_2$$

where

$$C = \frac{1}{R} \quad (= \text{conductivity})$$

and from Table 5.1

$$\sigma_C = \frac{\sigma_R}{(\bar{R})^2}$$

The end results are the same as above.

Example 5.9
Population density v. infant survival.

The first step is to rewrite Table 5.4 with population densities in descending order (Table 1).

Table 1.

Locality	Population density (x)	% Infant survival (y)	Agreement	Disagreement
Beverley	27·5	10·8		
Berelay	23·6	17·8		—
Wilmslow	14·8	17·1	+	
Langside	14·6	23·8		—
Lidingo	12·2	28·3		—
Ormholm	7·2	33·6		—
Wallington	7·0	38·2		—
Neasden	6·3	30·8	+	
Wellboro	5·2	33·6		—
Bilstow	5·0	46·3		—
Denham	4·5	37·0	+	
Longtown	4·2	42·2		—
Kirkside	3·1	33·3	+	
Tilwell	2·5	37·4		—
Dockland	2·2	43·4		—
Sandhamn	2·0	38·8	+	
Washton	1·4	40·0		—
Harlem	1·3	42·9		—
	No. of pairs = 18		$\Sigma = (+5)$	$\Sigma = (-12)$

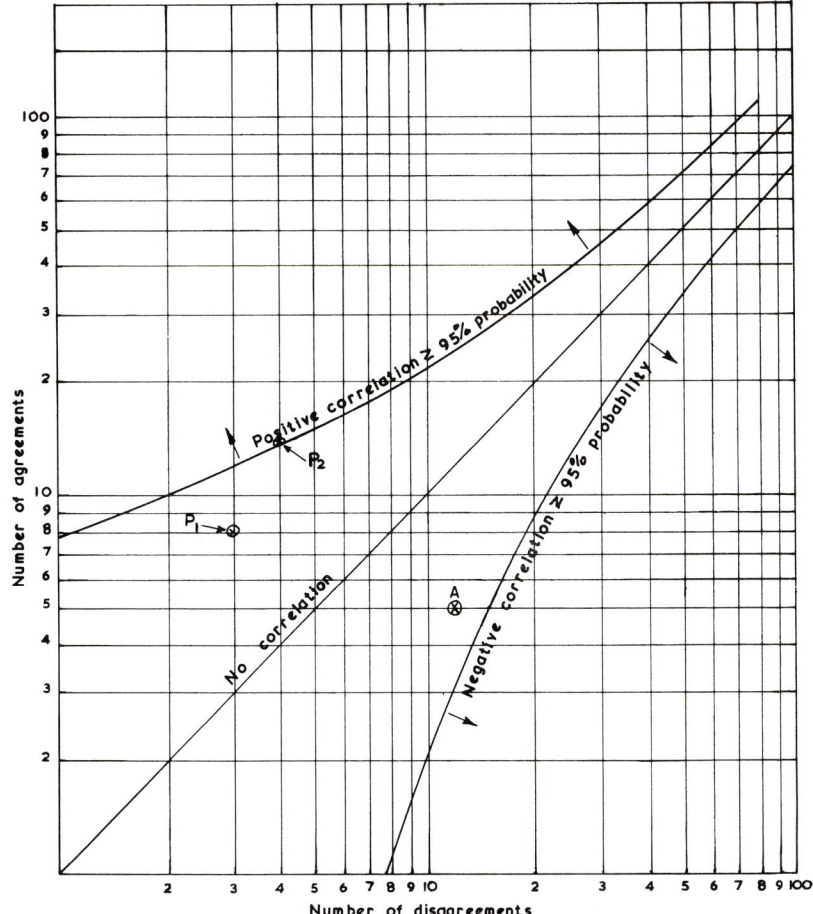

By plotting $\sum(+\text{ve})$ v. $\sum(-\text{ve})$ on correlation chart point A is achieved. As can be seen this gives a ($-$ve) correlation of less than 95% (or 0·95). This point is confirmed by the calculation of the correlation coefficient as below, the result for r being -0.9119 (or -91.19%):

$$r = \frac{\sum[(x-\bar{x})(y-\bar{y})]}{[\sum(x-\bar{x})^2 \cdot \sum(y-\bar{y})^2]^{1/2}}$$

or

$$r = \frac{n \cdot \sum(x \cdot y) - (\sum x \cdot \sum y)}{\{[n \cdot \sum x^2 - (\sum x)^2][n \cdot \sum y^2 - (\sum y)^2]\}^{1/2}}$$

Table 2.

x	y	$(x-\bar{x})$	$(y-\bar{y})$	$(x-\bar{x}) \cdot (y-\bar{y})$	$(x-\bar{x})^2$	$(y-\bar{y})^2$
27·5	10·8	19·47	-22.272	-433.676	379·081	496·042
23·6	17·8	15·57	-15.272	-237.785	242·425	233·234
14·8	17·1	6·77	-15.972	-108.13	45·833	255·105
14·6	23·8	6·57	-9.272	-60.917	43·165	85·970
12·2	28·3	4·17	-4.772	-10.899	17·389	22·772
7·2	33·6	-0.83	0·528	-0.438	0·6889	0·279
7·0	38·2	-1.03	5·128	-5.282	1·0609	26·296
6·3	30·8	-1.73	-2.272	3·931	2·993	5·162
5·2	33·6	-2.83	0·528	-1.494	8·009	0·279
5·0	40·3	-3.03	13·228	-40.08	9·181	174·98
4·5	37·0	-3.53	3·928	-13.866	12·461	15·429
4·2	42·2	-3.83	9·128	-34.96	14·669	83·32
3·1	33·3	-4.93	0·228	-1.124	24·305	0·052
2·5	37·4	-5.53	4·328	-23.934	30·581	18·732
2·2	43·4	-5.83	10·328	-60.212	33·989	106·668
2·0	38·8	-6.03	5·728	-34.54	36·361	32·810
1·4	40·0	-6.63	6·928	-45.933	43·957	48
1·3	42·9	-6.73	9·828	-66.142	45·293	96·590
$= 144.6$	$= 595.3$					
$\bar{x} = 8.033$	$\bar{y} = 33.072$			-1184.44	991·442	1701·72

$$\therefore r = \frac{-1184.441}{(991.442 \times 1701.72)^{1/2}} = -0.9119.$$

Chapter Seven

Example 7.1
Braking reliability of a motor car. Some parts of the system are in series and others in parallel. We shall proceed in a systematic manner from left to right of the diagram, (Fig. 7.17, p. 161).

(a) Front brakes

$$R_{(1)\to(2)} = (p_1 \cdot p_2 \cdot 0{\cdot}6)$$

$$R_{(3)\to(4)} = (p_3 \cdot p_4)$$

Now since $R_{(1,2)}$ is stand-by to $R_{(3,4)}$ their combined reliability is

$$R^{(1,2)}_{(3,4)} = 1 - [(1 - p_1 \cdot p_2 \cdot 0{\cdot}60) \cdot (1 - p_3 \cdot p_4)]$$

Unit (5) is in series with $R^{(1,2)}_{(3,4)}$ and, therefore,

$$R^{(1,2)}_{(3,4)\to(5)} = p_3 \cdot R^{(1,2)}_{(3,4)} \equiv R_{ss} \tag{a}$$

(b) Rear brakes

$$R_{(6)} = 0{\cdot}50 p_3 \text{ and is a stand-by to } R_{(3)} = p_3$$

$$\therefore R^{(6)}_{(3)} = 1 - [(1 - 0{\cdot}50 \cdot p_3) \cdot (1 - p_3)]$$

Thence

$$R_{|\frac{6}{3}|\to(7)} = p_7 \cdot \{1 - [(1 - 0{\cdot}50 \cdot p_3)(1 - p_3)]\}$$

and

$$R_{|\frac{6}{3}|\to(7)\to(8)} = p_8 \cdot p_7 \{1 - [(1 - 0{\cdot}50 \cdot p_3)(1 - p_3)]\} \equiv R_{s_8} \tag{b}$$

Now R_{ss} and R_{s_8} make up an active redundant system, i.e. both operate simultaneously and each, it may be assumed, can effectively brake the car by itself.

Using Poisson's distribution to assess the probability of both sub-systems failing simultaneously, we obtain:

probability of both sub-systems failing

$$F_2 = \frac{z^2}{2!} \cdot e^{-z} \qquad \text{(iii)}$$

where z is the average probability of failure of either subsystems.
Now, for

(a) $z_a = (1 - R_{s_5})$

(b) $z_b = (1 - R_{s_8})$

We may take

$$z = \frac{z_a + z_b}{2} \qquad \text{(iv)}$$

and inserting this into (iii) we obtain

$$F_2$$

and consequently

$$R_{s_9} = (1 - F_2) \qquad \text{(v)}$$

Subject Index

ACCRETION, 215
ACCURACY, MEASURE OF, 17
AGEING, 215
APPROXIMATE COMPUTATIONS, 1
  finite differences, 22, 29
  first order, 9, 10
  of functions, 29
  iterative methods, 8
  least square method, 19–22
  linear interpolation, 4
  Newton's method, 4
  root mean square, 21
APPROXIMATE NUMBERS
  inverse problem, 38
  influence on design, 44–48
  operations with, 32

BATH-TUB CURVE, see Failure
BRITTLE FRACTURE, see also Fracture
  crack lengths, 218
  intensity factor, 217, 219
  stress concentrations, 229
  toughness, 218
  transition point, 224
  velocity of, 216

CASE STUDIES
  machine failures, 203
    aero propeller, 204
    car steering, 203
    compressor, 204
  modelling,
    base plate, 313
    free-wheel spring coupling, 304
    machine reversibility, 308
    poppet valve, 321
    shaft frequency, 326

  reliability, 166
    block and hook, 177
    gears, high-performance, 184
    industrial fan, 168
    pumps, thermal station, 190
    stores lift, 170
CENTRAL LIMIT THEOREM, 80
CONFIDENCE, LIMITS OF, 118
CONVERGENCE
  checking for, 9
  inversion technique, 12
CORRELATION
  coefficient, 111
  confidence limits, 118
  continuous functions, 114
  discrete events, 112
  paired comparison, 112
CORROSION
  effect on fatigue, 283
  galvanic series, 282
  prevention of, 285, 291
  rates of, 286
  reliability and, 292
  surface treatments, 291
  types of, 279
    caustic embrittlement, 279
    chemical attack, 284
    crevice, 279
    fatigue, 283
    fretting, 285
    stress, 285
CREEP
  critical temperature, 241
  curves of, 242
  Larsen–Miller parameter, 245, 249, 250
  rupture strength, 244
  strain, 242
  stress relaxation, 241

SUBJECT INDEX

DEFECTS
  analysis, 210
  surveys, 210–214
DESIGN
  conforming surfaces, 267
  criteria for creep, 243
  damage-tolerant, 240
  data for wear, 269
  defect analysis, 210
  fracture safe, 238
  joint wear, 255
  "non-zero" wear, 268
  statistical factors, 50
  "zero" wear, 266
DETERMINISTIC ANALYSIS, 139, 333
DEVIATION FACTOR, 53
DISCRETE EVENTS
  arithmetic mean, 81
  correlation coefficients, 112, 114
  distributions of, 69, 72
  standard deviation of, 81
DISTRIBUTIONS
  binomial, 69
  central limit theorems, 80
  continuous, 73
  definition of terms, 74
  grouped frequency, 90
  mean of, 81
  normal (Gaussian), 80
  Poisson's, 71
  standard deviation, 81
  Weibull, 76

ERRORS
  absolute, 31
  defintion of, 31
  distribution of, 96
  estimation of, 14
  of functions, 36
  iteration procedure, 14
  relative, 15, 31
  rounding off, 32
  sample size, effect of, 98
  statistical, 50

FACTOR OF SAFETY
  actual, 135
  apparent, 135
  deterministic, 135, 139
  for fatigue, 229
  old-fashioned, 133
  probabilistic, 137, 140
  "safety margin", 149
  statistical, 149
FAILURE, see also Machine failure and Case studies
  bath-tub curve, 76, 134
  burnishing roller, 311
  of component, 215
  density, 86, 89, 90
  pattern of, 75
  probability of, 75, 136
  rates of, 75, 86
  static loading, 215
FATIGUE
  of burnishing roller, 311
  combined stresses, 227
  corrosion effects, 283
  endurance limit, 224, 231
  life, 228
  stress concentration, 228
  variation of strength, 128
FIELD INFORMATION, ORGANIZATION OF, 180, 209
FINITE DIFFERENCES
  equations, 22
  single variable, 22
  two variables, 23
FRACTURE
  critical flaw size, 239
  fatigue life, 237
  flaw parameter, 231
  "leak before break", 241
  mechanics, 230
  safe design, 238
  toughness, 231
FREQUENCY, GROUPED, 90

HARDNESS
  tests, various, 263
  v. shear strength, 262
HISTOGRAM, 90
  cell interval, 91

IMPACT LOAD, 215
INEQUALITY CRITERIA FOR SHAFT FREQUENCY, 236

SUBJECT INDEX

INTERPOLATION, *see also* Approximate computations
  linear, 5
INVERSE PROBLEM, 38
INVERSION TECHNIQUE, 12
ITERATIVE METHOD, 9

JOINTS, SLIDING, *see also* Wear *and* Design
  conforming surfaces, 255, 267
  groups of, 255
  non-conforming surfaces, 255
  pressure and velocity, 276
  types of, 255

LARSEN–MILLER PARAMETER, *see* Creep
"LEAK BEFORE BREAK", 241
LEAST SQUARE METHOD, 19
  R.M.S., 21
LOADING "SMOOTH" AND "ROUGH", 149, 159

MACHINES
  design defect analysis, 210
  failure, 199
    definition of, 199
    diagnosis of, 203
    diagnostic procedures, 209
    causes of, 200
    origins of, 199
    processes causing, 200
  performance and tolerances, 333
  reversibility of, 308
MATERIAL PROPERTIES
  abrasive wear, 272
  cavitation wear, 278
  compatibility for wear, 253
  corrosion resistant, 285
  cost for bearings, 270
  data sources, 129
  erosive wear, 277
  micro hardness, 256, 261
  variabilities of, 123
    coefficients of variation, 123, 124
    fatigue strength, 128
    helical springs, 127
    modulus of elasticity, 125
    static strength, 130
  strength of metals, 124
  tolerances, 125, 126
  welded joints, 127
  wear resistance, 256, 261
MEAN, ARITHMETICAL
  continuous functions, 81
  discrete numbers, 81
MODELLING
  burnishing rollers, 311
  case plate bolt loads, 313
  channel section, 300
  diamond section, 297
  free-wheel coupling, 304
  geometrical, 294
  iconic, 294
  ink-roller assembly, 332
  interpretation of, 295
  machine reversibility, 308
  mathematical, 294
  poppet valve, 321
  shaft frequency, 326
  spring clip, 303
  thin tubular cylinder, 295
MONOTONIC AND NON-MONOTONIC FUNCTIONS, 343
MONTE CARLO SIMULATION, 342

PAIRED COMPARISONS
  chart for, 113
  method of, 112
PARETO CURVE, 163
PARTIAL DERIVATIVE
  method, 106
  parameter sensitivity, 341
POPPET VALVE, *see also* Modelling
  seal resistance, 324
PROBABILITY
  density function, 74
  of failure, 74
  of survival, 74
  tolerance influence on performance, 332

QUALITY, *see also* "Zero defect"
  responsibility for, 163

RANDOM NUMBERS
  algebra of, 102, 104, 105

388 SUBJECT INDEX

definition of, 102
partial derivative method, 106
RECURRENCE EQUATION, 8
REDUNDANCY
 active, 156
 optimum, 157
 stand-by, 153
RELIABILITY, *see also* Case studies *and Failure*
 of assemblies and systems, 151, 159, 162, 165
  combined systems, 155
  in parallel, 153
  in series, 152
 of components, 133
 cost v., 146, 157
 fatigue, 52
 role of design for, 165
 rolling element bearings, 53
ROLLING ELEMENT BEARINGS
 life of, 53
 misalignment effects, 56
 radial clearance, 65
 reliability, 55

SAMPLE
 confidence level, 119
 size effect, 98
SERIES
 of function, 3
 power, 3
 Taylor's, 1
SHEAR CENTRE, CHANNEL SECTION, 300
SPRING
 clip pressure on shaft, 303
 force for poppet valve, 321, 325
STANDARD DEVIATION
 continuous function, 82
 discrete numbers, 81
 tolerance v., 50
STRESS
 concentration factors, 228
 corrosion, 285

in diamond section, 297
intensity factor, 217, 219–223
relaxation, 241
in thin tubular cylinder, 295

TOLERANCES
 manufacturing, influence on performance, 332
 standard deviations of, 50
 sure-fit, 339, 342, 378

VARIABLES
 continuous, 69
 discrete, 69
VARIANCE, 82
VIRTUAL DEVIATION, 47

WEAR
 abrasive, 252
 adhesive, 251
 cavitation, 278
 compatibility for, 253
 conforming surfaces, 255, 267
 erosive, 254, 277
 factor of, 275
 fatigue, 252
 fretting, 254
 micro-hardness, 262
 "number of passes", 260
 profiles, 256, 259
 rates of, 256, 274
 "sliding distance", 260
 sliding joints, life of, 255
 specific resistance, 256
 surface treatments, 280
 "unit of operation", 260
 "zero" and "non-zero", 260, 268

"ZERO DEFECT" production system, 163

RAYMOND H. FOGLER LIBRARY
DATE DUE